FLORA OF TROPICA

VERBENACE

BERNARD VERDCOURT

Herbs, shrubs, trees or woody climbers (varying from small annual herbs to large forest trees), sometimes thorny. Leaves predominantly opposite, sometimes whorled, rarely alternate, simple or digitately compound, entire, serrate or variously lobed, frequently aromatic; stipules absent. Flowers mostly ± irregular and 2-lipped, sometimes almost regular, usually 4–5-merous, usually ⚥ in often bracteate panicles, cymes or spikes. Calyx tubular or campanulate, lobed or subentire, sometimes 2-lipped. Corolla usually with narrow tube and spreading limb, minute to quite large and showy; lobes imbricate in bud. Stamens 4, didynamous, rarely 2 or, in a few genera, the same number as the corolla-lobes, inserted in the corolla-tube and alternate with lobes; filaments free; anthers free or connivent, dorsifixed, 2-thecous, the thecae mostly parallel, opening by longitudinal slits, introrse. Disk usually present, sometimes conspicuous. Ovary superior, sessile, entire or slightly 4-lobed, 2(–9)-locular, usually soon 4(or more)-locular by development of false septa, with axile or free central (in *Avicennia*) placentation; ovules 2 in each true locule (2 per carpel), erect or pendulous (in *Avicennia*); style terminal, simple and entire or shortly 2(rarely 4–5)-lobed. Fruit a drupe, a capsule (infrequent) or dividing at maturity into 2 or 4 nutlets. Seeds without or with much reduced endosperm or present and fleshy (in *Avicennia*); testa membranous.

About 70–80 genera and 3000–3500 species, widely distributed in tropical and temperate regions of both Old and New Worlds.

Cladistic studies of Verbenaceae and Labiatae strongly suggest that those parts of the former family other than the Verbenoideae are not logically separable from the Labiatae, also that *Cyclonema* and *Kalaharia* are distinct from *Clerodendrum*.

Teak is probably the best known member (see below) and the family contains numerous well-known garden plants, many of which are grown in East Africa (see below). For the purpose of this Flora and following Hutchinson (Fam. Fl. Pl., ed. 3: 487 (1973)) and Cronquist (Integr. Syst. Class. Fl. Pl.: 920 (1981)) *Avicennia* has not been recognised as a separate family but is so recognised by Moldenke and many other botanists. Moldenke has produced a vast amount of literature, most of it summarized in "A fifth summary of the Verbenaceae ... Synonymy", 2 vols., New Jersey (1971) and many similar earlier works. A very interesting historical account is given by H. & A. Moldenke in their 'A brief historical account of the Verbenaceae and related families', Plant Life, Jan–July 1946: 13–98 (1946). His account of the family in the Revised Flora of Ceylon 4: 196–487 (1983) is the latest treatment and is a mine of detailed information and synonymy. Also useful is his account for Flora of Panama (Ann. Missouri Bot. Gard. 60: 41–148 (1973) and Troncoso's keys to S. American genera (Darwiniana 18: 296–412 (1974)). The pollen morphology has been investigated by Bhoj Raj (Rev. Palaeobot. and Palynol. 39: 343–422 (1983)).

The family contains many herbs and shrubs used as ornamentals and trees for timber. Those belonging to genera not otherwise dealt with in the text are mentioned below; all are included in the generic key.

Aloysia triphylla (L'Hérit.) Royle** (*Lippia triphylla* (L'Hérit.) Kuntze, *Lippia citriodora* (Lam.) Kunth, *Aloysia citriodora* (Lam.) Pers.); Jex-Blake, Gard. E. Afr., ed. 4: 118 (1957)). The 'sweet-scented or lemon-scented Verbena' has been cultivated in East Africa (*Gillett*

---

*Marais has separated Cyclocheilaceae and Nesogenaceae (both completed for the Flora, 1984); see K.B. 35: 797–812 (1981). The genera concerned are mentioned in the generic key of this present part.
** Usually given as (L. Hérit.) Britton.

1

19825, Nairobi, 8 July 1972). It is an aromatic shrub 1–5 m. tall with narrowly oblong-elliptic lemon-scented leaves, 2.5–9.5 × 0.5–2.5 cm., in whorls of 3 and panicles of small purplish flowers 4–5 mm. long. Its properties are detailed by Moldenke in Rev. Fl. Ceylon 4: 234 (1983).

Citharexylum spinosum L. (*C. quadrangulare* Jacq.) has been grown at Karen near Nairobi (Apr. 1960, *Gardner* in *E.A.H.* H95/60/1! & 18 Jan. 1966, *Gillett* 17041!). It is a shrub or small tree to 15 m., with ovate-lanceolate, elliptic or elliptic-oblong leaves 3.5–29 × 1–11.5 cm. and very fragrant white, cream or reddish flowers in simple or compound axillary and terminal racemes 2.5–35 cm. long; corolla ± 5 mm. long; fruits blackish, fleshy, oblong, ± 8 mm. long.

Species of the genus *Petrea* are very frequently cultivated in East Africa and popular for planting by grassy boulevards in cities as well as in gardens. In gardening literature it is usually called *P. volubilis* L. (Jex-Blake, Gard. E. Afr., ed. 4: 139, 332 & 356 (1957); U.O.P.Z.: 406 (1949)) but Moldenke in his revision (F.R. 43: 1–48 (1938)) recognises 29 species, many of which seem very closely allied and more deserving of infraspecific rank. All E. African material appears to key to *P. kohautiana* Presl (T.T.C.L.: 640 (1949)); Jex-Blake mentions that "a second species smaller in every way with leaves harsh as sandpaper and dainty trusses of paler flowers is now to be had" but what this may be I do not know. About 7 sheets of *P. kohautiana* have been seen, all from Amani and Nairobi, e.g. Tanzania, Lushoto District, Amani Nursery, 23 Apr. 1930, *Greenway* 2231! & Amani, 28 Oct. 1969, *Ruffo* 268!; Kenya, Nairobi Arboretum, 26 May 1952, *G.R. Williams Sangai* 412!. *P. volubilis* has more distinctly petiolate leaves and more densely spreading hairy calyx, whereas *P. kohautiana* has subsessile or shortly petiolate leaves and ± sparsely adpressed pubescent calyx. East African material is a shrub, small tree or woody climber 2.5–6 m. tall, with coriaceous rather scabrid ovate, oblong or elliptic leaves 3.5–16.5 × 1.5–10 cm., mostly obtuse and with prominent nerves, and trusses of purple, blue or mauve flowers, the sepals accrescent, linear-oblong and venose, ± 2 cm. long and 5–8 mm. wide, persisting after the corolla has fallen.

Callicarpa rubella Lindl. (T.T.C.L.: 630 (1949)) has been grown at Amani over a period of 40 years (Amani Nursery, 25 Mar. 1931, *Greenway* 2908! & 15 Mar. 1973, *Ruffo* 613!) and probably is grown elsewhere. It is a shrub or small tree to 6 m., with oblong to oblong-lanceolate or -oblanceolate leaves 6–18 cm. long, 1.5–4.5 cm. wide, acuminate at the apex, cordate at the base, pubescent above, velvety stellate-tomentose beneath, crenulate, small greenish or purple flowers in dense axillary many-flowered cymes, and small purple very ornamental drupes.

Tectona grandis L.f., teak (T.T.C.L.: 641 (1949); U.O.P.Z.: 463 (1949); Descr. List. Introd. Trees Uganda: 67 (1937)), has been cultivated in many places in all three territories (Uganda, Bunyoro District, Budongo Forest, 24 Sept. 1962, *Styles* 61!, also at Maracha, Mbale, Soroti, Buyaga [Buwaga] and Kityerera; Kenya, Kisumu-Londiani District, Muhoroni to Songhor, June 1956, *Bally* 10563!; Tanzania, Lushoto District, Korogwe, Ugawe, 18 Aug. 1951, *Hughes* 96!; also at Tanga, Maweni, Longuza, Morogoro and Kilosa; U.O.P.Z. records it from Zanzibar and Pemba). It is a large tree at maturity, up to 50 m., with large elliptic leaves 10–100 cm. long, 5–50 cm. wide, mostly about 30 × 25 cm., stellate-tomentose beneath and often with red coloration on rubbing; inflorescences massive, ± 40 cm. long, 35 cm. wide with stellate-tomentose axes and small white flowers, the corolla-tube ± 1.5–3 mm. long; fruit subglobose, ± 1.5 cm. long and wide, enclosed in the inflated bladdery calyx 2.5 cm. long and wide. Information on teak has been summarised by Krishna Murthy (Bibliography on teak, Dehra Dun, 1981) and Moldenke gives much information (Phytologia 1: 154–164 (1935) & 5: 112–120 (1954)).

Three species of *Gmelina* (all Asian) have been grown in the Flora area separable as follows:

1. Short spine-like branches at each node of at least flowering
    shoots; shrubs or small trees with mostly elliptic leaves   2
   Branches not spiny; tree with distinctly ovate leaves   *G. arborea*
2. Leaves glabrous beneath save for sparse hairs on midrib;
    bright yellow flowers subtended by conspicuous purple
    bracts 2.5–3 cm. wide   . . . .   *G. philippensis*
   Leaves velvety tomentose beneath; yellow flowers
    subtended by narrower bracts 5 mm. wide; calyx with
    large glands outside   . . . .   *G. villosa*

Gmelina arborea Roxb. (T.T.C.L.: 639 (1949), Introd. Trees Uganda: 43 (1953)) has been grown in all three territories (Uganda, Bunyoro, Budongo, Nyabeya [Nyabyeya] Arboretum, 3 Oct. 1962, *Styles* 118!; also planted in Northern and Eastern Provinces; Kenya, Kilifi District, Jilore Forest Station, 4 Dec. 1969, *Perdue & Kibuwa* 10167!; Tanzania, Lushoto District, Sigi, bank of Kwamkuyo [Kwamkuyu] R., 26 Oct. 1948, *Greenway* 8318! and also at Rungwe, Iringa (Musekera Estate) and Tukuyu and doubtless many other places). It is a deciduous tree to 20 m., with broadly ovate entire leaves up to 20 cm. long and wide, acuminate at the apex, subcordate, rounded, or broadly cuneate at base and with 2 glands, velvety stellate-pubescent or -tomentose beneath; corolla orange-brown and yellow, 2–4 cm. long, very irregular, velvety hairy outside. Much used as a 'nurse' tree.

Gmelina philippensis Cham.* (*G. hystrix* Kurz; U.O.P.Z.: 276 (1949); Jex-Blake, Gard. E. Afr., ed. 4: 135 (1957)), grown as an ornamental, may be quite widely cultivated (Kenya, Masai District, Namanga Hotel, 10 Apr. 1955, *Verdcourt* 1254A!; Tanzania, Uzaramo District, Dar es Salaam, State House Grounds, 29 Sept. 1972, *Ruffo* SH/129!; Zanzibar, Mbweni, Kirk's old garden, 4 Feb. 1929, *Greenway* 1317!). It is a shrub about 3 m. tall or sometimes more or less climbing, with elliptic leaves ± 8 cm. long, 4.5 cm. wide, shortly acuminate at the apex, cuneate at the base into a distinct petiole, glabrous save for a few hairs on midrib beneath but with dense minute glands; corolla yellow, ± 4 cm. long, very narrowly tubular at base, irregularly cup-shaped above, ± glabrous; bracts large, purplish, ± 3.5 cm. long, 2.5–3 cm. wide.

Gmelina villosa Roxb. (U.O.P.Z.: 277 (1949)), cultivated in Zanzibar (Migombani [Mgombani], Feb. 1929, *Greenway* A.D.Z.12!) is a shrub or small tree similar to *G. philippensis* but with the leaves velvety beneath and bracts green and narrower, ovate to lanceolate.

Oxera pulchella Labill. (Jex-Blake, Gard. E. Afr., ed. 4: 138 (1957)), a native of New Caledonia, has been grown at Amani (Amani Nursery, 25 Sept. 1948, *Greenway* 8317!) and presumably in Kenya. It is a liane or ± climbing shrub to 6 m. with subacute or narrowly obtuse elliptic-oblong glabrous leaves ± 11 cm. long, 4 cm. wide, entire or coarsely shallowly crenate; flowers in profuse axillary inflorescences but appearing terminal when clustered; pedicels well developed; calyx ± 2 cm. long, deeply divided into elliptic lobes; corolla white, curved trumpet-shaped, 5–6.5 cm. long, narrowed below and with irregular limb; stamens and style well exserted.

Nyctanthes arbor-tristis L. (U.O.P.Z.: 381 (1949); Jex-Blake, Gard. E. Afr., ed. 4: 121 (1957)) is often placed in a separate family Nyctanthaceae or sometimes in the Oleaceae; I have seen no specimens of this native of India but it has obviously been grown in Kenya and Zanzibar and is a small tree or shrub; leaves ovate, up to 15 cm. long, 8.5 cm. wide, acuminate at the apex, rounded at the base, minutely hairy or glabrescent, rough above; flowers in small terminal clusters; calyx ± 7 mm. long; corolla very sweetly scented, white with an orange eye and tube salver-shaped, ± 2 cm. long, the 5–8 lobes ± obtriangular.

Caryopteris odorata (D. Don) Robinson (*C. wallichiana* Schauer) has been grown in Nairobi (City Park, 1 Aug. 1961, *Verdcourt* 3203!) and there is also a sheet from Kitale (Aug. 1968, *Tweedie* 3565!). Mrs. Tweedie gives the habitat as wet savanna at 1860 m., which suggests it might be naturalised, but without more evidence I have refrained from treating the species in full. It is a shrub 1.5–4.5 m. tall, with ovate-lanceolate to lanceolate leaves ± 8 cm. long, 2–5 cm. wide, long-acuminate at the apex, cuneate at the base, distinctly serrate, very densely glandular-punctate and pubescent, particularly on the venation beneath; inflorescences axillary, shorter than leaves but running together terminally and spike-like; calyx 3–7 mm. long, deeply divided into narrowly triangular teeth; corolla-tube 5 mm. long, pink or white and lobes about equal, pink or white with distal part of lower lip blue, magenta or carmine, pubescent outside; anthers and style well exserted.

Caryopteris paniculata C.B.Cl. has been grown in Tanzania (Lushoto District, E. Usambaras, near Amani, Kihuhwi, 24 July 1906, *Braun* in *Herb. Amani* 1349!) but no recent material has been seen. It is a shrub 1.2–3.6 m. tall, with elliptic to lanceolate petiolate acuminate toothed to ± entire leaves up to 12 × 5.5 cm.; cymes small, axillary, mostly not running into terminal thyrses; calyx 2 mm. long, deeply divided into narrow lobes; corolla much smaller than in last species, pale purple to crimson; tube 3 mm. long, marked carmine inside, the lobes 3 mm. long. Capsule 2.5–3.5 mm. in diameter.

Congea tomentosa Roxb., sometimes put in a separate family Symphoremataceae** (Jex-Blake, Gard. E. Afr., ed. 4: 135 (1957); Nairobi, Spring Valley, Nov. 1954, *Jex-Blake*), is a striking climbing shrub, with elliptic-ovate leaves up to 18.5 cm. long and 9.5 cm. wide, ±

---

* *G. philippensis* Cham. in Linnaea 7: 109 (1832) was published in the form *Gmelina* (*asiatica?*) *philippensis* but I am assured by nomenclatural experts that this does not invalidate it. The type at LE has been examined.

** Moldenke spells this Symphoremaceae

acuminate at the apex, subcordate at the base, entire, velvety pubescent; cymes in axillary and terminal inflorescences, each cyme 5–9-flowered and surrounded by 3 violet elliptic-oblong emarginate to mucronate involucral bracts, 1.5–3 cm. long, 0.5–1.2 cm. wide, tomentose and with long hairs above at base; calyx violet, 5–7 mm. long, densely hairy outside, with triangular teeth; corolla white tinged pink, ± 8 mm. long.

Holmskioldia sanguinea Retz., a native of India and Pakistan, is now widely cultivated for ornament (U.O.P.Z.: 300 (1949); Jex-Blake, Gard. E. Afr., ed. 4: 115 (1957)). It is a straggling shrub or small tree 3–10 m. tall, with ovate entire or ± crenate-serrate leaves 3–12 cm. long, 1.5–8.5 cm. wide; fruiting calyx yellow, orange or red, 1.5–2.5 cm. diameter and the corolla yellow, orange, brick red, scarlet or red-brown (Kenya, Nairobi, National Museum Grounds, Aug. 1963, *Verdcourt* 3709H!; Tanzania, Lushoto District, Amani Nursery, 5 July 1940, *Greenway* 5958!)

I have a note that the genus *Aegiphila* Jacq., has been grown in East Africa but can trace no source for the information; it has been included in the key.

1. Shrubs or small trees of mangrove swamps; pneumatophores present; flowers in terminal and axillary globose heads; placentation free-central with 4 pendulous orthotropous ovules; endosperm fleshy (*Avicennioideae*) . . . . 13. **Avicennia**
   Shrubs, trees or herbs not growing in mangrove associations; pneumatophores absent; placentation axile; endosperm absent or almost so 2
2. Leaves compound with (1–)3–5 leaflets in E. Africa (*Viticoideae* in part) . . . . 9. **Vitex**
   Leaves simple but sometimes deeply lobed 3
3. Flowers actually axillary and solitary but appearing fascicled; bracteoles conspicuous, broad and flat . . . . **Cyclocheilaceae — Asepalum**

   Flowers in several–many-flowered inflorescences or if solitary and axillary then without conspicuous bracteoles 4
4. Slender annual herb with 1–several flowers in the leaf-axils . . . . **Nesogenaceae — Nesogenes**

   If an annual herb then flowers in compound or many-flowered inflorescences 5
5. Each cyme with an involucre of 3–4 coloured bracts (cultivated) (*Symphorematoideae*) . . . . . . **Congea**, p. 3
   Bracts, if present, usually small and not forming a coloured involucre 6
6. Inflorescence spike-like but sometimes short and subcapitate, at least in flower (*Verbenoideae*) 7
   ·Inflorescences raceme-like or cymose; flowers pedicellate (save in *Nyctanthes*) 14
7. Fertile stamens 2; rhachis hollowed out beneath the flowers . . . . 3. **Stachytarpheta**
   Fertile stamens 4; rhachis not hollowed out beneath the flowers 8
8. Fruit of 4 1-seeded schizocarps; naturalized or cultivated, some now weeds . . . . 1. **Verbena**
   Fruit of 2 pyrenes 9
9. Ovary 4-locular; pyrenes normally 2-locular and 2-seeded, rarely 1-locular and 1-seeded; fruiting calyx enveloping the echinate pyrenes . . . . 4. **Priva**
   Ovary 2-locular; pyrenes normally 1-locular and 1-seeded 10
10. Calyx tubular and conspicuous . . . . 2. **Chascanum**
    Calyx very small and inconspicuous 11
11. Fruit fleshy with hard endocarp; calyx truncate or shallowly toothed, corolla 4–5-lobed . . . . . 7. **Lantana**
    Fruit ± dry; calyx 2–4-fid or conspicuously toothed; corolla 4-lobed 12

12. Herbaceous, usually creeping, with medifixed hairs; inflorescence ± subcapitate, often elongating in fruit   **5. Phyla**
    Shrubby or sometimes small woody-based pyrophytes 13
13. Spikes elongate, axillary, 3–4 per node or aggregated in terminal panicles; flowers ± spaced (cultivated)   **Aloysia**, p. 1
    Spikes short and subcapitate or if elongate then the flowers dense . . . . **6. Lippia**
14. Inflorescences racemose (*Verbenoideae*) 15
    Inflorescences cymose, forming false umbels, umbel-like panicles, thyrses, etc. 17
15. Fruiting calyx-lobes enlarged, wing-like, violet, net-veined (cultivated) . . . . **Petrea**, p. 2
    Fruiting calyx-lobes not enlarged nor coloured 16
16. Ovary 8-locular; calyx shrunk over fruit and beaked; fruit with 4 pyrenes, each 2-locular and 2-seeded (cultivated and naturalized) . . . . **8. Duranta**
    Ovary 4-locular; calyx not enveloping fruit; fruit with 2 pyrenes, each 2-locular and 1–2-seeded (cultivated)   **Citharexylum**, p. 2
17. Flowers sessile in 3-flowered bracteate heads arranged in terminal trichotomous cymes; corolla-tube cylindric, orange; limb flat, (5–)6–7(–8)-lobed, white; anthers 2 (cultivated) (*Nyctanthoideae*) . . . . **Nyctanthes**, p. 3
    Inflorescences of different structure and other characters not combined 18
18. Flowers ± regular; stamens 4–6 19
    Flowers ± irregular; stamens 2–4 21
19. Fruits with 1 4-locular pyrene; fruiting calyx conspicuously inflated; leaves often huge, 10–100 × 5–50 cm. (cultivated)   **Tectona**, p. 2
    Fruits with 4–5 1-locular 1-seeded pyrenes; fruiting calyx not inflated; leaves not as above, 6–18 × 3.5–4.5 cm. (*Viticoideae* in part) 20
20. Stigma deeply bifid with long subulate branches; fruiting calyx enlarging and indurated; stamens inserted at or about the middle of the corolla-tube; ovary mostly glabrous (cultivated) . . . . **Aegiphila**, p. 4
    Stigma depressed capitate or peltate; fruiting calyx not enlarging; stamens inserted near the base of the corolla-tube; ovary ± pubescent (cultivated) . . . **Callicarpa**, p. 2
21. Stamens 2 (cultivated) (*Viticoideae*) . . . . **Oxera**, p. 3
    Stamens 4 22
22. Fruit a 4-valved capsule; lowest and largest corolla-lobe fimbriated (cultivated) (*Caryopteridoideae*) . . . . **Caryopteris**, p. 3
    Fruit not dehiscent; corolla-lobes not fimbriate (*Viticoideae*) 23
23. Fruits with 1 4-locular pyrene or 2 2-locular pyrenes; corolla-tube 2–4 mm. long save in cultivated species with yellow corollas over 2 cm. long 24
    Fruits with up to 4 1-locular pyrenes or 2 2-locular pyrenes; corolla-tube rarely as short as 4 mm. and then pyrenes 1-locular 25
24. Corolla short, tube 2–4 mm. long, cylindric . . . . . **10. Premna**
    Corolla large, over 2 cm. long, mostly funnel-shaped and widened at apex (cultivated) . . . . **Gmelina**, p. 2
25. Fruiting calyx tubular, cupular or inflated but never flat and accrescent, usually toothed or lobed . . . . **12. Clerodendrum**
    Fruiting calyx accrescent, ± flat and saucer-shaped or only slightly convex, entire or shallowly obtusely lobed 26
26. Corolla-tube tubular, slightly curved but scarcely zygomorphic, relatively long and narrow, not split; stamens inserted near the middle of the tube, prolonged below their insertion as raised pubescent lines; fruit somewhat fleshy, verrucose but not hairy, dividing into 4 pyrenes (cultivated) . . . . . . **Holmskioldia**, p. 4

Corolla-tube ± short, gibbous and fissured; stamens
inserted near the mouth and not prolonged below
their insertion; fruit hard and dry, pubescent above,
tardily splitting into 2?–4 pyrenes . . . .   11. **Karomia**

1. VERBENA

L., Sp. Pl.: 18 (1753) & Gen. Pl., ed. 5: 12 (1754)

Annual or perennial herbs or subshrubs. Leaves opposite or sometimes in whorls of 3, simple and dentate or deeply divided, rarely entire. Flowers in simple spikes or panicles of spikes, often dense and compact or more elongate and lax, bracteate, terminal. Calyx tubular, distinctly and ± unequally (4–)5-toothed, 5-ribbed. Corolla funnel-shaped or hypocrateriform; tube straight or ± curved, ± pubescent within; limb 5-lobed, ± spreading, feebly to distinctly 2-lipped. Stamens (2–)4(–5), included, usually inserted near the middle of the tube but sometimes higher, didynamous. Ovary entire or 4-lobed, 4-locular with a single ovule in each locule; style slender but ± short; stigma unequally 2-lobed. Fruit splitting into 4 nutlets at maturity, enveloped in the persistent calyx. Nutlets with inner face often scabridulous or granulate.

About 200–250 species in tropical and temperate America with a few in Europe, Asia and N. Africa. The five species occuring in East Africa are all either naturalised and/or cultivated.

The common garden Verbena, × *V. hybrida* Groenl. & Rümpl., with white, pink, red, blue, purple or variegated flowers, has I think been grown, but only one sterile specimen has been seen from an unlikely unpopulated area — Tanzania, Buha District, Birira to Nisusi, 27 Feb. 1926, *Peter* 37891. It is easily distinguished by the corolla-limb up to 2 cm. wide and leaves lobed or toothed. Jex-Blake (Gard. E. Afr., ed. 4: 83 (1957)) mentions *V. venosa* Gillies & Hook. 'var. *lilacina*' (now called *V. rigida* Sprengel) as "perhaps one of the most dangerous weeds we have introduced. A tremendous root system, running for many yards and forming dense mats, is a real danger in a garden". *Glandularia* Gmelin is usually recognised as a genus in S. American literature (see N.S. Troncoso in Burkart, Fl. Ilustr. Entre Rios (Argentina) 5: 244 (1979)). It is distinguished by the more cylindrical stems, nutlets ± prolonged at the base and anthers sometimes with a connective-gland. For the purpose of this Flora however I have retained species 5 in *Verbena*; I feel *Glandularia* is no more than a section of *Verbena;* the facies is exactly similar.

1. Leaves deeply divided 2
   Leaves distinctly toothed but not deeply divided 3
2. Ultimate leaf-lobes oblong, oblanceolate or elliptic . .   1. *V. officinalis*
   Ultimate leaf-lobes filiform . . . .   5. *V. aristigera*
3. Corolla-limb 5–7 mm. wide; tube 8–10 mm. long; spikes
     solitary to few; tips of leaf-serrations more spreading   2. *V. rigida*
   Corolla-limb mostly much narrower; tube 2.5–7 mm. long;
     spikes numerous; tips of leaf-serrations more
     ascending 4
4. Corolla-limbs 3.5–4.5(–5.5) mm. wide, borne above the
     apices of spikes; tube distinctly exserted from the
     calyx . . . .   3. *V. bonariensis*
   Corolla-limbs 2.7–3.7 mm. wide, not exceeding spike-
     apices and forming a circlet just below its conical apex;
     tube scarcely exserted from the calyx . . . . .   4. *V. brasiliensis*

1. **V. officinalis** *L.*, Sp. Pl.: 20 (1753); Bak. in F.T.A. 5: 286 (1900); Ross-Craig, Drawings Brit. Pl. 23, t. 38 (1966); Franco in Fl. Europaea 3: 123 (1972); Townsend in Fl. Iraq 4: 651 (1980) & in Fl. Turkey 7: 33, t. 117/8–10 (1982); Meikle, Fl. Cyprus: 1251 (1985). Types: Mediterranean, specimens in *Hort. Cliff.* 11.6 (BM, 3 syn.) & *Linnaean Herbarium* 35.15 (LINN, syn.)

Erect perennial, rarely annual, laxly branched herb 0.3–1.6 m. tall; stems ± woody at base, 4-angled, sulcate, sparsely scabridulous or ± glabrous. Leaves more or less ovate in outline, 2–7(–11) cm. long, 0.8–8.5 cm. wide, pinnatisect or pinnatifid with broad acute bluntly serrate lobes, adpressed hispidulous above, scabridulous beneath, acute or ± obtuse at the apex, cuneate at the base into an indistinct winged petiole 1–2 cm. long; upper stem-leaves often lanceolate or oblanceolate in outline, less lobed and more

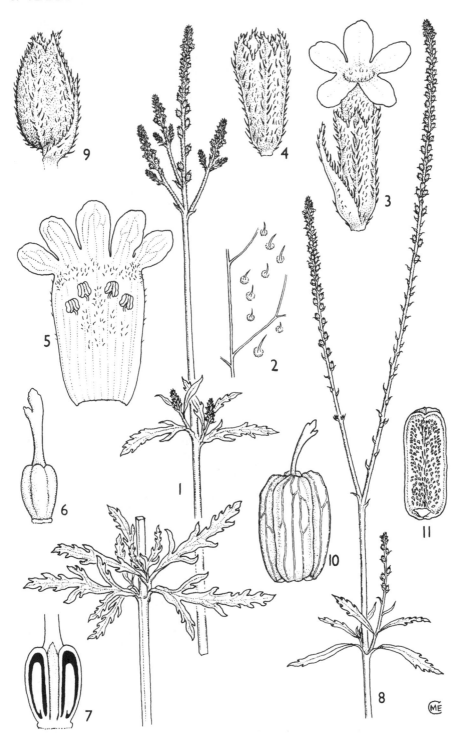

FIG. 1. *VERBENA OFFICINALIS* subsp. *AFRICANA* — **1**, habit, × ²/₃; **2**, part of upper leaf surface, × 8; **3**, flower, × 14; **4**, calyx, × 14; **5**, corolla, opened out, × 14; **6**, ovary, × 20; **7**, longitudinal section of ovary, × 30; **8**, fruiting stem, × ²/₃; **9**, fruit, enclosed by calyx, × 12; **10**, fruit, × 16; **11**, nutlet, × 16. 1–7, from *Gillett* 16559; 8–11, from *Tweedie* 1529. Drawn by Mrs. M.E. Church.

bluntly serrate. Flowers minute in numerous elongate spikes 5–30 cm. long, dense above, lax below; rhachis and narrowly ovate-acuminate bracts glandular-pubescent. Calyx ± 2–2.5 mm. long, glandular-pubescent outside; teeth triangular, ± 0.4 mm. long. Corolla pale lilac-purple, pink or mauve, often with yellow throat, glabrous; tube ± 3 mm. long; limb subequally 5-lobed; lobes ± oblong, 0.75–2 mm. long. Stamens inserted at 2 levels near or above the middle of the tube. Style 0.5–1 mm. long. Nutlets brown, oblong, 1.5–2 mm. long, 0.6–0.8 mm. wide, longitudinally rugulose-reticulate outside, whitish scabridulous-papillose inside.

subsp. **africana** R. *Fernandes & Verdc.* in Bol. Soc. Brot., sér. 2, 62: 305, fig. 1 (1990). Type: Zimbabwe, Harare, between Avondale West and Mabelreign, *Drummond* 4858 (K, holo.!, B, BR, LISC, S, SRGH, iso.!)

Leaf-lobes more lanceolate, together with serrations distinctly more acute than in typical subsp. *officinalis*. Floral bracts lanceolate to ovate-lanceolate, attenuate-acuminate, up to 2.5 mm. long, equalling the calyx. Corolla-tube cylindric, constricted just below the limb, 2.5 mm. long; lower lobe 1.75 mm. wide; upper lobe 1.25 mm. wide with lobes 0.75 × 0.5 mm. Stigma 1 mm. long. Fig. 1.

UGANDA. Toro District: Ruwenzori, Ruimi [Wimi] valley, June 1884, *Scott-Elliot* 7800!; Ankole District: Mitoma County (Saza), Oct. 1938, *Purseglove* 497!; Kigezi District: Nyarusanje, 7 May 1952, *Norman* 124!

KENYA. NE. Elgon, Apr. 1958, *Tweedie* 1529!; Naivasha, 1 May 1932, *Napier* 1853!; Nanyuki, 100 m. E. of Sportman's Arms Hotel, 18 Mar. 1968, *Mwangangi & Fosberg* 546!

TANZANIA. Masai District: Ngorongoro Crater, inner NE. slopes, W. of Oljoro Nyuki, 10 Oct. 1977, *Raynal* 19530! & Ngorongoro Crater, near Siedentopf's, Apr. 1941, *Bally* 2344!; Arusha, Feb. 1958, *Ivens* 1091!

DISTR. U 2; K 2–4, 6; T 2; Zaire, Sudan, Ethiopia, Zambia, Zimbabwe, South Africa; possibly also Nepal and Taiwan

HAB. Short volcanic grassland, long grassland by water, hillside fallow, bushland, *Acacia lahai* woodland; 1500–2100 m.

SYN. *V. officinalis* L. var. *natalensis* Krauss in Flora 28: 68 (1845); Moldenke in Phytologia 36: 279 (1977), *nomen.* Authentic specimen: South Africa, Natal, R. Umlaas, *Krauss* (K!, TUB)
[*V. officinalis* sensu U.K.W.F.: 611 (1974), *non* L. sensu stricto]

NOTE. My long-held suspicions that material in East Africa always called *V. officinalis* did not resemble specimens I had collected in Britain were confirmed when R. Fernandes considered they were not conspecific. I have, however, been unable to match the East African plant with any American taxon. An added difficulty is that in South Africa two taxa occur, one undoubtedly true *V. officinalis* the other (to which Hochstetter gave an unpublished varietal name) identical with the plant growing in the Flora Zambesiaca and Flora of Tropical East Africa areas and in Ethiopia where some specimens are, however, not clear cut. Throughout the Old World there is a great deal of variation in what is called *V. officinalis* and it is possible that two extreme variants have been introduced into different parts of Africa at different times. The differences, however, are quite marked. The possibility of there being a truly native African subspecies seems unlikely but nevertheless subspecific rank has been employed as a solution. The typical taxon occurs in Europe, Middle East, Asia to China and Indonesia, Canary Is. and N. Africa; now introduced almost throughout the world (see Fl. Iraq for more detailed listing). S. Barber (Syst. Bot. 7: 433–456 (1982)) has made a detailed numerical study of some N. American species and decided that *V. halei* Small could not be separated specifically from *V. officinalis* and treated it as *V. officinalis* subsp. *halei* (Small) S. Barber. She accepts that subsp. *officinalis* is introduced from the Old World and that hybrids between the two have been reported. Presumably the possibility that *V. officinalis* was introduced from America into the Old World very early is not acceptable and at least it is native to America in the broad sense. Typical *V. officinalis* has more obtuse leaf-segments and serrations, shorter ovate floral bracts and the corolla-tube not constricted below the limb.

2. **V. rigida** *Sprengel*, Syst. Veg. 4, cur. post.: 230 (1827); Moldenke in Phytologia 11: 62, 80 (1964); Franco in Fl. Europaea 3: 123 (1972); Troncoso in Burkart, Fl. Ilustr. Entre Rios (Argentina) 5: 233, fig. 106 (1979). Type: Uruguay, Rio Grande, *Sellow* (B, holo.†)

Rhizomatous perennial herb forming mats with slender branched or ± unbranched ascending stems 15–60 cm. tall, decumbent at the base, ± square and grooved, with dense short spreading tubercle-based pubescence, later tuberculate only. Leaves narrowly oblong or oblong-elliptic, 3–11 cm. long, 0.8–3.5 cm. wide, acuminate at the apex, subcordate to amplexicaul at the base, sessile, strongly incised-serrate with rather distant ± straight teeth or some upper leaves ± entire, very scabrid with short tubercle-based hairs and similar longer hairs on margins and venation beneath. Inflorescence of simple spikes or 3–5 in dichotomous cymes, 1–2.5(–5 in fruit) cm. long; peduncle up to 6 cm. long; bracts lanceolate-acuminate, 4–5 mm. long, 0.8 mm. wide, setulose on the margin. Calyx greenish purple, cylindrical, 3–4 mm. long, glandular-pubescent, with short acute ±

subulate lobes. Corolla mauve-purple or bluish, tube 8–10 mm. long, shortly spreading pubescent outside; limb 5–7 mm. diameter. Stamens inserted near middle of tube. Style 2.5–3 mm. long. Nutlets brown, ovoid-trigonous to oblong, 2–2.5 mm. long, 0.8 mm. wide, reticulate-ridged outside, white scabridulous scaly inside.

KENYA. Elgeyo Forest Reserve, near Singore Mission, 27 Apr. 1968, *Dyson* 568!; Nairobi, Chiromo Estate, 5 Jan. 1970, *Mathenge* 517! & Nairobi Arboretum, 18 May 1953, *G.R. Williams* 552!
TANZANIA. Lushoto District: Magamba Forest Reserve, 22 Nov. 1957, *Mgaza* 162!
DISTR. K 3, 4; T 3; eastern S. America from Brazil to N. Argentina, also cultivated and widely naturalised in Europe, Africa, Australia, N. America, etc.
HAB. Cultivated but now often as an escape, e.g. in forest plantation edges, grassy roadside verges, etc.; 1650–2400 m.

SYN. *V. venosa* Gillies & Hook. in Hook., Bot. Misc. 1: 167 (1829); Hook., Bot. Mag. 59, t. 3127 (1832). Type: Argentina, "inter pampas prov. Bonariae", *Gillies* (K, holo.!)
V. bonariensis L. var. *rigida* (Sprengel) Kuntze, Rev. Gen. Pl. 3(2): 255 (1898)

3. **V. bonariensis** *L.*, Sp. Pl.: 20 (1753); Burkart in Darwiniana 11: 383, t. 16 (1957), pro parte; Moldenke in Phytologia 8: 246 (1962), pro parte; Franco in Fl. Europaea 3: 123 (1972), pro parte; Troncoso in Burkart, Fl. Ilustr. Entre Rios (Argentina) 5: 241, fig. 110 (1979), pro parte; Yeo in K.B. 45: 105, fig. 2 (1990). Type: Argentina, Buenos Aires*, *Linnaean Herb.* 35.11 (LINN, lecto.)

Stiffly erect branched mostly annual herb 0.6–1.8(–2.5) m. tall; stems scabrid to hispid or hairy. Leaves ovate, obovate or ovate-lanceolate below to oblong, elliptic or elliptic-lanceolate or linear-lanceolate above, 3–15 cm. long, 0.2–3 cm. wide, acuminate at the apex, cuneate, rounded or ± amplexicaul, sessile, regularly serrate or doubly serrate, the base sometimes entire, scabrid with tubercle-based hairs; venation impressed above, raised and closely reticulate beneath. Inflorescences of numerous spikes 0.5–1.5(–4.5) cm. long, 5–7 mm. wide, arranged cymosely in mostly dense groups of triads, the central ones subsessile, the whole forming a dense flat-topped or rounded cluster; bracts lanceolate, 2.75–4 mm. long, setulose-ciliate. Calyx ± 3.5 mm. long. Corolla purple or magenta, the limb sometimes bluish; tube 5.5–7 mm. long, exserted for about ⅛ its length or 1.5 mm. (in dry state), pubescent outside; limb (3.5 dry –)4.2–5.5 mm. wide. Anthers inserted near the middle of the corolla-tube. Nutlets brown, narrowly oblong, ± 8 mm. long, 0.6 mm. wide, reticulate-ridged outside, scabridulous white scaly inside.

KENYA. Nakuru, May 1963, *Tweedie* 2600!; Nairobi, bank of Nairobi R. near National Museum, 12 Aug. 1953, *Verdcourt* 1006!; Londiani, *Gardner* in F.D. 919!
DISTR. K 3–6; Argentina and Uruguay (region of Parana and Uruguay rivers and their common gulf, Rio de la Plata), also Chile (Valparaiso) but now widely naturalised in Europe, Asia, Africa and Australia
HAB. Grassy roadsides, river-banks, grassland with scattered trees, etc.; 1650–2650 m.

NOTE. Troncoso gives range as Argentina, Chile, Uruguay, Paraguay, Bolivia and Brazil but presumably due to confusion with the next species.

4. **V. brasiliensis** *Vell.*, Fl. Flum.: 17 (1825) & 1, t. 40 (1835); Yeo in K.B. 45: 111, fig. 3 (1990). Type: Brazil, no specimen cited, near Rio de Janeiro, Fl. Flum., t. 40 (lecto.!)

Annual or short-lived perennial herb, very similar to the last species, (0.3–)0.9–2(–?3) m. tall, branched above, sparsely to densely scabrid with tubercle-based hairs. Leaves oblong-elliptic to lanceolate at base and ± linear-lanceolate above, (1.5–)3–14 cm. long, 0.4–4.5 cm. wide, acute at the apex, cuneate to amplexicaul at the base, sessile, evenly to ± unevenly serrate, usually entire towards base, scabrid with tubercle-based hairs on both surfaces; venation impressed above, raised beneath. Inflorescences much branched, of numerous spikes 1.5–4.5(–10) cm. long, 5.5–6.5 mm. wide, dense or sometimes ± lax at base, arranged as in last species in dense heads; bracts lanceolate, (2.25–)4–5 mm. long, 1 mm. wide, hispidulous-ciliate. Calyx green, tinged purple, 2.5–3.5 mm. long. Corolla

---

* Linnaeus cites Dillenius, Hort. Elth., t. 300/387 (1732), and according to Druce & Vines (The Dillenian Herb.: 182 (1907)) the specimen collected in 1726 by Mr Milan in Buenos Aires is extant; this is the locality cited by Linnaeus.

bluish purple, mauve or violet; tube 2.75–3.3 mm. long, scarcely exserted or not by over ⅓ its length, pubescent; limb 2.7–3.7 mm. wide. Anthers situated in upper ⅓ of tube. Nutlets brown, oblong, 1.2–1.5 mm. long, 0.5 mm. wide, ridged-reticulate outside, white scabridulous-scaly inside.

KENYA.   Near Naivasha, 21 May 1961, *J. G. Williams* in *E.A.H.* 12350!; Kiambu District: Limuru, 31 Oct. 1947, *Bogdan* 1413! & Kikuyu Escarpment, Naivasha road, Upper Ngubi Forest Reserve guard post, 15 Dec. 1966, *Perdue & Kibuwa* 8245!
TANZANIA.   Arusha District: Kilinga Forest, 20 Sept. 1971, *Arasululu* in *Richards* 27256!; Lushoto District: W. Usambaras, 6.4 km. NE. of Lushoto, Mkuzi, 11 Apr. 1953, *Drummond & Hemsley* 2092! & Lushoto, 18 July 1959, *Semsei* 2873!
DISTR.   K 3, 4, ?5; T 2, 3; similar to last species but also northwards to Brazil (Minas Gerais) and NW. to Bolivia, Colombia and Peru, now widely naturalised in Africa, Malay Peninsula, Indonesia, tropical Australia, etc.
HAB.   Grassland, bushland, e.g. *Acacia, Rhus,* etc., disturbed ground and as a roadside weed; 1050–2220 m.

SYN.   *V. quadrangularis* Vell., Fl. Flum.: 16 (1825) & 1, t. 39 (1835). Type: 'in campis maritimis' (ubi?); Fl. Flum., t. 39 (lecto.!)

NOTE.   All the E. African material (and much else besides) had been named *V. bonariensis* and but for Yeo's determinations I would probably not have thought it any more than a form of that and indeed some specimens seemed identical; after study Yeo's distinctions became more evident but I am still doubtful if specific rank is correct. The species has also been much confused with *V. litoralis* Kunth but that name has never been used in East Africa.

5. **V. aristigera** S. *Moore* in Trans. Linn. Soc., Bot., ser. 2, 4: 439 (1895); Moldenke in Phytologia 8: 189 (1962). Type: Brazil, Matto Grosso, near Mt. Pão d'Assucar, between Coimbrá and R. Apa, *Moore* 1083 (BM, holo.)

Perennial herb, often woody at the base, 10–40 cm. tall; stems decumbent below, sparsely pubescent but soon glabrescent. Leaves ± triangular in outline, pinnatisect, 2–3.5 cm. long, 2–4 cm. wide; segments linear, up to 1 cm. long, 0.5–1(–3) mm. wide, ± acute or obtuse at the apex, substrigose-pubescent or ± glabrous, entire or toothed. Inflorescences terminal, dense and ± ovate, at first 1.5 × 2 cm. but soon elongating to 5–12 cm.; peduncles 1.5–5 cm. long, elongating to 10 cm. or more in fruit. Bracts subovate at base, lanceolate-subulate, 2–4 mm. long, 1.2–1.5 mm. wide, canescent-puberulous. Calyx tubular, 6–7 mm. long, densely adpressed strigose-pubescent and with some sparse dark glands; teeth ovate at base, contracted into a long seta, 0.5–2 mm. long, unequal. Corolla blue or purple; tube 1–1.1 cm. long, glabrous outside, hairy inside at throat; limb ± 1 cm. wide, the lobes ± broadly obcordate, 3–4 mm. long and wide, emarginate for 1–1.5 mm. Stamens included or with glands just exserted; style ± included, 0.9–1.3 cm. long, persistent. Nutlets subcylindrical, 3 mm. long, 0.8 mm. wide, reticulately ribbed outside, white verrucose inside.

KENYA.   Uasin Gishu District: near Eldoret, 15 Aug. 1963, *Heriz-Smith* 938!; Kiambu District: Muguga, staff quarters, 6 Oct. 1976, *Mungai* 106!; Nairobi, Dec. 1948, *Bally* 6554!
TANZANIA.   Just E. of Arusha, 25 Jan. 1970, *Wendelberger* 217!
DISTR.   K 3, 4; T 2; Zaire, Mozambique, Malawi, Zambia, Zimbabwe, Botswana, Brazil, Paraguay, Uruguay and Argentina
HAB.   Cultivated but also as a weed locally on cleared land; 1650–2060 m.

SYN.   *V. tenuisecta* Briq. in Ann. Conserv. Jard. Bot. Genève 7–8: 294 (1904); Moldenke in Phytologia 11: 280 (1965). Type: Paraguay, La Trinidad, *Balansa* 1025 (G, holo.)
      Glandularia aristigera (S. Moore) Tronc. in Darwiniana 14: 636 (1968)
      [*Verbena tenera* sensu U.K.W.F.: 611 (1974) *non* Sprengel]

NOTE.   Kenya material had been wrongly determined at Kew as *V. tenera* Sprengel (*Glandularia tenera* (Sprengel) Cabrera) over 50 years ago and that name was applied to material from other areas. I am grateful to R. Fernandes for arousing my suspicions; she suspected Flora Zambesiaca area material was wrongly identified and this proved to be the case in East Africa also. The plant is often known in Kenya as Burkitt's Blue.

2. CHASCANUM

E. Mey., Comm. Pl. Afr. Austr.: 275 (1838); Moldenke in F.R. 45: 113–160, 300–319 (1938), 46: 1–12 (1939) & Rev. Sudam. Bot. 6: 15–24 (1939); Gillett in K.B. 10: 131 (1955); Saxena in Indian Journ. Bot. 3: 95–102 (1980) (pollen); Moldenke in Phytologia 52: 323–329 (1983); Gilbert et al. in Taxon 35: 391 (1986); Sebsebe & Verdc. in K.B. 43: 519 (1988), *nom. conserv.*

Plexipus Raf., Fl. Tellur. 2: 104 (1837); R. Fernandes in Bol. Soc, Brot., sér. 2, 57: 265 (1984)
    Chascanum E.Mey. sect. *Pterygocarpum* Walp., Repert. 4: 39 (1845)
    Svensonia Mold. in F.R. 41: 129 (1936); Gillett in K.B. 10: 131 (1955)

Erect or spreading perennial herbs or undershrubs, often quite woody at base and often densely pubescent. Leaves opposite or subopposite, mostly ovate or oblong, toothed or incised-dentate, rarely ± entire, ± petiolate. Flowers sessile or shortly pedicellate in terminal spikes or spike-like racemes; rhachis not excavated; bracts small and narrow; bracteoles minute or absent. Calyx narrowly tubular, 5-ribbed and 5-angled, somewhat oblique, 5-toothed, dilating and splitting longitudinally at maturity. Corolla white, yellowish or pink, infundibuliform or hypocrateriform; tube slender, straight or curved, often widened at apex, glabrous outside, pilose inside between stamens; limb ± oblique, with 5 subequal or unequal lobes, ± 2-lipped. Stamens 4, didynamous, inserted in upper part of the corolla-tube, included or slightly exserted. Ovary cylindrical or narrowly oblong, glabrous, 2-locular, each locule with one basal anatropous ovule; style terminal, filiform, elongate; stigma ± oblique, sub-bilobed. Fruit straw-coloured to black, oblong, of 2 1-seeded linear-oblong nutlets ribbed on outer surface, sometimes winged at apex, minutely verruculose-asperulous on seed-part inside. Seeds linear, exalbuminous.

A genus of about 27 species, mainly African, extending to Madagascar, Arabia, Pakistan (Sind) and India.
    The two African species referred to *Svensonia* are undoubtedly not separable from *Chascanum* at generic level, despite the winged nutlets. There is, however, a complication. The distinctly different looking pink- or white-flowered Indian species, C. *hyderabadense** (Walp.) Mold. (*Svensonia hyderabadensis* (Walp.) Mold.), is undoubtedly nearer to *Bouchea* Cham. than any other Old World species. At one time most *Chascanum* and *Svensonia* species were placed in *Bouchea* Cham. but this is now restricted to the New World by Moldenke. I am not convinced that it should not be reinstated as a pantropical genus. If the type of *Svensonia* had been the Indian species it could have been maintained for it but a naïve original type selection renders this solution impossible. Saxena's work on pollen (overlooked by Sebsebe and I in 1988) supports the amalgamation of *Svensonia* and *Chascanum*.

1. Nutlets without wings, usually truncate or at the most
        slightly produced or pointed 2
    Nutlets with distinct apical wing up to 2.5 mm. long 8
2. Leaves linear, 1.5–5 cm. long, 1.5–2 mm. wide . . . .   7. *C. rariflorum*
    Leaves not linear, much wider 3
3. Venation of leaves impressed above and distinctly closely
        raised reticulate beneath 4
    Venation of leaves not so impressed and only main nerves
        raised beneath 5
4. Leaves ovate, broadly elliptic or ± round, 0.5–4.5 cm. long,
        0.5–3 cm. wide, rounded cuneate or subtruncate at the
        base (**K** 1, 2) . . . .   1. *C. marrubiifolium*
    Leaves ± obovate in outline, 0.6–2 cm. long, 0.4–1 cm. wide,
        attenuate-cuneate at the base (**K** 6) . . . . . .   2. *C. sp. A*
5. Inflorescences ± densely shortly pubescent, sometimes
        with long and short hairs but without additional much
        longer spreading hairs; corolla cream, yellowish,
        brownish yellow, apricot or purplish, ± 1.4–1.5 cm.
        long; leaves coarsely to more finely toothed or ±
        subentire 6

___
* Walpers' original spelling *hyderobadensis* followed by Moldenke was probably deliberate but since it has never been employed for the well-known state, district and city the corrected spelling used in most Indian floras and the Index Kewensis is perfectly justified.

Inflorescences pubescent and with additional much
    longer spreading hairs; corolla white or cream, 1.3–3
    cm. long; leaves toothed save at base 7
6. Pubescence less dense and more equal; leaves more
    obovate and less coarsely toothed; corolla cream and
    yellow, never brownish or purplish . . . . . .   3. *C. obovatum*
    Pubescence denser and of longer and shorter hairs; leaves
    elliptic-oblong, subentire to coarsely toothed; corolla
    brownish yellow, apricot or purplish . . . . .   4. *C. gillettii*
7. Leaves attenuate at the base into a usually distinct petiole,
    narrowly oblong, elliptic-oblong or obovate; corolla
    1.3–2.9 cm. long (the most widespread and
    commonest species) . . . .   5. *C. hildebrandtii*
    Leaves less attenuate, almost sessile, narrowly elliptic-
    oblong; corolla ± 3 cm. long (still known only from the
    type from **T** 5 or 6) . . . .   6. *C. hanningtonii*
8. Corolla under 2 cm. long; calyx ± 8–9 mm. long; nutlet-
    wings erect . . . .   8. *C. laetum*
    Corolla 3–3.5 cm. long; calyx ± 1.1 cm. long; nutlet-wings
    shorter and eventually ± curving outwards . . . .   9. *C. moldenkei*

1. **C. marrubiifolium** *Walp.*, Repert. 4: 38 (1845); Moldenke in F.R. 45: 132 (1938);
Gillett in K.B. 10: 133 (1955); Hepper, F.W.T.A., ed. 2, 2: 437 (1963); Jaffri & Ghafoor in Fl.
W. Pakistan 77: 12, fig. 3/A–C (1974); Collenette, Fl. Saudi Arabia: 444 (1985). Type:
Sudan, Kordofan, Abu Gerad, *Kotschy* 32 (ubi? holo., B†, BM, BR, G, K!, L, LD, M, NY, P, S,
US, W, iso.)

Much-branched perennial herb or subshrub 10–40(–90) cm. tall, woody at the base
with ± stout corky stems; young stems slender, densely white pubescent. Leaves opposite
or nearly so, sometimes appearing fasciculate due to abbreviated axillary branchlets;
blades ovate, broadly elliptic or ± round, 0.5–4.5 cm. long, 0.5–3 cm. wide, rounded or
obtuse at the apex, rounded, cuneate or subtruncate at the base, conspicuously coarsely
crenate-serrate, densely adpressed whitish-pubescent on both surfaces, rather greyish;
nervation impressed above and raised and reticulate beneath; petioles 0.1–2.5 cm. long.
Spikes terminal, solitary, numerous, 3–3.5 cm. long, dense, many-flowered but more
scattered towards base; flowers closely adpressed to the rhachis, peduncles 0.5–1.7 cm.
long; bracts lanceolate, (3–)4–5 mm. long. Calyx narrowly tubular, 5–10 mm. long,
5-plicate and ribbed, ± densely pubescent with short ± tangled white hairs, minutely
5-toothed. Corolla white or yellowish; tube very slender, 1–1.3 cm. long, straight or slightly
curved, glabrous; lobes narrowly elliptic-oblong, subequal, 2.1–2.2 mm. long, 0.8 mm.
wide, obtuse. Stamens included. Ovary oblong, 1 mm. long; style filiform, ± 9 mm. long.
Nutlets black, linear-oblong, 3–4 mm. long, 0.8 mm. wide, ribbed on the back and rather
obscurely reticulate at the apex, glabrous.

KENYA. Northern Frontier Province: SW. Lake Turkana [Rudolf], near Mugurr, 6 June 1970, *Mathew*
    6652! & near Lake Turkana [Rudolf], Laridabach [Lardabash], 2 Sept. 1944, *J. Adamson* 43!;
    Turkana District: 8 km. S. of Lodwar, 3 July 1954, *Hemming* 281!
DISTR. **K** 1, 2; Mauritania, Guinée, Mali, Sudan, Arabia and Pakistan (Sind)
HAB. *Aristida* grassland with scattered *Acacia*, *Commiphora* and *Cordia*, watercourses with *Boswellia*,
    Commiphora and *Lannea* on laterite, lava or sandy soils, also crevices in basement rocks; 200–
    1110 m.

SYN. *Bouchea marrubiifolia* (Walp.) Schauer in DC., Prodr. 11: 558 (1847); Wight, Ic. Pl. Ind. Or., t. 1461
    (1849); Clarke in Hook.f., Fl. Brit. Ind. 4: 564 (1885); Bak. in F.T.A. 5: 282 (1900); Schwartz in
    Mitt. Inst. Bot. Hamb. 10: 216 (1939)
    Plexipus marrubiifolius (Walp.) R. Fernandes in Bol. Soc. Brot., sér. 2, 57: 272 (1984)

var. **A**

Spikes stouter, up to 1 cm. wide at flowering part; bracts forming cone at apex. Calyx with dense
spreading rather longer white hairs.

KENYA. Northern Frontier Province: 12 km. from Baragoi on road to South Horr, 3 June 1979, *Gilbert*
    et al. 5490!
DISTR. **K** 1; not known elsewhere
HAB. *Acacia tortilis* – *A. nilotica* woodland with *Duosperma* on granite hill; 1260 m.
NOTE. This is a rather distinctive variant but more is needed to assess its status.

2. C sp. A

Subshrubby herb to 15 cm., woody and corky at the base; young shoots densely shortly white-pubescent. Leaves ± obovate in outline, 0.6–2 cm. long, 0.4–1 cm. wide, ± rounded in outline at apex but with central acute tooth, attenuate-cuneate at the base into a 1–2 mm. long petiole, densely covered with short white ± curled pubescence; venation impressed above, raised and reticulate beneath. Spikes 3.5–9 cm. long; bracts lanceolate-acuminate, 4.5 mm. long, 1 mm. wide. Calyx 7 mm. long, 5-ribbed, shortly toothed. Corolla white; tube ± 1.2 cm. long, widened at the throat; lobes obovate, ± 4 mm. long, 2 mm. wide. Anthers just exserted. Style 1.2–1.5 cm. long. Nutlets straw-coloured but probably black at maturity, linear-oblong, ± 4 mm. long, 1 mm. wide, strongly reticulate-ribbed outside at apex, long-ribbed beneath.

KENYA. Masai District: Athi Plains, km. 140.8, Namanga–Nairobi road, 26 June 1961, *Lind* 3100!
DISTR. **K** 6; known only from the above collection
HAB. Grassland; 1500 m.

NOTE. This is similar to *C. obovatum* and *C. gillettii* in all but the reticulately veined leaves in which it approaches *C. marrubiifolium* but differs in leaf-shape. It is strange that more material is not available from so well known a locality close to the capital.

3. **C. obovatum** *Sebsebe* in Nordic Journ. Bot. 8: 629, fig. 2 (1988) & in K.B. 45: 139 (1990). Type: Ethiopia, Bale Region, 66 km. from Ginir to Imi, *Puff et al.* 870513-4/2 (ETH, holo.)

A perennial woody herb up to 50 cm. tall; young stems angled and green, pubescent with short hairs; older stems terete, brown or grey. Leaf-blades obovate to obovate-oblong, 0.8–2 cm. long, 0.5–1.5 cm. wide, truncate in outline at the apex, cuneate at the base, coarsely crenate-serrate at the margin with 5–9 spreading teeth, puberulous to pubescent with short hairs on both surfaces; lateral nerves 2–3, visible above and beneath; petiole 2–6 mm. long, pubescent. Flowers spreading, sessile in dense terminal spikes 2–13 cm. long; bracts 1.5–3 mm. long, acuminate. Calyx tubular, 6–8 mm. long, 5-ribbed, hyaline between the ribs, shortly toothed, pubescent. Corolla cream and yellow; tube 1.4–1.6 cm. long; limb 5–6 mm. wide, the lobes somewhat unequal, 1.5–3 mm. long, 1–2 mm. wide. Nutlets black, 3.5–4.5 mm. long, reticulate.

subsp. **obovatum**; Sebsebe in K.B. 45: 139 (1990)

Leaves concolorous, green but not glaucous; petioles 2–4 mm. long. Corolla-lobes 1.5–3 × 1–2 mm. Filaments under 1 mm. long.

KENYA. Northern Frontier Province: 20 km. SSW. of El Wak, 29 May 1952, *Gillett* 13383! & Dandu, 14 May 1952, *Gillett* 13196! & Mandera, War Gedud, 1 May 1978, *Gilbert & Thulin* 1306!
DISTR. **K** 1; Ethiopia, ? Somalia
HAB. *Acacia, Commiphora* open scrub and rich bushland, cracks in granite rocks; 400–1020 m.

SYN. [*C. gillettii* sensu Gillett in K.B. 10: 133 (1955); E.P.A.: 792 (1962), pro parte, *non* Mold. sensu stricto]

NOTE. *C. gillettii*, with which this species has been confused, has the stems and leaves covered with mixed long and short hairs, elliptic-oblong leaves, acute at the apex, bracts 3–4 mm. long, corolla-tube yellow but the lobes yellowish brown to purplish red. *C. obovatum* subsp. *glaucum* Sebsebe occurs in the South Yemen and Somalia.

4. **C. gillettii** *Mold.* in Rev. Sudam. Bot. 6: 18 (1939); Gillett in K.B. 10: 133 (1955); E.P.A.: 792 (1962). Type: Somalia N., near Buramo, Debrawen, *Gillett* 4934 (S, holo., EA, K, iso.!)

Perennial herb 13–50 cm. tall; stems branched and woody at the base, densely pubescent with short greyish spreading hairs and usually some longer ones also. Leaf-blades ± grey-green, elliptic, oblong-elliptic or obovate-oblong, 1.5–2.7 cm. long, 0.8–1.3 cm. wide, acute at the apex, attenuate at the base, coarsely dentate or subentire, sparsely to densely pubescent with short hairs; nerves ± raised beneath but venation not or only slightly reticulate; petioles 1–4 mm. long or obsolete, pubescent as the stems. Flowers in terminal ± dense many-flowered spikes 4.5–12 cm. long; rhachis densely pubescent; bracts narrowly lanceolate, 3–4 mm. long, 0.5–1 mm. wide, densely shortly pubescent. Calyx tubular, 6.5–7 mm. long, 5-ribbed, densely shortly pubescent, shortly 5-toothed, splitting in fruit. Corolla deep yellow, yellow brown, apricot or purplish, hypocrateriform; tube cylindric, curved, 1.4–1.5 cm. long, slightly widened at the apex, glabrous; limb 5-lobed; lobes oblong, 2–3.4 mm. long, 1.2–2 mm. wide, obtuse. Stamens included or anther-tips slightly exserted. Ovary 1.5 mm. long; style filiform, ± 1.4 cm. long. Nutlets straw-coloured or black, 4 mm. long, 1 mm. wide, longitudinally ribbed at base outside, reticulate at the apex, obtuse.

KENYA. Northern Frontier Province: Mandera, 5–10 km. SSE. of Ramu, 2 May 1978, *Gilbert & Thulin* 1339!
DISTR. **K** 1; Somalia (N.), Ethiopia
HAB. *Acacia, Commiphora* bushland with grasses; 400–500 m.

SYN. [*C. africanum* sensu Mold. in F.R. 45: 136 (1938), quoad *Thomson* 41 & 46, & Hutch. & Bruce in K.B. 1941: 176 (1941), quoad *Gillett* 4934, *non* Mold. sensu stricto]
[*C. adenostachyum* sensu Mold. in Rev. Sudam. Bot. 6: 16 (1939), quoad *Lort Phillips* s.n., *non* Mold. sensu stricto]

NOTE. The sheet cited above differs from typical *C. gillettii* in having the leaves almost entire but similar specimens occur in Somalia. I had at first considered they might deserve varietal rank but further material is needed.

5. **C. hildebrandtii** *(Vatke) Gillett* in K.B. 10: 134 (1955); U.K.W.F.: 614 (1974); Blundell, Wild Fl. E. Afr.: 397, t. 130 (1987); Sebsebe & Verdc. in K.B. 43: 520, fig. 1/B (1988). Type: Kenya, Kitui District, Ukambani, *Hildebrandt* 2737 (B, holo.†)

Perennial herb or subshrub, ± much branched and often spreading, ± woody at base, 20–40 cm. tall but often appearing to be a less branched annual, perhaps flowering in the first year, and frequently in such plants the lower stems are very distinctly corky; stems usually densely white-pubescent and with longer shaggy white hairs as well. Leaf-blades oblong-elliptic, narrowly oblong or obovate, 0.8–6.5 cm. long, 0.3–3.5 cm. wide, obtuse or rounded in outline at the apex, cuneate-attenuate at the base, ± uniformly toothed save at base, sparsely to densely pubescent on both sides with ± curled white hairs or sometimes glabrescent; main nerves prominent beneath but intermediate areas plane, the venation not reticulate; petiole 0.5–3 cm. long, usually shaggy white-pubescent with long and short hairs. Flowers in dense terminal spikes 7–33(–100) cm. long, hairy like the stems when young with bracts comose at the apex; bracts lanceolate, ± 4–5 mm. long, acuminate, pilose with scattered white hairs. Calyx narrowly tubular, 0.8–1.3 cm. long, conspicuously 5-plicate and ribbed, with sparse to dense short and longer white hairs, minutely 5-toothed, eventually splitting. Corolla white, sometimes at least with a yellow throat, ± hypocrateriform, very variable in size; tube slender, distinctly curved, 1.3–2.9 cm. long, distinctly widened at the throat; lobes oblong or ± round, 2.2–6 mm. long, 1.5–5.8 mm. wide, broadly rounded at the apex. Stamens included, the anthers bright yellow. Ovary oblong, ± 1 mm. long; style filiform, 1–2.6 cm. long, often ± persistent. Nutlets straw-coloured but eventually usually black, linear-oblong, 5–6.5 mm. long, ± 1 mm. wide, ribbed outside and quite strongly reticulate near apex which is a little variable, truncate, slightly produced or pointed but not distinctly winged. Fig. 2.

UGANDA. Karamoja District: Karita [Karitha], 22 May 1940, *A.S. Thomas* 3509! & Amudat, 11 June 1959, *Symes* 538! & Moroto, 5 Oct. 1952, *Verdcourt* 759!
KENYA. Northern Frontier Province: Mathews Range, Ngeng, 10 Dec 1958, *Newbould* 3164!; W. Suk District: below Morobus [Moribus] Pass, May 1932, *Napier* in *C.M.* 2069!; Kitui District: km. 174.4 on Nairobi–Garissa road, near Mwingi, 25 Jan. 1958, *Verdcourt* 2088!
TANZANIA. Masai District: foot of Longido Mt., on track to Moshi, 30 Dec. 1968, *Richards* 23580!; Pare District: Kisangiro [Kisangerio], 15 May 1955, *Greenway* 9537!; Morogoro District: N. Uluguru Reserve, above Morningside, June 1953, *Semsei* 1250!
DISTR. **U** 1; **K** 1, 2, 4, 6, 7; **T** 2, 2/5, 3, 6; SW. Ethiopia and S. Somalia*
HAB. Usually *Acacia* grassland or *Acacia, Commiphora*, etc., sparse to dense mixed bushland, *Brachystegia* and *Hyphaene, Acacia* woodland, rarely (**T** 6) "in forest"; in volcanic areas, often in rocky places, also black-cotton soils; 15–1650 m.

SYN. *Stachytarpheta hildebrandtii* Vatke in Linnaea 43: 529 (1882); Gürke in P.O.A. C: 338 (1895); Bak. in F.T.A. 5: 284 (1900); W.F.K.: 107 (1948)
Chascanum africanum Mold. in F.R. 45: 136 (1938). Type: Kenya, Machakos District, Kiteta [Kateta], *Gardner* in *F.D.* 3372 (K, holo.!)
Plexipus hildebrandtii (Vatke) R. Fernandes in Bol. Soc. Brot., sér. 2, 57: 270 (1984)

NOTE. Some small-leaved and small-flowered specimens, e.g. *Gilbert & Thulin* 1075 (Kenya, Northern Frontier Province, Isiolo–Wajir road, 85 km. from the turning near Isiolo), seem to be correctly placed in this species but there is no correlation between geography and variation.

---

* Gillett demonstrates that the records from N. Somalia refer to other species but Chiovenda (Fl. Somala 2: 359 (1932)) records it from S. Somalia, Kismayu (*Senni* 519) (E.P.A.: 792 (1962)) and it is now known to occur there.

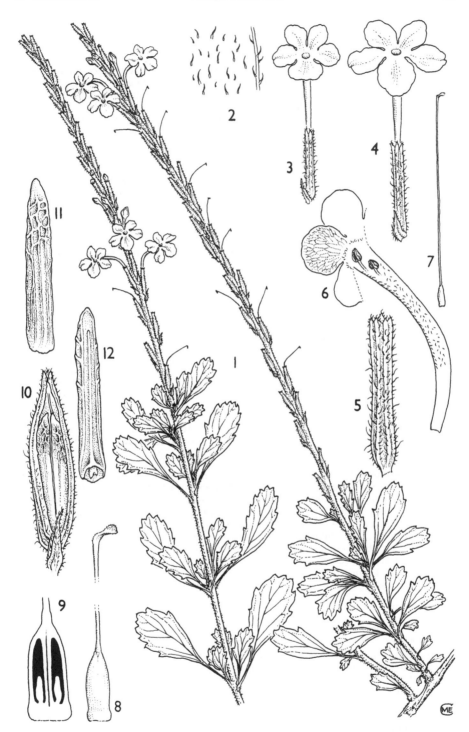

Fig. 2. *CHASCANUM HILDEBRANDTII* — **1**, habit, × ⅖; **2**, part of upper leaf surface, × 10; **3**, **4**, flowers, × 2; **5**, calyx, × 4; **6**, longitudinal section of corolla, × 3; **7**, ovary and style, × 3; **8**, ovary and stigma, × 10; **9**, longitudinal section of ovary, × 14; **10**, dehisced calyx, showing 2 seeds, × 4; **11**, **12**, 2 views of seed, × 6. 1, 2, 4, from *Symes* 538; 3, from *Polhill & Paulo* 597; 5–9, from *Verdcourt* 759; 10–12, from *Leippert* 5059. Drawn by Mrs. M.E. Church.

6. **C. hanningtonii** (*Oliv.*) *Mold.* in Torreya 34: 9 (1934) & in F.R. 45: 155 (1938); Gillett in K.B. 10: 135 (1955). Type: Tanzania, area between 6°10' and 6°20'S. and 36°25' and 37°35'E., *Hannington* (K, holo.!, NY, fragment and photo., Z, photo.)

Probably a subshrubby herb but lower parts not preserved; stems becoming woody, with spreading hairs and shorter ± adpressed pubescence. Leaves narrowly elliptic-oblong, 1.5–3.5 cm. long, 3–8 mm. wide, acute at the apex, long-attenuate at the base into 1–5 mm. long petiole (fide Moldenke) or ± sessile, sparsely toothed along the upper half with rather small distant teeth, adpressed pubescent, particularly on the nerves. Spike 15–18 cm. long, many-flowered, dense save at base; peduncle 1.5–2.5 cm. long; pedicels obsolete; bracts ovate-lanceolate, 6 mm. long, 2 mm. wide, long-acuminate. Calyx narrowly tubular, ± 1.5 cm. long, 5-ribbed and plicate, with scattered but conspicuous long white setae 1–2 mm. long; teeth unequal, 0.2–1.7 mm. long. Corolla probably pale but colour unknown, hypocrateriform, the slender tube 3–4 cm. long, widening at the apex, glabrous; lobes unequal, oblong-elliptic or obovate, 4–4.5 mm. long, 2.5–3 mm. wide, obtuse. Ovary cylindric, ± 1.5 mm. long; style filiform, ± 3.5 cm. long. Fruit oblong, 4.5–5 mm. long, 1–1.5 mm. wide, the two cocci not separating, reticulate on the upper ¼, ribbed beneath, rounded at apex and in no way winged, glabrous.

TANZANIA. Mpwapwa/Kilosa/Morogoro Districts: on the route Kwa Chiropa, Mamboya [Mamboia], Mpwapwa and Chisokwe [Kisokwe], *Hannington*!
DISTR. **T** 5 or 6; not known elsewhere
HAB. Not known

SYN. *Bouchea hanningtonii* Oliv. in Hook., Ic. Pl. 15: 37, t. 1446 (1883); Bak. in F.T.A. 5: 283 (1900)
Plexipus hanningtonii (Oliv.) R. Fernandes in Bol. Soc. Brot., sér. 2, 57: 269 (1984)

NOTE. It is extraordinary that no further material of this distinct species has been collected in an area even if not very well known certainly by no means neglected. The possibility that some plant from another part of the world had become mixed with the collection is discounted since nothing remotely like it has been found. The nearest African material I have seen is a specimen from S. Malawi named *Plexipus angolensis* (Mold.) R. Fernandes subsp. *zambesiacus* R. Fernandes. It is undoubtedly close to *C. hildebrandtii* which sometimes has flowers almost as large and similar comose bracts at the top of young spikes which are, however, narrower, but the ± sessile narrower less toothed leaves of *C. hanningtonii* are different.

7. **C. rariflorum** (*A. Terracc.*) *Mold.* in Phytologia 1: 167 (1935) & in F.R. 45: 129 (1938); Gillett in K.B. 10: 135 (1955); E.P.A.: 793 (1962). Type: Ethiopia, Rer Amaden, *Baudi di Vesme & Candeo* (FT, holo.)

Subshrubby herb 25–60 cm. tall; stems erect, glabrous or with virtually invisible scaly emergences, woody at the base and with nodules resulting from leafless reduced shoots. Leaves linear, 1.5–5 cm. long, 1.5–3 mm. wide, opposite or on lower stems, fasciculate on very short shoots, acute at the apex, cuneate at the base, sessile, almost imperceptibly puberulous. Spikes 10–14 cm. long, fairly densely flowered above, laxer below, slender; bracts lanceolate, 2.5–3 mm. long, 1 mm. wide, obscurely ciliolate; pedicels obsolete. Calyx tubular, 5–6 mm. long, glabrous, scarcely oblique, with very short teeth. Corolla cream or yellow; tube curved, slender, 1–1.2 cm. long, glabrous; lobes ovate or rounded, ± 1.5 mm. long, ± 1 mm. wide, obtuse. Style ± 3 mm. long. Fruit black, of 2 linear nutlets, ± 6 mm. long, 0.7 mm. wide, rounded at the apex, reticulately nerved.

KENYA. Northern Frontier Province: 11 km. Mandera–Ramu road, 3 May 1978, *Gilbert & Thulin* 1379!
DISTR. **K** 1; Ethiopia, S. Somalia
HAB. Open *Commiphora, Boswellia, Acacia* bushland on low rocky hills; 300 m.

SYN. *Hebenstreitia rariflora* A. Terracc. in Bull. Soc. Bot. Ital. 1892: 424 (1892)
Bouchea rariflora (A. Terracc.) Chiov. in Ann. Bot. Roma 9: 127 (1911)
Plexipus rariflorus (A. Terracc.) R. Fernandes in Bol. Soc. Brot., sér. 2, 57: 273 (1984)

NOTE. This very distinctive species is virtually unknown. As Moldenke surmised freshly collected material shows it to have 4 stamens.

8. **C. laetum** *Walp.*, Repert. 4: 39 (1845); Sebsebe & Verdc. in K.B. 43: 521, fig. 1/C (1988). Types: Sudan, Kordofan, between Mt. Kohn and Tekele, *Kotschy* 230 (ubi?, syn., BM!, BR, G, K!, L, M, NY, O, P, S, US, W, isosyn.) & Ethiopia, Tigray, Adeganna, *Schimper* 1012* (ubi?, syn., BM!, BR, G, K!, L, M, NY, O, P, S, UPS, W, isosyn.)

* Incorrectly cited as *Kotschy* 1012 by Walpers.

Herb or erect shrublet 0.35–1.2 m. tall, base quite woody with grey bark. Leaf-blades ± ovate, 1–4.7 cm. long, 0.7–2.5 cm. wide, rounded at apex, rounded, truncate or cuneate at the base, coarsely crenate, very finely pubescent or glabrous save on the nerves; petiole 0.3–2 cm. long, densely short spreading pubescent. Spikes dense, 6–40 cm. long; bracts narrowly lanceolate, 2.5–4 mm. long. Calyx 9(–10.5) mm. long, densely very shortly spreading velvety puberulous; teeth 0.1–0.25 mm. long. Buds often tinged pink or purple. Corolla greenish yellow, yellow or yellowish cream; tube 1.2–1.5 cm. long; lobes 1.5–2.5 mm. long and wide, obtuse. Style 1.1–1.3 cm. long, ± persistent. Nutlets linear-oblong, 4.5–6 mm. long, rounded and ribbed outside, with apical wing 1.8–2.5 mm. long, ± reticulate outside near the wing, excavated and minutely verruculose inside.

KENYA. Northern Frontier Province: 25 km. NE. of Mado Gashi [Modo Gash] on road from Habaswein [Habas Wen], 15 May 1974, *Gillett & Gachathi* 20692!; Masai District: Ol Orgasailie, 10 June 1956, *Bally* 10568!; Teita District: near Voi on road to Mackinnon Road, 21 Dec. 1953, *Verdcourt* 1109!
TANZANIA. Moshi District: Kilimanjaro, Himo steppe, 22 Dec. 1932, *Geilinger* 4894!; Pare District: Ngulu, May 1928, *Haarer* 1327!; ?Lushoto District: steppe by Pangani R., July 1893, *Volkens* 450!
DISTR. K 1, 4, 6, 7; T 2, 3; Sudan, Ethiopia, Somalia (N.), also Arabia (Yemen)
HAB. Dry grassland with scattered *Acacia*, *Commiphora*, etc., also *Acacia*, *Commiphora* and mixed bushland on lava and basalt soils and black cotton soil boundaries; 60–1380 m.

SYN. *Bouchea pterygocarpa* Schauer in DC., Prodr. 11: 558 (1847); A.Rich.. Tent. Fl. Abyss. 2: 166 (1850); Engl., P.O.A. A: 57 (1895); Gürke in P.O.A. C: 338 (1895); Bak. in F.T.A. 5: 282 (1900); Grenzebach in Ann. Missouri Bot. Gard. 13: 78 (1926); Schwartz in Mitt. Inst. Bot. Hamb. 10: 216 (1939), *nom. illegit.* Types as for *C. laetum* but syntypes at G, all rest isosyntypes
Chascanum arabicum Mold. in F.R. 45: 138 (1938); Gillett in K.B. 10: 133 (1955). Type: Yemen, Hodjeilah, *Deflers* 72 (P, holo.!)*
Svensonia laeta (Walp.) Mold. in F.R. 46: 5 (1939); Gillett in K.B. 10: 132 (1955); F.P.S. 3: 199 (1956); E.P.A.: 793 (1962)
NOTE. A specimen in advanced fruiting stage, *Ohta* 192 (Turkana District, July 1982) with very small ± round leaves about 6–8 mm. long and wide and petioles short, 2–3 mm. long, is probably a state of this species and if so supplies a K 2 record.

9. **C. moldenkei** (*Gillett*) *Sebsebe & Verdc.* in K.B. 43: 521 (1988). Type: Kenya, 53 km. SW. of Mandera on road to El Wak, *Gillett* 13397 (K, holo.!, EA, iso.!)

Subshrub to ± 2 m. with densely adpressed puberulous stems, ± woody at the base. Leaves opposite or subalternate; blades ovate-oblong, up to 3 cm. long, 2 cm. wide, coarsely serrate-dentate, subacute at apex, narrowed into the petiole from a ± subtruncate base, shortly puberulous on both sides; nerves prominent beneath; petiole 6–8 mm. long. Inflorescences terminal, spicate, up to 35 cm. long, densely puberulous; bracts narrowly deltoid-subulate, ± 5 mm. long. Calyx ± 1.1 mm. long, densely puberulous with short spreading hairs, splitting in fruit; teeth unequal, 0.2–0.8 mm. long. Corolla white; tube cylindrical, ± 3.5 cm. long, 1.5 mm. diameter, pubescent inside; lobes subequal, oblong, 5 mm. long, 2.2–2.8 mm. wide, rounded at apex. Style 3.7 cm. long; stigma ovoid, 0.5 mm. long, papillose, oblique. Nutlets narrowly linear-subcylindric, 7.5 mm. long including the short 1.3 mm. long wing, finely ribbed outside and ± reticulate near the wing, minutely papillate inside.

KENYA. Northern Frontier Province: 53 km. SW. of Mandera on road to El Wak, 30 May 1952, *Gillett* 13397!
DISTR. K 1; Somalia
HAB. *Commiphora*, *Acacia* open scrub on red sandy soil; 600 m.

SYN. *Svensonia moldenkei* Gillett in K.B. 10: 132 (1955), pro parte; E.P.A.: 794 (1962), pro parte
NOTE. Material from the Ogaden with very different nutlets, some of which was cited by Gillett, is a distinct species.

3. STACHYTARPHETA

Vahl, Enum. Pl. 1: 205 (1804); Moldenke in Rev. Fl. Ceylon 4: 246–267 (1983); R. Fernandes in Bol. Soc. Brot., sér. 2, 57: 87–111 (1984), *nom. conserv.*

Annual or perennial herbs or shrubs, glabrous to densely hairy. Leaves opposite or

* Moldenke also cites *Deflers* 1038 (S. Yemen, Bilad Fodhli, Mt. Areys) from which some of the description is derived and ironically this is a new species (see Sebsebe in K.B. 45: 137 (1990)).

alternate, entire or toothed. Flowers sessile, crowded in elongate terminal spikes, the rhachis hollowed beneath each flower; bracts present. Calyx persistent, closely adpressed, not enlarging in fruit; tube narrow, 4–5-lobed or -toothed or with 2 2-fid lobes. Corolla coloured, hypocrateriform; tube cylindrical; limb equally or usually subequally 5-lobed, ± irregular. Stamens 2, anterior plus 2 staminodes inserted in the upper part of the tube; fertile stamens with divergent anther-thecae, ± included. Ovary 2-locular with 1 erect anatropous ovule in each locule; style filiform with capitate or subcapitate stigma. Fruit included in the calyx, splitting into 2 1-seeded linear-oblong mericarps. Seeds linear, exalbuminous.

About 100 species, nearly all American, a few naturalised or native in the tropics of the Old World.
Four species are naturalised or wild in East Africa and a fifth, *S. mutabilis* (Jacq.) Vahl, is cultivated and easily known from the others by its much larger flowers (Uganda, Bunyoro District, Budongo Forest, Busingiro Rest House, edge of colonizing forest beside road, 21 May 1971, *Synnott* 582!; Kenya, Nairobi Arboretum, 4 Feb. 1952, *G.R. Williams Sangai* 326!; Tanzania, Amani, 4 July 1950, *Verdcourt* 281!); it is a shrub 2–2.5 m. tall, with densely pubescent crenate-dentate leaves to ± 14 × 10 cm. and long spikes to 30(–50) cm. long and ± 6 mm. wide (excluding corollas); corolla rose-coloured, with tube tinged blue and throat often violet; stigma blue. Species in this genus are reported to hybridise freely but only one possible hybrid specimen has been seen (see under *S. urticifolia*).

1. Corolla-limb 1–2 cm. wide; inflorescence-rhachis coarse,
      over 5 mm. wide when dry (up to 1 cm. wide after
      flowering, *fide* Moldenke) (cultivated) . . . . .   *S. mutabilis*
   Corolla-limb usually under 1 cm. wide; inflorescence-
      rhachis more slender, ± 3 mm. wide (up to 7 mm. wide
      after flowering *fide* Moldenke) 2
2. Leaves narrowly oblong or oblanceolate with large rather
      distant teeth; calyx seen from outside appearing bifid
      at the apex, the 2 central teeth reduced to small
      inconspicuous denticles on the side of the much
      larger laterals . . . .   4. *S. indica*
   Leaves more ovate or broadly elliptic with closer more
      regular teeth; calyx seen from outside with 4 ±
      prominent teeth 3
3. Inflorescence usually with some pubescence in African
      specimens; corolla-tube 4–5 mm. long; calyx-teeth 4, ±
      equal; usually shrubby . . . .   1. *S. cayennensis*
   Inflorescence glabrous; corolla-tube ±± 8–11 mm. long 4
4. Leaves essentially obtuse in outline, the serrations short,
      bluntish, with apices excentric . . . .   2. *S. jamaicensis*
   Leaves sharply acute, the teeth very distinctly spreading,
      the acute apices central . . . .   3. *S. urticifolia*

1. **S. cayennensis** (*Rich.*) *Vahl*, Enum. Pl. 1: 208 (1804), as "*cajanensis*"; Schauer in DC., Prodr. 11: 562 (1847), as *Stachytarpha*, & in Mart., Fl. Bras. 9: 199 (1851), as *Stachytarpha*; Danser in Ann. Jard. Bot. Buitenz. 40: 2 (1929); Moldenke in Fl. Suriname 4(2): 274 (1940); Brenan in K.B. 5: 223 (1950); Hepper, F.W.T.A., ed. 2, 2: 434, fig. 305/G–L (1963); Adams, Fl. Pl. Jamaica: 632 (1972); Jafri & Ghafoor in Fl. W. Pakistan 77: 14, fig. 4/F (1974). Type: French Guiana, *Le Blond* (G, holo.)

Shrubby, 0.9–2.5 cm. tall; stems woody at the base, glabrous or shortly pubescent and with a few longer hairs. Leaves ovate or elliptic, 1.8–8 cm. long, 0.5–4 cm. wide, shortly acute at the apex, attenuate at the base into a 1–1.5 cm. long petiole, closely shortly serrate, the teeth ± rounded with apex excentric, slightly scabrid with short pubescence. Spikes slender, up to 20–25(–34) cm. long, mostly sparsely but distinctly pubescent or almost glabrous; bracts linear to triangular-subulate, 4–5 mm. long, acuminate. Calyx 4–5 mm. long, with 4 ± equal teeth. Corolla white or mostly pale blue with white centre; tube 4–5 mm. long, scarcely exceeding the calyx; limb ± 5 mm. across, the lobes 1.5–2 mm. long; style included. Mericarps blackish, oblong, 2.5–3(–5) mm. long, 1.2 mm. wide, reticulately ridged.

UGANDA. Mengo District: W. Mengo, 0.5 km. E. of Port Bell pier, 16 Jan. 1969, *Lye* 1186! & same locality, 25 Oct. 1968, *Lye* 49! & near same locality, *Rwaburindore* 222!
DISTR. U 4; W. Africa from Sierra Leone to Cameroon, Mozambique, Zimbabwe, widespread in tropical America from Mexico to W. Indies and S. America to Peru and Argentina, now widely naturalized throughout the tropics; not previously reported from E. Africa

HAB. Grassland; 1140–1200 m.

SYN. *Verbena cayennensis* Rich. in Act. Soc. Hist. Nat. Paris 1: 105 (1792)
[*Stachytarpheta indica* sensu Bak. in F.T.A. 5: 284 (1900), pro parte, *non* (L.) Vahl]
[*S. jamaicensis* sensu F.W.T.A. 2: 277 (1931), pro parte, *non* (L.) Vahl]

NOTE. Brenan (K.B. 5: 225 (1950)) claimed that Vahl's spelling of the specific epithet was so different from the original that it should be taken as a new name but it is clearly only an orthographic variant.

2. **S. jamaicensis** (*L.*) *Vahl*, Enum. Pl. 1: 206 (1804); Sims in Bot. Mag. 44, t. 1860 (1817); Schauer in DC., Prodr. 11: 564 (1847), pro parte; Briquet in E. & P. Pf. 4, 3a: 154 (1895), pro parte; F.W.T.A. 2: 277 (1931), pro parte; Moldenke in Lilloa 4: 300 (1939) & in Fl. Madag., Fam. 174: 22, fig. 3/1–2 (1956); Adams, Fl. Pl. Jamaica: 632 (1972); Fosberg & Renvoize, Fl. Aldabra: 225, fig. 35/7 (1980); Moldenke in Rev. Fl. Ceylon 4: 253 (1983) (extensive information and synonymy); R. Fernandes in Bol. Soc. Brot., sér. 2, 57: 96, 100 (1984). Type: specimen in *Linnean Herbarium* (S, lecto., microfiche 7.13)

Perennial well-branched herb 0.6–1.2 m. tall, woody at base; stems glabrous or very slightly pubescent, rarely more distinctly hairy on the young shoots, often purplish. Leaves ovate, elliptic or ± oblong, 1.5–11 cm. long, 0.8–5 cm. wide, ± obtuse in outline at apex, long-attenuate into ± 1 cm. long petiole at base. Spikes long and narrow, 14–45(–50) cm. long, glabrous; bracts lanceolate or oblong-lanceolate, 5–8 mm. long, 2 mm. wide, acuminate. Calyx tubular, 5–6.5 mm. long, bifid at apex, 4-toothed, the 2 central teeth rather shorter. Corolla pale blue to deep or royal blue or purple, sometimes with a white centre; tube 0.8–1.1 cm. long, slightly curved; limb ± 8 mm. diameter, the lobes ± 3 mm. long. Stigma included. Mericarps dark, linear-oblong, 4(–?7) mm. long, 1 mm. wide, dorsally ribbed.

TANZANIA. Uzaramo District: Dar es Salaam University Campus, 8 July 1973, *Wingfield* 2189/A!; Zanzibar I., Mazizini [Massazine], 2 Aug. 1959, *Faulkner* 2321, in part! & Pemba I., Banani, 23 Dec. 1930, *Greenway* 2776!
DISTR. T 6; Z; P; S. U.S.A. to Ecuador and Brazil and W. Indies, introduced to and naturalized in Africa, Madagascar, Indian Ocean Is., tropical Asia, Australasia and Oceania
HAB. Seasonally wet grassland, near villages and reported as a very common weed in Zanzibar, occurring in pure stands over large areas; ± 0–90 m.

SYN. *Verbena jamaicensis* L., Sp. Pl.: 19 (1753) & ed. 2: 27 (1762); Jacq., Obs. Bot. 4: 6, t. 85 (1771)
[*Stachytarpheta indica* sensu Schauer in DC., Prodr. 11: 564 (1847), pro parte; Bak. in F.T.A. 5: 284 (1900), pro parte; Danser in Ann. Jard. Bot. Buitenz. 40: 5 (1929); U.O.P.Z.: 453 (1949); Brenan in K.B. 5: 225, fig. B (1950); Hepper, F.W.T.A., ed. 2; 2: 434, fig. 305/M–R (1963), *non* (L.) Vahl]

3. **S. urticifolia** *Sims* in Bot. Mag. 43, t. 1848 (1816); Fosberg & Renvoize, Fl. Aldabra: 225 (1980); Moldenke in Rev. Fl. Ceylon 4: 257 (1983), as "*urticaefolia*" Sims (extensive information and synonymy). Type: specimen cultivated near London collected by Isaac Swainson (BM, lecto.)

Herb or slender subshrub 0.5–2 m. tall, with ± erect branched glabrous or very sparsely pubescent stems from a long taproot, woody at the base. Leaves broadly elliptic to ovate, 4–12.5 cm. long, 2–7 cm. wide, distinctly acute at the apex, cuneate and attenuate into 0.5–2 cm. long petiole, glabrous, regularly sharply serrate with spreading teeth, the sharply acute apices central. Spikes slender, elongate, many-flowered, 14–45 cm. long, ± 2.5 mm. diameter; bracts lanceolate, 7 mm. long, 2 mm. w:de, ± glabrous, long-acuminate. Calyx cylindrical, 7 mm. long, 2 mm. wide, 5-ribbed and shortly 5-toothed. Corolla dark purple-blue, mauve or royal blue with light or white throat; tube 7–8 mm. long; limb ± 1 cm. wide, the lobes 2–3 mm. long, 3–5 mm. wide. Stigma exserted. Mericarps dark, linear-oblong, 3–4 mm. long, 1 mm. wide, reticulately ribbed.

UGANDA. Mengo District: 6 km. N. of Mukono, 16 Oct. 1974, *Katende* 2299! & Kampala, Makerere University Hill, 30 Oct. 1968, *Lye* 115!
KENYA. Kwale District: Diani Beach, 5 Apr. 1974, *Bock* in E.A.H. 15433! (possibly a hybrid with *S. indica*) & 4.8 km. S. of Diani, 10 July 1961, *Church* 73! & Shimba Hills Forestry Expt. area, 15 Apr. 1968, *Magogo & Glover* 892!
TANZANIA. Lushoto District: Amani, 19 Apr. 1968, *Renvoize & Abdallah* 1600!; Tanga District: Sawa, 3 Sept. 1964, *Faulkner* 4923!; Uzaramo District: Pugu Hills, Minaki, 12 June 1972, *Flock* 346!; Zanzibar I., Dole Ridge, 13 Apr. 1952, *R. O. Williams* 166!
DISTR. U 4; K 7; T 3, 6; Z; P; widespread in Africa from Sudan to South Africa, Madagascar, Seychelles, etc.; Moldenke considers it is probably a native of tropical Asia whence it extends into the Pacific; also in S. America, W. Indies and Florida

HAB.   Roadside weed, grassland, bushland and forest; 0–1200 m.

SYN.   *Cymburus urticaefolius* Salisb., Parad. Lond., t. 53 (1806), excl. synonymy, *nom. illegit.*
        [*Stachytarpheta indica* sensu Bak. in F.T.A. 5: 284 (1900), pro parte quoad *Hildebrandt* 997, *non* (L.)
        Vahl]
        [*S. jamaicensis* sensu Danser in Ann. Jard. Bot. Buitenz. 40: 7, t. 1/3, 4 (1929); U.O.P.Z.: 453
        (1949) and auctt. mult. *non* (L.) Vahl]

NOTE.   R.O. Williams (U.O.P.Z.: 453 (1949)) noted that two species grew in Zanzibar which he called
        S. jamaicensis and *S. indica*, the former with deep blue flowers and the latter with pale-blue flowers
        and more rounded leaf-apices; Vaughan notes the same.
            It is much used as a hedge plant and has been cultivated in Nairobi (e.g. City Park, 29 July 1953,
        Verdcourt 1002) at 1650 m. Since Sims particularly excludes the synonymy cited by Salisbury the
        name has been treated as a new one dating from Sims' publication. Moldenke continues to give
        the authority as (Salisb.) Sims.

4. **S. indica** (*L.*) *Vahl*, Enum. Pl. 1: 206 (1804); Moldenke, Rev. Fl. Ceylon 4: 265 (1983);
R. Fernandes in Bol. Soc. Brot., sér. 2, 57: 88, 95 (1984); Type: Ceylon*, plant grown at
Uppsala from seed sent by David van Royen, specimen 35/1 (LINN, lecto.)

Annual sometimes somewhat subsucculent herb 9–60(–120) cm. tall, with glabrous
unbranched or branched stems. Leaves oblanceolate to narrowly oblong in outline,
2–9(–13) cm. long, 0.3–3.5(–6) cm. wide, subacute at the apex, attenuate at the base into a 1
cm. long petiole with a few hairs at base and on junction across stems, ± remotely coarsely
serrate. Spikes 4–35 cm. long; bracts elliptic to lanceolate, 5–5.5 mm. long, 1–1.5 mm. wide,
sometimes obscurely ridged, acuminate, often ± spreading in older inflorescences. Calyx
4–4.5 mm. long, the apex appearing bifid from outside. Corolla deep to light blue, mauve
or lavender, the tube and throat often white; tube 5 mm. long; limb 7.5 mm. wide, the
lobes rounded, 2.5–3 mm. long and wide. Style 5 mm. long, just exserted; stigma green.
Fruit pale brown turning black, oblong, 3.5 mm. long, 1.5 mm. wide, reticulately ridged
towards apex, longitudinally ribbed beneath, slightly beaked at apex.

UGANDA.   W. Nile District: Metu, Oct. 1959, *E.M. Scott* in *E.A.H.* 11765! & Koboko, Mar. 1938, *Hazel*
        437! & Amua, 7 June 1936, *A.S. Thomas* 1954!
KENYA.   Northern Frontier Province/Ethiopia: Lake Turkana [Rudolph], *Wellby*!
TANZANIA.   Rufiji District: Mafia I., Kikuni cultivation area, 14 Aug. 1937, *Greenway* 5095! & Utete, by
        R. Rufiji, 2 Dec. 1955, *Milne-Redhead & Taylor* 7525! & Rufiji, 6 Dec. 1930, *Musk* 8!; Kilwa District:
        Matandu R., 6 Aug. 1975, *Vollesen* in *M.R.C.* 2637!
DISTR.   U 1; K ?1; T 6, 8; widespread in tropical Africa and in tropical America
HAB.   Bare earth, grassland, weed in heavy cultivated soils, also in rice-fields; 10–1050 m.
SYN.   *Verbena indica* L., Syst. Nat., ed. 10: 851 (1759) & Sp. Pl., ed. 2: 27 (1762); Murr., Syst. Veg., ed. 14:
        66 (1784); Willd., Sp. Pl., ed. 4, 1: 115 (1797)
        V. angustifolia Mill., Gard. Dict., ed. 8, no. 15 (1768). Type: Mexico, La Vera Cruz, *Houston* (BM,
        holo.)
        Stachytarpheta angustifolia (Mill.) Vahl, Enum. Pl. 1: 205 (1804); Schauer in DC., Prodr. 11: 563
        (1847); Briquet in E. & P. Pf. 4, 3a: 154, fig. 59/C, D (1895) as "*augustifolia*"; Bak. in F.T.A. 5: 284
        (1900); F.W.T.A. 2: 277 (1931); Moldenke in Pulle, Fl. Suriname 4(2): 275 (1940); Brenan in
        K.B. 5: 226 (1950); F.P.S. 3: 199 (1956); Hepper, F.W.T.A., ed. 2, 2: 434, fig. 305/A–F (1963);
        Adams, Fl. Pl. Jamaica: 632 (1972); Vollesen in Opera Bot. 59:83 (1980)
        S. jabassensis Winkler in E.J. 41: 284 (1908). Type: Cameroon, Yabassi, *Winkler* 927 (WRSL, holo.)

NOTE.   R. Fernandes gives additional synonymy and references. It must be emphasized that the
        name *S. indica* has been almost universally applied to the species correctly known as *S. jamaicensis*
        which name has itself been equally misapplied to the species correctly to be called *S. urticifolia*. It is
        likely that this confusion will persist for a long while.

4. PRIVA

Adans., Fam. Pl. 2: 505 (1763); Moldenke in F.R. 41: 1–76 (1936); R. Fernandes in Bot.
Helv. 95: 33–45 (1985)

Erect or ascending mostly almost subscabrid pubescent perennial herbs. Leaves
opposite or subopposite, simple, mostly ± toothed, sessile or petiolate. Inflorescences
raceme-like or subspicate, terminal or axillary; bracts small; flowers ± small, arranged in a
spirally alternate or pseudo-secund manner on the mostly elongate rhachis. Calyx
tubular, 5-ribbed, with 4–5 mostly short equal or subequal teeth, persistent, enlarging and

* Apparently an erroneous locality as it is not otherwise recorded.

covering the fruit, usually contracted and ± beaked at the mouth. Corolla white or coloured, ± irregular, funnel-shaped or hypocrateriform; tube cylindric, straight or slightly curved, widened at the throat; limb ± 2-lipped, with 5 unequal lobes. Stamens 4, didynamous, inserted near the middle of the tube, included; a minute staminode is sometimes present also. Ovary 2(-4)-locular, with each locule 1-2-ovuled; style terminal, filiform, 2-lobed at apex with a longer curved or erect stigmatiferous lobe and a very minute tooth-like lobe; ovules basal, erect, anatropous. Fruit eventually splitting into 2 1-2-locular subglobose cocci; pericarp hard, the dorsal surface echinate, scrobiculate or ridged, the inner excavated, flat or concave. Seeds without endosperm.

A genus of about 20 species in tropical and subtropical Asia, Asia Minor, Africa and America (Florida and Texas to Paraguay and W. Indies).

1. Leaves essentially sessile or petioles 2(-5) mm. long, very
      rarely longer; cocci densely echinate with short stout
      straight puberulous spines on backs and sides, save on
      the margins of the commissures where the spines are
      much reduced or reduced to ridges . . . . . .   4. *P. curtisiae*
   Leaves petiolate; cocci with spines more restricted 2
2. Cocci glabrous with spines arranged in 2 combs, one on
      either side projecting almost at right-angles and
      forming a shelf-like structure, the spines in each comb
      united for ¼-¾ their length by a membrane   . . .   1. *P. flabelliformis*
   Cocci glabrous or puberulous to pubescent with spines
      rather differently arranged and never united by a
      membrane   3
3. Cocci usually glabrous or even glossy, distinctly beaked, the
      elevated part strongly reticulate-ridged with 2 rows of
      7-11(-18) similar glabrous or ± puberulous spines;
      back reticulate-alveolate without spines; sides
      reticulate, the lower part strongly ribbed; fruiting calyx
      distinctly obcordate   . . . .   2. *P. adhaerens*
   Cocci usually densely pubescent all over, ± beaked, with
      several rows of straight or curved puberulous spines or
      sometimes reduced to tubercles; back ribbed and
      sometimes with scattered tubercles or spines; sides
      reticulate, the lower part smooth or ribbed   . . .   3. *P. tenax*

1. **P. flabelliformis** (*Mold.*) *R. Fernandes* in Bot. Helv. 95: 37 (1985). Type: Tanzania, NW. Uluguru Mts., above Morogoro, *Schlieben* 3231 (S, holo., BM, BR, iso.!, NY, fragm.)

Perennial, rather weak erect or straggling herb 0.5-1.5 m. tall; stems square, pubescent with fine uncinulate hairs or glabrescent. Leaves ovate, triangular or elliptic, 2-10 cm. long, 0.8-6 cm. wide, acute or subobtuse at the apex, rounded then cuneate at the base into a 0.8-2.5 cm. long petiole, ± thin, serrate save on basal margin, puberulous and with longer tubercle-based bristly hairs, very adherent. Spikes slender, 9-17(-28) cm. long; bracts linear-lanceolate, 2-3 mm. long; pedicels up to 1.5 mm. long in fruit. Calyx-tube 5-6 mm. long, densely uncinulate-pubescent. Corolla white or white tinged purple, often with 2 blue or purple lines on the lower lip; tube slender, 8 mm. long; lobes oblong, 3 mm. long, 1.2 mm. wide. Fruit ± round in outline, compressed, 5 mm. long, 4.5 mm. wide, beaked; cocci glabrous with 2 combs of spines, one on either side and projecting almost at right-angles to the sides and forming a shelf-like structure; the spines in each comb are united for ¼-¾ their length by a membrane; the back of the coccus between the two combs is strongly wrinkled; the sides are strongly reticulate above and the lower faces bordering the commissure are ribbed.

UGANDA. Mbale District: W. Bugwe Forest Reserve, 25 Apr. 1951, *G.H. Wood* 69!; Mengo District; Entebbe, Oct. 1922, *Maitland* 266! & Entebbe, *Linder* 2669!
TANZANIA. Lushoto District: Korogwe, Magunga Estate, 26 Nov. 1956, *Faulkner* 1269! & W. Usambaras, Baga–Bumbuli road, 2.4 km. NE. of Sakarani, 6 May 1953, *Drummond & Hemsley* 2413!; Morogoro, 7 July 1969, *Batty* 579!; Zanzibar I., Ndagaa, 27 Sept. 1963, *Faulkner* 3277!
DISTR. U 2-4; T 2, 3, 6; Z; Zaire (Ruwenzori), S. Ethiopia, Mozambique, Malawi, Zimbabwe, Botswana, South Africa (Transvaal, Natal) and Madagascar
HAB. *Pennisetum purpureum* and *Themeda* grassland, bushland, forest edges and clearings, old cultivations and also as a weed; 0-1400 m.

SYN. [*P. meyeri* sensu Mold. in F.R. 41: 19 (1936), quoad *Snowden* 344 et spec. mult., et sensu Hutch.,
    Botanist in S.Afr.: 356 (1946), *non* Jaub. & Spach]
    [*P. cordifolia* (L.f.) Druce var. *abyssinica* sensu Mold. in F.R. 41: 46 (1936), quoad *Dummer* 30 et
    spec. mult., *non* (Jaub. & Spach) Mold. sensu stricto]
    P. cordifolia (L.f.) Druce var. *flabelliformis* Mold. in F.R. 41: 47 (1936); Brenan in Mem. N.Y. Bot.
    Gard. 9: 37 (1954)
    P. cordifolia (L.f.) Druce var. *australis* Mold. in F.R. 41: 47 (1936). Type: South Africa, Natal,
    Richmond, R. Umkomaas, *Krook* in Penther 1776 (W, holo.)
NOTE. In this case I have no doubt that Mme. R. Fernandes is entirely correct in considering it a
distinct and very widely distributed species. *Peter* 12501 (Tanzania, E. Usambaras, Ngambo-
Magunga, 5 Aug. 1915) has been annotated *P. meyeri* Jaub. & Spach and the almost spineless cocci
are indeed similar but I suspect it is a specimen of *P. flabelliformis* with ± undeveloped fruits.

2. **P. adhaerens** (*Forssk.*) *Chiov.* in Bull. Soc. Bot. Ital. 1923: 115 (1923); Moldenke in
F.R. 41: 39 (1936) & in Phytologia 5: 62, 63 (1954) & 43: 330, 331 (1979) & 49: 60 (1981);
E.P.A.: 794 (1962); R. Fernandes in Bot. Helv. 95: 36 (1985), pro parte quoad typos solum;
Verdc. in K.B. 43: 671 (1988). Types: Yemen, Hadîe, *Forsskål* (C, syntypes!)

Subshrub 0.75–1.8 m. tall, branched, ?rarely an annual (in Arabia); stems woody at the
base, pubescent and with some larger hairs. Leaves ovate, 1.5–13 cm. long, 1–9 cm. wide,
acute or acuminate at the apex, subtruncate to slightly cordate at the base or subcordate
then ± cuneate or distinctly cuneate, coarsely serrate, finely pubescent with tubercle-
based bristly non-uncinulate hairs, also micropuberulous and with longer hooked
pubescence rendering leaves very adherent; petiole up to 4 cm. long. Spikes 15–50 cm.
long; bracts linear-elliptic, 2–4 mm. long; pedicels 1–2 mm. in fruit. Calyx-tube 7–8 mm.
long including ± distinct 1.5 mm. long lobes, densely covered with hooked hairs. Corolla
white, ? rarely reddish (with lilac stripes on lobes in Arabia), usually drying dark (at least in
East Africa but pale in Arabia, etc.); tube 0.9–1.1 cm. long; limb ± 6 mm. wide, with
rounded lobes ± 3 mm. long, 2.5–3 mm. wide. Fruiting calyx distinctly subobcordate, 3–4
mm. long, 5–6 mm. wide, with an additional beak 1.5–2.5 mm. long; cocci compressed,
semi-globose, distinctly beaked at one end, the elevated part strongly reticulate, ridged,
with 2 rows of quite separate glabrous or puberulous spines 0.1–2 mm. long, 4–11(–18) on
each side, the intervening back of the coccus ± broad and reticulate-alveolate, without
spines; lower part of sides with 4–5 strong ribs down to commissural margin, thus dividing
into 2 distinct zones; surface mostly glabrous or even glossy.

KENYA. Northern Frontier Province: Mathews Range, Mandasion, 7 Dec. 1960, *Kerfoot* 2556!;
    Machakos District: km. 222.5 Mombasa–Nairobi road, Kanga, 10 Jan. 1964, *Verdcourt* 3876!; Teita
    District: Voi, 6 May 1931, *Napier* 931!
TANZANIA. Mwanza District: Massanza II, Kalemera, 14 Jan. 1954, *Tanner* 1890! (calyx only 4 mm.
    long); Masai District: Mto wa Mbu, 10 May 1952, *Tanner* 838!; Mbulu District: Lake Manyara
    National Park, Chem Chem, 3 Dec. 1963, *Greenway & Kirrika* 11119! (in absence of fruit)
DISTR. **K** 1, 4, 7; **T** 1, 2; Yemen, Egypt, Ethiopia (including Eritrea) and Sudan
HAB. Grassland, *Acacia*, *Commiphora*, *Adenia* bushland, *Acacia* woodland and forest undergrowth;
    200–2250 m.
SYN. *Phryma* Forssk., Fl. Aegypt.-Arab.: 114 (1775), *nomen*. Authentic material as type above
    Ruellia adhaerens Forssk., Fl. Aegypt.-Arab.: 114 (1775)
    Verbena forskaolaei Vahl, Symb. Bot. 3: 6 (1794), as *"forskalii"*, nom. illegit. Type as above
    Tamonea arabica Mirb., Hist. Nat. Pl., ed. 2, 15: 233 (1805), *nom. illegit.* Type as above
    Priva dentata Juss. in Ann. Mus. Hist. Nat. Paris 7: 70 (1806); Pers., Syn. Pl. 2: 139 (Nov. 1806);
    A.Rich., Tent. Fl. Abyss. 2: 165 (1850), *nom. illegit.* Type as above
    P. abyssinica Jaub. & Spach, Ill. Pl. Or. 5: 57, 58, t. 453, 454 (1855); R. Fernandes in Bot. Helv. 95:
    36 (1985). Type: Ethiopia, *Schimper* [565, number not actually mentioned by authors] (P, holo.
    & iso.!)
    P. forskaolaei (Vahl) Jaub. & Spach, Ill. Pl. Or. 5: 59 (1855), as *"forskalii"*, quoad typos solum excl.
    t. 455
    [*P. leptostachya* sensu Bak. in F.T.A. 5: 285 (1900), pro parte, *non* Juss.]
    P. adhaerens (Forssk.) Chiov. var *forskaolaei* (Vahl) Chiov. in Bull. Soc. Bot. Ital. 1923: 115 (1923),
    as *"forskalii"*, nom. superfl.
    P. cordifolia (L.f.) Druce var. *abyssinica* (Jaub. & Spach) Mold. in F.R. 41: 45 (1936), pro parte, & in
    Phytologia 5: 68–70 (1954), pro parte, & 49: 62 (1981) & Fifth Summ. Verbenac.: 905 (1971),
    pro parte
    [*P. cordifolia* sensu U.K.W.F.: 613 (1974) *non* (L. f.) Druce]
NOTE. Forsskål's names had always been assumed to refer to 3, *P. tenax*, but examination of the
material shows them to refer to the taxon mostly previously known as *P. abyssinica*. *Peter* 51780
(Tanzania, Meru, Oldonyo Sambu–Engare Nanyuki, 5 Mar. 1914) has cocci similar to *P. meyeri*
Jaub. & Spach and is annotated as such by Peter but does I think belong here; the fruits are
probably not mature.

3. **P. tenax** *Verdc.* in K.B. 43: 672 (1988). Type: Ethiopia, Eritrea, vicinity of Saati, *Schweinfurth & Riva* 490 (K, holo.!)

Annual or perennial herb 10–50(–70) cm. tall, simple or branched at the base; stems uncinulate-pubescent. Leaves elliptic, elliptic-ovate or ovate-triangular, often ± narrower in form than other species, 1.5–8.5(–?10) cm. long, 1–4(–?5.5) cm. wide, obtuse to acute at the apex, cuneate or truncate then cuneate at the base, crenate-serrate, puberulous and with tubercle-based hairs above, puberulous and with additional uncinulate-pubescence beneath; petiole 0.5–3 cm. long. Spikes 12–43 cm. long, lax below; bracts filiform to linear-lanceolate, 2–4 mm. long; pedicels 1.5 mm. long. Calyx 6–8 mm. long, densely uncinulate-pubescent. Corolla white, usually drying dark in Flora area; tube ± 1 cm. long; limb ± 9 mm. wide, with lobes 2–3.5 mm. long, 2–3 mm. wide. Fruits more rounded than obcordate, compressed subglobose or ovoid, 5 mm. long, 6 mm. wide, the beak short or up to 1.5(–2) mm. long; mericarps ± 5 mm. long, 2.5 mm. wide, usually densely pubescent all over, ± beaked, with several rows of straight or curved spines to ± 1.5(–2) mm. long or sometimes only 0.1–0.5 mm. long tubercles, totalling 13–30 on each side, the back ribbed and less reticulate, sometimes with scattered tubercles or spines; sides reticulate; the lower part smooth or ribbed.

KENYA. Northern Frontier Province: Dandu, 6 May 1952, *Gillett* 13094!; Meru District: Isiolo–Wajir road, 6 km. E. of junction with Marsabit road, 7 Dec. 1977, *Stannard & Gilbert* 803! (fruit with short tubercles); Teita District: Tsavo National Park, Voi, 15 Apr. 1966, *Hucks* 793!
TANZANIA. Lushoto District: Mkomazi–Buiko km. 167, 7 June 1926, *Peter* 41007!
DISTR. **K** 1, 4, ?6, 7; **T** 3; Yemen, Somalia, Ethiopia
HAB. *Acacia* bushland, *Acacia, Commiphora* scrub, on sandy and black cotton soils; 200–1080 m.

SYN. [*P. forskaolaei* sensu Jaub. & Spach, Ill. Pl. Or. 5: 59, t. 455 (1855), as "*forskalii*", *non* Vahl]
[*P. leptostachya* sensu Bak. in F.T.A. 5: 285 (1900), pro parte, *non* Juss.]
[*P. adhaerens* sensu Mold. in F.R. 43: 39 (1936), etc., pro parte; R. Fernandes in Bot. Helv. 95: 36 (1985), pro parte et auctt. et adnot. mult., *non* (Forssk.) Chiov.]

NOTE. See note after previous species. A number of specimens which are not in fruit cannot be named with absolute certainty. Two specimens seem to represent a variety with narrowly elliptic leaves 4.5–9.5 cm. long and 1.5–4 cm. wide, narrowly cuneate into the petiole, *Stannard & Gilbert* 803 (Meru District, Isiolo–Wajir road, 6 km. E. of junction with Marsabit road, 7 Dec. 1977) and *Lindsay* 84 (Masai District, Amboseli South, Embaringoi, 21 Apr. 1978). The former was on black cotton soil with dominant *Acacia seyal* var. *fistula* at 1080 m. and the latter in open *A. drepanolobium* bushland. Despite the rather distinctive appearance other specimens are somewhat intermediate. The nutlets differ in these two, those of 803 having short tubercles rather than spines but those of 84 are more typical.

4. **P. curtisiae** *Kobuski* in Ann. Missouri Bot. Gard. 13: 7, t. 2 (1926); Moldenke in F.R. 41: 52 (1938); U.K.W.F.: 613, fig. on p. 612 (1974). Type: Kenya, Masai District, Loita plains, *Curtis* 499 (GH, holo., MO, fragm.)

Perennial branched herb with procumbent and ascending stems 20–60 cm. tall, sometimes even ± mat-forming; young stems pubescent with sparse to dense greyish or brownish ± viscid hairs. Leaves ovate-triangular or ovate-oblong to oblong, 0.5–6 cm. long, 0.4–4.5 cm. wide, mostly ± rounded at apex or shortly acute, truncate and abruptly cuneate or gradually cuneate at the base, mostly sessile or petioles ± 2(–5) mm. long, crenate-serrate with obtuse or ± acute teeth or rarely somewhat lobed at base, ± scabrid with tubercle-based hairs and ± viscid pubescent above; similar beneath with longer hairs mostly on the ± prominent nerves. Inflorescences 8–24(–35) cm. long, many-flowered but ± lax at base; bracts subulate to ± rhomboid, 1–3 mm. long, up to 1.2 mm. wide, acuminate, with white hairs; pedicels ± obsolete or up to 1.5–2 mm. long in fruit. Calyx tubular, 5–8 mm. long, ± 3 mm. wide, 5-ribbed, obscurely 5-toothed, densely pubescent with short hooked viscidulous hairs and interspersed longer tubercle-based hairs. Corolla white, pink or magenta-pink; tube cylindric, 0.8–1.2 cm. long, slightly curved, funnel-shaped near apex, glabrous outside, hairy inside, ± twisted at base; limb ± 2-lipped, 5-lobed, the lobes ± 2–4 mm. long and 2–3 mm. wide. Ovary 1.3 mm. long, obscurely 4-lobed, 4-locular; style ± 6 mm. long. Fruit compressed-subglobose, including the completely enveloping calyx, 5–6 mm. long, 5–7 mm. wide, the calyx indumentum persisting but now more distinctly uncinulate rendering the fruits burr-like; schizocarp flattened, oblate-spheroid, composed of 2 joined dry 2-locular cocci, each 4 mm. long, 3 mm. wide, the backs and sides densely echinate with short stout straight puberulous spines 0.4–1 mm. long save on the margins of the commissures where the spines are much reduced or reduced to ridges; ventral surface excavated. Fig. 3.

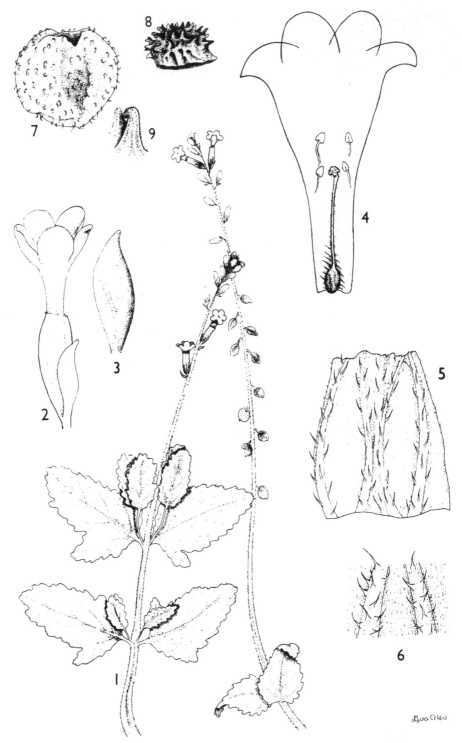

FIG. 3. *PRIVA CURTISIAE* — **1**, habit, × ⅔; **2**, flower, × 4; **3**, bract, × 6; **4**, longitudinal section of corolla, × 6; **5**, calyx, opened out, × 6; **6**, detail of calyx, × 12; **7**, fruit, × 4; **8**, nutlet, × 4; **9**, detail of nutlet, × 18. 1, from *Polhill & Paulo* 2314 and *Richards* 24833; 2–4, from *Hepper & Jaeger* 6681; 5–9, material used not recorded. Drawn by Lisa Coles.

UGANDA. Karamoja District: Lodoketemit [Lodoketeminit] catchment, 12 July 1958, *Kerfoot* 428! & Timu Forest, Apr. 1960, *J. Wilson* 959!
KENYA. Nairobi/Masai Districts: Athi Plains, 24 May 1971, *Kokwaro* 2579! & S. Ngong Hills, 19 July 1952, *Verdcourt* 686!; Teita District: Tsavo National Park, Murka Camp, 6 Aug. 1968, *Willy* 3838!
TANZANIA. Musoma District: Togoro Plains, 13.7 km. on return to Seronera, 13 Mar. 1962, *Greenway* 10520!; Masai District: 50 km. Arusha to Dodoma, 10 May 1965, *Leippert* 5739!; Mbulu District: Tarangire National Park, 12.8 km. from Tarangire Camp, 29 Nov. 1969, *Richards* 24833!; Mpwapwa District: Kongwa, 15 Feb. 1950, *Anderson* 629!
DISTR. U 1; K 1, 3–7; T 1–3, 5; Ethiopia
HAB. Grassland, grassland with scattered trees, e.g. *Acacia drepanolobium*, also mixed thicket and bushland, and *Acacia-Commiphora* woodland, frequently by roadsides; 915–1860(–2250) m.

SYN. [*P. leptostachya* sensu Bak. in F.T.A. 5: 285 (1900), pro parte quoad *Scott Elliot* 6127 & *Volkens* 2152, non Juss.]

NOTE. The variation in flower-colour needs examination; magenta forms are particularly frequent in the Machakos, Nakuru, Ngong, Laitokitok and N. Kavirondo areas; certainly in some areas only white forms are found but no collector mentions if the two occur together. *Hepper & Jaeger* 7168 (Kenya, Northern Frontier Province, Mt. Kulal, below Gatab lower airstrip) seems closest to this species despite petioles to 1.1 cm.; the fruit-shape and pyrenes are correct and it does not seem to be a hybrid, but all other *P. curtisiae* seen, a considerable amount, has sessile leaves.

5. **PHYLA**

Lour., Fl. Cochinch.: 66 (1790); Greene in Pittonia 4: 45–48 (1899)

Trailing perennial herbs with angular stems rooting at the nodes. Leaves opposite, simple, subentire to toothed, mostly tapering into an obscure petiole; lamina with sparse to dense adpressed medifixed hairs. Inflorescences ± long-pedunculate axillary short ovoid to cylindric very dense bracteate spikes. Calyx membranous, 2-lobed, compressed. Corolla with short tube and spreading somewhat irregularly 4-lobed white or purple limb. Stamens 4, included. Fruit of 2 nutlets enveloped by the persistent calyx.

About 6 species, mainly American, one now widely dispersed throughout the warmer parts of the world. *P. filiformis* (Schrad.) Meikle (*P. canescens* auctt. *non* (Kunth) Greene) has been cultivated in Mombasa and at the Grasslands Research Station, Kitale (28 Sept. 1959, *Bogdan* 4898) presumably as a grass-like ground-cover; it is very similar to *P. nodiflora* but differs in the abaxially less lobed calyx, longer corolla-tube, 2.5 mm. long, much exceeding the bracts and the corolla-limb 2.5–3 mm. in diameter.

P. nodiflora (*L.*) *Greene* in Pittonia 4: 46 (1899); Brenan in Mem. N.Y. Bot. Gard. 9: 37 (1954); Meikle in F.W.T.A., ed. 2, 2: 437 (1963); Jafri & Ghafoor in Fl. W. Pakistan 77: 11, fig. 2/C – E (1974); U.K.W.F.: 613 (1974); Vollesen in Opera Bot. 59: 82 (1980); Townsend in Fl. Turkey 7: 32 (1982); Meikle, Fl. Cyprus 2: 1249 (1985). Type (see note): U.S.A., Virginia, *Clayton* 448 (BM, lecto.)

Prostrate creeping herb, with slender often purplish stems 20–45 cm. long covered with fine medifixed hairs; rootstock woody. Leaves often purplish, oblanceolate to obovate, 0.5–5 cm. long, 0.2–3 cm. wide, acute and sharply serrate at the apex, cuneate and entire at the base, glabrous or thinly pubescent with adpressed medifixed hairs. Spikes purplish, cylindrical, 0.3–2.5 cm. long; bracts broadly obovate to oblate-cuspidate, 2–3 × 3–5 mm.; peduncles 1.3–8 cm. long. Calyx dorsiventrally flattened, 1.4 mm. long, split almost to base abaxially and to about the middle adaxially. Corolla mauve, pink or white with yellow centre, often white and purple in one inflorescence, and sometimes purple in bud, white when open; tube ± 1.5 mm. long; lobes unequal, 0.6–0.8 mm. in diameter. Stamens 4, didynamous. Stigma obliquely capitate. Fruit rounded obovoid, 1.2 × 1–1.2 mm., divided into 2 pyrenes. Fig. 4.

UGANDA. W. Nile District: Maracha, 12 June 1936, *A.S. Thomas* 2010!; Bunyoro District; Lake Albert, June 1943, *Purseglove* 1592!; Kigezi District: Lake Edward, Rwensama [Rwenshama], Mar. 1951, *Purseglove* 3596!
KENYA. Lake Baringo, W. side, 28 July 1976, *Timberlake* 232!; Teita District: Tsavo National Park East, Sabaki R., Sala Gate, 29 Dec. 1966, *Greenway & Kanuri* 12904!; Kilifi District: Malindi, June 1962, *Tweedie* 2370!
TANZANIA. Arusha District: Nyumba ya Mungu, reservoir lake, 1 Sept. 1969, *Batty* 596!; Ufipa District: Lake Tanganyika, Kasanga, 15 June 1957, *Richards* 10110!; Singida Lake, 27 Apr. 1962, *Polhill & Paulo* 2208!; Zanzibar I., Mnazi Moja, 26 May 1964, *Faulkner* 3383!

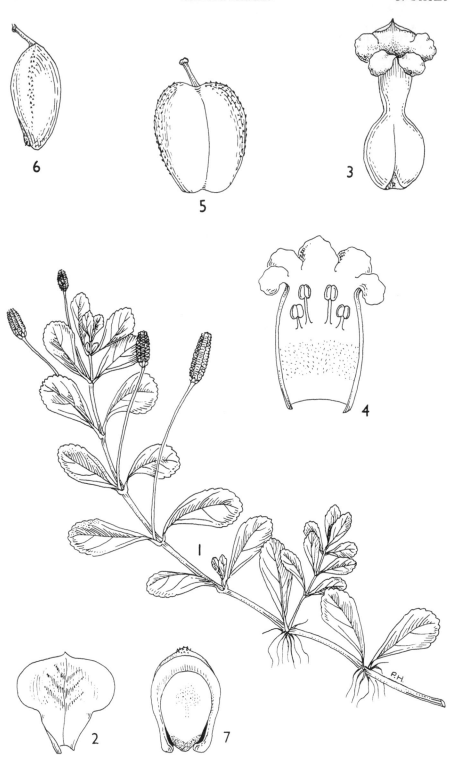

FIG. 4. *PHYLA NODIFLORA* — **1**, habit, × ⅔; **2**, floral bract, × 10; **3**, flower, × 12; **4**, corolla, opened out, × 12; **5**, fruit, × 18; **6**, fruit, side view, × 18; **7**, seed, embedded in half fruit, × 18. 1, from *A.S. Thomas* 2010; 2–7, from *Whyte* s.n. Drawn by P. Halliday.

DISTR. U 1, 2; **K** 3, 5, 7; **T** 2–6, 8; **Z**; now widespread in tropics, subtropics and warmer temperate
regions; throughout tropical Africa from Senegal, Central African Republic, Sudan, Ethiopia and
Somalia to South Africa and also in North Africa.
HAB. Sandy lake shores, shore mud, lava boulders by lakes, lakeside grassland, also roadsides and
Avicennia salt marshes; 0–1170 m.

SYN. *Verbena nodiflora* L., Sp. Pl.: 20 (1753)
    Lippia nodiflora (L.) Michx., Fl. Bot.-Amer. 2: 15 (1803); Bak. in F.T.A. 5: 279 (1900); F.W.T.A. 2:
    270 (1931)

NOTE. Townsend gives the type as 'a cultivated specimen (holo. BM–Hb Cliff.!)' but Linnaeus cites
    5 syntypes and gives the locality as Virginia. I have therefore accepted the specimen forming the
    basis of the *Gronovius* reference as lectotype.

6. LIPPIA*

L., Sp. Pl.: 633 (1753) & Gen. Pl., ed. 5: 282 (1754); Moldenke in Phytologia 12, 13
(1965–1966) (numerous contributions); R. Fernandes in Bol. Soc. Brot., sér. 2, 59: 245–272
(1987)

Erect shrubs, subshrubs or perennial herbs, mostly pyrophytes from large woody
rootstocks, pubescent or hairy; dioecious in species 11. Leaves opposite or in whorls of
3–4, rarely alternate, entire or variously crenate, toothed or lobed, often scabrid above
and rugose. Flowers small, sessile in mostly pedunculate dense heads or spikes which
sometimes elongate in fruit, 1–several heads per axil and sometimes forming terminal
panicles but often in numerous lower axils; each flower supported by a bract; eventual
bare rhachis with floral scars very closely placed. Calyx campanulate or compressed and
laterally 2-keeled or 2-winged, undulate, truncate, 2–4-fid or 4-toothed. Corolla white,
greenish or creamy yellow, often darker at the throat, more rarely magenta; tube cylindric
or funnel-shaped, straight or curved; limb ± 2-lipped, oblique, the anterior lip larger than
the posterior. Stamens 4, didynamous, inserted in the corolla-tube, included or slightly
exserted. Ovary 2-locular, each locule with 1 ovule; style usually short with oblique or
recurved stigma. Fruit dry, surrounded by the adpressed calyx, separating into 2 pyrenes
at maturity; pericarp papery or hard. Seeds without endosperm.

A genus of about 200 species mainly in South and Central America but many also in Africa.

1. Inflorescences essentially terminal, with additional axillary
    elements in upper and sometimes middle axils 2
   Inflorescences essentially axillary, often in many axils,
    often middle and sometimes basal ones, but those
    from uppermost sometimes overtopping undeveloped
    short apex and appearing terminal or occasionally
    some truly terminal elements but never extensive
    panicles or spicate arrangements 4
2. Leaves narrowly oblong-lanceolate or narrowly oblong,
    broadest about the middle, smooth to rough above,
    variously pubescent beneath; individual flower-heads
    0.5–2 × 0.7–1.2 cm., bracts 3–5 × 2–3 mm. 3
   Leaves ovate or ovate-lanceolate, widest near to the base,
    very rough above, densely to sparsely pubescent
    beneath with scarcely adpressed hairs and obviously
    gland-dotted; individual flower-heads 1.3–5 × 1.5–2
    cm.; bracts ovate-lanceolate, 5–10 × 3–5 mm. . . .   9. *L. plicata*

* The work of R. Fernandes for Flora Zambesiaca is gratefully acknowledged, some of which has
)een used with little alteration.

3. Leaves ± smooth above (although tuberculate-based hairs
   are present), very densely white pubescent-tomentose
   beneath with completely adpressed hairs (in East
   Africa); heads 0.5–2 × 0.7–0.8 cm., arranged in more
   paniculate inflorescences; bracts ovate, 3–4 × 2–3 mm.,
   acuminate, the upper pale yellowish hirsute (U 1)   1. *L. multiflora*
   Leaves rough above, sparsely to ± densely pubescent
   beneath with pale scarcely adpressed hairs; heads 1–2
   × 0.8–1.2 cm., arranged in more spicate inflorescences;
   bracts ovate-lanceolate, 3–5 × 2–3 mm., narrowly
   acuminate, shortly pubescent with few stiff hairs or
   more densely pubescent and glandular (widespread)   2. *L. abyssinica*
4. Leaves distantly almost lobulate-crenate, the lobulae
   rounded, densely gland-dotted and with a few sparse
   hairs on the venation (K 1) . . . .   5. *L. dauensis*
   Leaves more closely crenate-serrate and usually more
   densely pubescent 5
5. Mature inflorescences usually shorter than the leaves 6
   Mature inflorescences usually longer than the leaves 8
6. Bracts very accrescent in functionally ♀ inflorescences
   becoming membranous, 0.8–1.2 × 1.1–1.5 cm., and
   venose; ♂ inflorescences small, up to 5 × 7 mm. in
   flowering state; leaves scarcely rugose, 0.7–4.8 × 0.4–3
   cm. (K 1) . . . .   11. *L. carviodora*
   Bracts much less accrescent 7
7. Inflorescences 1–5 per axil, each conical or oblong head
   0.45–1.5(–2.2) × (0.3–)0.5–0.8 cm.; peduncles 0.3–3 cm.
   long; bracts 1.8–4 × 1.2–2 mm.; corolla-tube 1.2–2 mm.
   long; leaves slightly to distinctly scabrid above   3. *L. javanica*
   Inflorescences 1–2(–3) per axil, each globose to oblong-
   ovoid head 0.5–1.5 × 0.7–1.5 cm.; peduncles 2–6.5 cm.
   long; bracts 5–8 × 2–4 mm.; corolla-tube 3–4 mm. long;
   leaves distinctly scabrid above . . . .   10. *L. kituiensis*
                                                              (in part)
8. Small pyrophyte with basal inflorescences often from the
   lowermost node, 1–8(–18) cm. tall . . . . . .   7. *L. praecox*
   Taller plant with inflorescences from upper nodes 9
9. Leaves small, rounded ovate, 0.6–2.5 × 0.4–1.5 cm., very
   rugose above; corolla exceeding the bracts; tube
   (2.5–)3.5–4 mm. long; bracts ovate, 4–4.5 × 2–4.3 mm.;
   peduncles 2–5.5(–12) cm. long, about twice as long as
   the leaves (K 1, ?3) . . . .   4. *L. somalensis*
   Leaves larger and more ovate, 1–11.5 × 0.5–5 cm., rugose or
   scarcely so; corolla shorter or longer than the bracts;
   peduncle not twice as long as leaves 10
10. Corolla exceeded by narrowly triangular-lanceolate
    densely hairy bracts; mostly of small habit, (7–)15–60(–
    100) cm. tall, usually with several unbranched stems
    from a woody root . . . .   8. *L. woodii*
    Corolla usually exceeding the bracts, at least at top of
    inflorescence 11
11. Subshrub or herb 65(–100) cm. tall; indumentum on stems,
    leaves and bracts much longer (in E.A. variety);
    leaves sessile . . . .   6. *L. baumii*
    Shrub or subshrub 1.5–4.5 m. tall; indumentum on stems,
    leaves and bracts usually shorter; heads 1–3 per axil;
    leaves usually shortly petiolate . . . .   10. *L. kituiensis*

1. **L. multiflora** *Mold.* in Phytologia 3: 168 (1949); Meikle in F.W.T.A., ed. 2, 2: 437, fig.
306 (1963). Type: N. Nigeria, Borgu, *Barter* 768 (NY, holo., K, iso.!)

Robust woody perennial or shrub, aromatic, 0.9–3.6 m. tall; stems ridged, sparsely
adpressed strigillose-pubescent and glandular or subglabrous, brown or purplish, often ±

shining, striate. Leaves mostly in whorls of 3; blades narrowly oblong-lanceolate or narrowly elliptic, 5.5–13(–17) cm. long, 1–3.3(–5.5) cm. wide, acute at the apex, cuneate at the base, widely to very closely serrulate, with venation impressed when dry above, rugulose and strigillose with bulbous-based hairs but ± smooth to the touch, densely adpressed puberulous and ± resinous-glandular beneath, soft, with longer hairs on the prominent nervation or sometimes ± glabrous in West Africa; petioles 2–10 mm. long. Inflorescences terminal, up to 20 cm. long, 10 cm. wide, of numerous subcapitate to cylindric-oblong spikes 0.5–2 cm. long, 7–8 mm. wide; main branches 3–10 cm. long, ultimate stalks 0.5–2 cm. long, densely adpressed pubescent; main bracts similar to the leaves; secondary bracts narrowly ovate-lanceolate, acuminate, 6 mm. long, 3 mm. wide; lower bracts of actual spikes ovate, 3–4 mm. long, 2–3 mm. wide, the upper more rounded, smaller, densely imbricated and pale yellowish hirsute. Calyx subglobose, ± 1 mm. long, densely spreading pubescent; lobes very short. Corolla white or pale yellow; tube funnel-shaped from a narrow base, 2 mm. long, densely puberulous outside; limb ± 3 mm. across, 2-lipped, the largest lobe oblong ± 1.5 mm. long and wide. Fruits enclosed in calyx, breaking into 2 half-ovoid nutlets 1–1.5 mm. long, 0.8 mm. wide.

UGANDA. W. Nile District: Maracha, Dec. 1939, *Hazel* 399!
DISTR. U 1; Sierra Leone to Nigeria, Equatorial Guinea, Gabon, Zaire, Sudan, Central African Republic, Angola
HAB. 'Near river' [Nile], probably coarse grassland; ± 1200 m.

SYN. [*L. adoensis* sensu Bak. in F.T.A. 5: 280 (1900), pro parte; F.W.T.A. 2: 270 (1931), pro parte, *non* Walp.]

2. **L. abyssinica** (*Otto & Dietr.*) *Cuf.*, E.P.A.: XXXI (1969); Sebsebe in K.B. (in press). Type: a cultivated plant raised from seed derived from *Schimper* 305 (B, holo.†). Neotype: Ethiopia, Wellega, 20 km. E. of Lekemt, *W. de Wilde* 8795 (ETH, neo., K, WAG, isoneo.)

Stiffly erect branched or ± unbranched woody herb, subshrub or softly woody shrub 0.6–2(–3) m. tall, slightly aromatic; stems ± 6-sided, sparsely strigillose with white hairs to densely pubescent and with many longer hairs or sometimes very scabrid; small glands also present and visible on glabrescent parts. Leaves in whorls of 3–4, much smaller towards apices of shoots; blades elliptic to elliptic-oblanceolate, 2–16 cm. long, 0.5–6 cm. wide, acute or acuminate at the apex, cuneate at the base, above very scabrid with short white tubercle-based hairs and venation impressed, beneath with hairs on the main nervation and pubescent and glandular between or almost glabrescent but sometimes ± densely pubescent, closely ± shallowly crenate-serrulate; petioles 0–5 mm. long. Inflorescence variable but due to reduction or suppression of upper leaves always appearing terminal with groups of heads at up to 5 widely separate nodes and some at lower leafy nodes; heads subglobose to conic, 1–2 cm. long, 0.8–1.2 cm. wide, 1–2 in each axil; peduncles 0.3–2 cm. lengthening with maturity but sometimes much longer and up to 5 cm. long; bracts ovate-lanceolate, narrowly acuminate, the lowest ± 5 × 3 mm., the uppermost 3 × 2 mm., shortly pubescent and glandular outside with few stiff hairs or more densely pubescent; lowermost claw-part ± glabrous. Calyx 1–1.3 mm. long, very densely white spreading hairy with 2 boat-shaped lobes. Corolla creamy white or yellowish with a yellow throat; tube funnel-shaped, 2.2–4 mm. long, pubescent above; limb 2.2–4 × 1.8–3 mm., the upper lip short and square, the lower 3-lobed, ± 3 mm. long. Nutlets half-ovoid, 1.2 mm. long, 0.8 mm. wide.

UGANDA. W. Nile District: Maracha, 6 Aug. 1953, *Chancellor* 119!; Kigezi District: Ruhinda, Jan. 1951, *Purseglove* 3553!; Teso District: Serere, Nov. 1931, *Chandler* 12!
KENYA. Elgon, Oct.-Nov. 1930, *Lugard* 109! & Kitale, Milimani, Aug. 1969, *Tweedie* 3682!; Kericho District: Ngoina Forest Reserve, 24 Nov. 1971, *Magogo* 1502!
TANZANIA. Bukoba District: Bunazi, Sept.-Oct. 1935, *Gillman* 494!; Mpanda District: Mahali Mts., 6 Sept. 1958, *Newbould & Jefford* 2384! & below Pasagulu, 16 km. N. of Kasoge, 8 Aug. 1959, *Harley* 9216!
DISTR. U 1–4; K ? 2, 3, 4 (see note), 5; T 1, 4; Zaire, Rwanda, Burundi, Sudan and Ethiopia
HAB. Grassland, grassland with scattered trees, *Hyparrhenia* associations in *Terminalia*, *Combretum* woodland, abandoned cultivations, often in rocky places; 900–1950(–? 2250) m.

SYN. *Lantana abyssinica* Otto & Dietr. in Allg. Gartenzeit. 9: 379 (1841)
　　Lippia grandifolia Walp., Repert. 4: 55 (1845); A. Rich., Tent. Fl. Abyss. 2: 167 (1850); E.P.A.: 791 (1962); U.K.W.F.: 614 (1974). Type: Ethiopia, Feurfeura [Ferrfera], *Schimper* 734 (BM, K, iso.!)
　　L. sp. 2 sensu Oliv., App. Speke's Journ.: 644 (1863)
　　L. adoensis var.; Oliv. in Trans. Linn. Soc., Bot., 29: 132 (1875)

[*L. adoensis* sensu Bak. in F.T.A. 5: 280 (1900), pro parte *non* Walp.]
L. grandifolia Walp. var. *longipedunculata* Mold. in Phytologia 3: 271 (1950). Type: Uganda,
    Mengo District, Kipayo, *Dummer* 54 (K, holo.!)
L. adoensis Hochst. var. *pubescens* Mold. in Phytologia 8: 58 (1961). Type: Uganda, Masaka
    District, Katera, *Drummond & Hemsley* 4496 (B, holo., K, Z, iso.!)
L. abyssinica (Otto & Dietr.) Cuf. var. *pubescens* (Mold.) Mold. in Phytologia 12: 43 (1965)

NOTE.   *L. sp. 1* (App. Speke's Journ.: 644) could from Grant's description be the same species. I have
followed the delimitation of Meikle arrived at while in charge of naming this group for many years,
i.e. treating it as a separate species from *L. adoensis* Walp. Cufodontis has assumed that his
combination *L. abyssinica* (Otto & Dietr.) Cuf. is referable to *L. adoensis*, but Sebsebe Demissew
shows the description better fits *L. grandifolia* and has neotypified it in this sense.
    The inflorescences can be much condensed, e.g. *Drummond & Hemsley* 4496 (Uganda, Masaka
District, Katera, 1 Oct. 1953) has the heads aggregated into a cylindrical inflorescence 10 × 3.5 cm.
and *Gardner* in *F.D.* 3727 (Kenya, Cherangani Hills, Oct. 1937) has a spike of heads. The former
was made the type of *L. adoensis* var. *pubescens* Mold. Moldenke has described a var.
longipedunculata based on material from Uganda. At least one specimen seen from the Sudan, with
apparent peduncles 7 cm. long, has the heads supported by reduced leaves and is in fact an axillary
branch. Since there is a good deal of variation in length, even in one plant, the variety has not been
formally treated. Extremes can, however, be very distinctive, e.g. Busoga District, Jinja, *E. Brown* 84,
and mostly occur in Uganda. The name is available for those who wish to use it. *Hamilton* 422
(Meru National Park, Kindani–Kilimakieru track W. of Park boundary, 2 Apr. 1979) has long
peduncles and very narrowly long-attenuate inflorescence bracts; it is the basis of the **K** 4 record
and may be a distinct taxon. I have not seen *L. grandifolia* var. *angustispicata* Mold. in Phytologia 3:
419 (1951), based on Zaire, *van der Gucht* 364 (BR, holo.)

3. **L. javanica** (*Burm.f.*) *Sprengel*, Syst. Veg. 2: 752 (1825); Meeuse in Blumea 5: 68 (1942)
(much synonymy); Meikle in Mem. N.Y. Bot. Gard. 9: 36 (1954); K.T.S.: 588 (1961); E.P.A.:
790 (1962); F.F.N.R.: 370 (1962); Ivens, E. Afr. Weeds: 69 (1967); Kokwaro, Luo Bot. Dict.:
12, t. 28/c (1972); U.K.W.F. 613 (1974); Vollesen in Opera Bot. 59: 82 (1980). Type: "Hab.
in Java*", *Kleinhof* (G, syn.!)

Much-branched shrub 0.6–4.5(–?6) m. tall; stems brown with adpressed or spreading
short stiff tubercle-based hairs and small glands. Leaves opposite or in whorls of 3; blades
lanceolate, elliptic- or ovate-lanceolate or ± oblong-elliptic, 1–10(–13.5) cm. long, 0.3–
4.3(–5) cm. wide, obtuse to acute at the apex, rounded then cuneate at the base, closely
serrulate, densely pubescent with tubercle-based hairs and shorter hair above, slightly to
distinctly scabrid and rugulose bullate, softly and densely often ± velvety adpressed
pubescent beneath and nerves raised, also with small glands. Flowers in conical or
oblong spikes 0.45–1.5(–2.2) cm. long, (3–)5–8 mm. wide, 1–5 per axil, sessile or with
peduncles 0.3–3 cm. long; lower bracts of spikes ovate, 4 mm. long, 2 mm. wide, long-
acuminate, upper smaller, 1.8 mm. long, 1.2 mm. wide, acuminate, densely spreading
white pubescent and glandular. Calyx ± 1 mm. long, 2-lobed, densely spreading
pubescent. Corolla white, yellowish white or greenish (rarely pink), usually with a yellow
throat; tube narrowly funnel-shaped from a very narrow base, 1.2–2 mm. long, pubescent
and glandular above; limb 1.2–2 mm. wide, the largest lobe squarish, 0.5 mm. long, 0.8
mm. wide. Nutlets brown, half-ovoid, 1.3 mm. long, 1 mm. wide.

UGANDA.   Mbale District: Elgon, Mbale, Jan. 1918, *Dummer* 3736! & Bugisu [Bugishu], Bufumbo, Nov.
    1932, *Chandler* 1007! & Sebei, 9.6 km. Kapchorwa–Kaburoni, 12 Oct. 1952, *G.H. Wood* 433!
KENYA.   Trans-Nzoia District: Kitale, Milimani, Aug. 1969, *Tweedie* 3680!; 7.2 km. N. of Fort Hall, 4
    Aug. 1958, *Trapnell* 2413; Masai District: N. of Kipleleo Massif, Ol Joro Ole Soyet, 24 May 1962,
    Glover & Samuel 2873!
TANZANIA.   Moshi District: Lyamungu, Coffee Research Station, 27 Oct. 1943, *Wallace* 1098!; Singida
    District: 56 km. Singida–Babati, near Mt. Hanang, 29 Mar. 1965, *Richards* 19957!; Mbeya District:
    17.5 km. SW. of Mbeya, 12 May 1956, *Milne-Redhead & Taylor* 10189!; Zanzibar I., Chwaka, 28 July
    1959, *Faulkner* 2314!
DISTR.   **U** 3; **K** 1–6; **T** 1–8 (rare in 3 & 4); **Z**; **P** (fide Bojer & Greenway); Zaire, Ethiopia, Mozambique,
    Malawi, Zambia, Zimbabwe, Botswana, Angola and South Africa
HAB.   Scrub bushland, grassy rocky hillsides, *Brachystegia, Combretum*, etc. woodland, forest clearings
    and plantations and becoming a weed in derived range lands; often predominant species over
    wide areas; 0–200 m. in **T** 6, 8 & **Z**, elsewhere 450–2350 m.

* An error since the species is not found there; the type labelled *V. javanica* in Burman's hand
bears the note 'Kleinhof, Java' in an unknown hand (*fide* Meeuse)

SYN. *Verbena javanica* Burm.f., Fl. Ind.: 12, t. 6/2 (1768)
[*Lippia asperifolia* sensu Schauer in DC., Prodr. 11: 583 (1847), pro parte; Gürke in P.O.A. C: 338 (1895); Bak. in F.T.A. 5: 280 (1900); U.O.P.Z.: 332 (1949); T.T.C.L.: 640 (1949), *non* Marthe]
L. whytei Mold. in Phytologia 1: 428 (1940). Type: N. Malawi, *Whyte* (NY, holo., K!, prob. iso.)

NOTE. There is great variation in the number of inflorescences per axil and their peduncle-length. *Glover et al.* 2125 (Kenya, Masai District, Olodungoro, Entasekera, 13 July 1961) and *Starzenski* 36 (Kedong Valley, 2 Aug. 1962) are exceptional in having the peduncles up to 5.5 cm. long. R. Fernandes intended to recognise two varieties, var. *javanica* and var. *whytei*, one with pedunculate heads and the other with them ± sessile but finally decided they were not constant enough. They grow together in many places, e.g. both are on the one sheet *Meyerhoff* 129M (Kenya, W. Suk, Mwina [Muino], Mutunyo, 7 Jan. 1979) and the variation is ± continuous, so despite the very different appearance I have not formally separated two varieties either. Meeuse first synonymised *Lippia alba* (Mill.) N.E. Br. with *L. javanica* but they appear to be distinct. Modern users of the name *L. alba* in recent American flora's make no reference to *L. javanica*. The problem needs further study. The plant long established in India appears to be *L. alba*. Meikle claims that the two following are, judging by their illustrations, the same as *L. javanica*.
Verbena globiflora L'Hérit., Stirp. Nov. 23, t. 12 (1786). Type: 'America calidiore' (? G, syn.) & Sloane Jam. 65 hist. 1: 173 t. 108/1 (syn.)
Zappania odoratissima Scop., Delic. Insub. 1: 34, t. 15 (1786). Type: origin not known, seeds sent from Zappa (ubi ?)
Tweedie 3736 (N. Cherangani, Dec. 1969) had been compared with *L. ukambensis* auctt. but is clearly nearer *L. javanica*. The long peduncles up to 4 cm. and very rough upper leaf surfaces suggest it might be *L. javanica* × *L. kituiensis* but it has been left as a form of *L. javanica*.
L. javanica yields an essential oil.

4. **L. somalensis** *Vatke* in Linnaea 43: 527 (1882); Bak. in F.T.A. 5: 279 (1900); E.P.A.: 792 (1962); Kuchar, Pl. Som. Checklist: 253 (1986). Type: Somalia, near Mait [Meid] in region of Mt. Surud [Serrut], *Hildebrandt* 1443 (B, holo.†, BM, K, iso.!)

Very aromatic much-branched shrub 0.9–3.6 m. tall, with adpressed scabridly pubescent slender branches, later glabrescent. Leaves in pairs; blades ovate, elliptic, rhomboid or almost round, 0.5–2.5 cm. long, 0.3–1.5 cm. wide, rounded to acute at the apex, rounded to cuneate at the base, regularly and closely or sometimes more coarsely crenate or crenulate, rugose and with impressed venation above, scabrid with short tubercle-based hairs (very scabrid and with huge tubercles in most Somali material) and small glands, beneath scabrid with tubercle-based hairs on the raised venation or more softly pubescent and glandular; petioles 0–8 mm., mostly very short. Inflorescences solitary or paired in upper and often lower axils and much overtopping the leaves; spikes globose to oblong, 1–2.2 cm. long; peduncles 2–5.5 cm. long in East Africa, but up to 7(–12 in one variant) cm. in Somali populations, pubescent to scabrid; bracts ovate to broadly elliptic, round or ± obovate, 4–5 mm. long, 2–4.3 mm. wide, acuminate or narrowly cuspidate, mostly densely adpressed silky pubescent and ciliate. Calyx 1–1.2 mm. long, densely spreading pubescent. Corolla white or cream with yellow eye; tube narrowly funnel-shaped from a very slender base, (2.5–)3.5–5 mm. long, pubescent and glandular on upper half; limb ± 3 mm. wide, the lobes rounded squarish, pubescent outside. Fruits not seen.

KENYA. Northern Frontier Province: 6.4 km. from Marsabit on road to Isiolo, 14 May 1970, *Magogo* 1331! & Mt. Marsabit, near lodge, 13 Jan. 1972, *Bally & Radcliffe-Smith* 14776! & Ndoto Mts., Sirwan, 1 Jan. 1959, *Newbould* 3397!
DISTR. K 1, ? 3/6; Somalia (N.)
HAB. *Cadia* and *Aloe, Euphorbia-Rhus-Carissa*, etc. scrub and bushland, forest edges, grassland often in rocky places; 900–2010 m.

SYN. *Lantana somalensis* (Vatke) Engl. in Sitzb. K. Preuss. Akad. Wiss. 1904: 410 (1904)

NOTE. Agnew, U.K.W.F.: 613 (1974), records this species from Mt. Suswa (K 3/6) on the basis of a specimen *Glover & Samuel* 3371. The bracts in N. Somalia populations have distinctly longer cusps than those of the Kenya populations but despite the considerable disjunction the differences do not seem to justify subspecific distinction.

5. **L. dauensis** (*Chiov.*) *Chiov.* in Fl. Somala 2: 359 (1932); E.P.A.: 791 (1962). Type: Somalia, Daua R. valley, Ueldá, *Riva* 1613 (FT, holo.)

Shrub; stems with youngest parts densely covered with white hairs becoming more scattered later and with dense pale glands; older stems very pale brown, 4-ribbed, ± glabrous. Leaves aromatic, opposite; blades elliptic to lanceolate, 1.5–5 cm. long, 1–1.7 cm. wide, obtuse at apex or subacute and mucronulate, very narrowly cuneate at the base, densely glandular and with sparse white hairs, almost lobulate-crenate, the lobules 2.5–3 mm. long; petiole ± obsolete or 2–4 mm. long. Heads small, 1–3 per node, globose, at first 4–5 mm. long, later up to 10 × 6 mm., the fruiting axis usually 1 cm. long; peduncle 2.5–6 cm. long, glabrous; bracts rounded-elliptic, 1.5–2 mm. long, 1 mm. wide, densely adpressed pubescent. Calyx globose, bilobed; lobes rounded, 1.5 mm. long, densely pubescent and glandular. Corolla white; tube 1.5 mm. long, minutely furfuraceous; lobes round, 0.5 mm. diameter. Fruit compressed-globose, 1.5–2.5 mm. long, glabrous, splitting into 2 mericarps.

KENYA. Northern Frontier Province: Furroli, 17 Sept. 1952, *Gillett* 13928! & 9 km. from Banisa [Banissa] on road to Malka Mari, Lulis, 5 May 1978, *Gilbert & Thulin* 1486!
DISTR. K 1, Somalia; also cultivated at Nakuru in 1958
HAB. Lava plain, by water-hole in *Commiphora-Acacia* bushland on thin stony soil; 960 m.
SYN. *Lantana dauensis* Chiov in Ann. Bot. Roma 10: 402 (1912)
NOTE. Of possible value for its essential oils.

6. **L. baumii** *Gürke* in Warb., Kunene-Sambesi-Exped.: 350 (1903); R. Fernandes in Bol. Soc. Brot., sér 2, 59: 247 (1987). Type: Angola, between Ungombekike and Cuito, *Baum* 515 (B, holo.†, BM, iso.!)

Erect or ascending subshrub or perennial herb to 65 or more cm. long (probably over 1 m. when unburnt), densely covered with short to long hairs or ± adpressed pubescent; lowest internodes long, up to 12 cm. Leaves opposite; blades lanceolate, oblong-lanceolate, oblong or oblong-obovate to ovate, 1.7–10 cm. long, (0.8–)2–3(–5) cm. wide, obtuse to ± acute at the apex, rounded to ± cuneate or sometimes almost subcordate at the base, crenate–crenulate serrate or obsoletely crenate above but lower ½–⅓ entire; venation impressed above, prominent and reticulate beneath, densely hairy on both surfaces. Flowers in spike-like heads in upper axils, 1.5–3 cm. long, 1.3–1.8 cm. wide; peduncles mostly solitary (7–)11–14(–13.5) cm. long; lower bracts longer than the flowers, ovate or ovate-lanceolate, 6–12 mm. long, 4–6 mm. wide, attenuate or shortly acuminate at the apex, hairy and ciliate, or sometimes the bracts leaf-like and 1.2–2.5 cm. long, 0.7–1 cm. wide. Calyx ± 1.5 mm. long, 2-lobed, ± densely adpressed white hairy or puberulous outside. Corolla yellow or white with a yellow centre or in one variety deep magenta, 3.5–8 mm. long; tube pubescent outside save at base; limb ± 2-lipped, ± 4 mm. wide, the upper lip entire, rounded, the lower 3-lobed, the lobes entire, microgranulate inside. Fruit ovoid, with mericarps very convex, 2–2.5(–3) mm. long, 1.25–1.75(–2.2) cm. wide, brownish and slightly shining outside, flat and white inside.

SYN. [*L. wilmsii* H. Pearson var. *villosa* sensu Mold. in Phytologia 13: 175 (1966), pro parte, *non* Mold. (1953) neque *L. africana* Mold. var. *villosa* Mold. (1948)]

var. **nyassensis** *R. Fernandes* in Bol. Soc. Brot., sér 2, 59: 251 (1987). Type: Malawi, Viphya Plateau, Mzimba, *Salubeni* 653 (SRGH, holo.!, BR, K, LISC, iso.!)

Perennial many-stemmed herb with an indumentum of denser silvery white longer and more spreading hairs up to 2.5–3 mm. long. Leaves broadly ovate to rounded, usually longer than the internodes. Corolla yellowish or pale yellow, rarely white, 3.5–4.5 mm. long.

TANZANIA. Ufipa District: Malonje Plateau, Nsanga Mt., 13 May 1959, *Richards* 12127! & Sumbawanga, path to Kasapa village, 9 Mar. 1957, *Richards* 8600! & Nsanga [Nsangu] Highlands, Rukwa Escarpment, 2 Jan. 1962, *Richards* 15862!
DISTR. T 4; Malawi
HAB. Upland grassland, open bushland on stony hillsides; 1800–2100 m.
SYN. *Lantana primulina* Mold. in Phytologia 28: 402 (1974). Type: Malawi, Mafinga Hills, *Robinson* 4452 (MO, holo., BR, K, M, iso.!)
NOTE. *Richards* 15862 has large bracts supporting the inflorescences which are essentially leaves 2.5 cm. long, 1 cm. wide, the more normal bracts being 1.2–1.5 cm. long, 0.7–1 cm. wide. Fernandes (loc. cit. and F.Z. MS) does not record this variety from Tanzania but I am certain the three specimens cited above are nearer to this variety than to var. *baumii*.

7. **L. praecox** *Mold.* in Phytologia 4: 292 (1953). Type: Tanzania, Rungwe District, Kyimbila, *Stolz* 2210 (S, holo., B, BM!, G, P, S, Z, iso.)

Pyrophytic herb with numerous short stems 1–8(–18) cm. long from a thick woody stock 2–4 cm. wide with numerous heads giving rise to many shoots and tough fibrous roots;

stems densely pubescent. Leaves oblong-ovate, 0.5–1.2(–5.5) cm. long, 0.4–0.7(–3) cm. wide, obtuse to rounded at the apex, narrowed into the short petiole at the base, densely white pilose; petioles obsolete or up to 5 mm. long. Inflorescences from axils near ground-level and few above and not from the uppermost, hemispherical, 0.7–1.5 cm. long and wide; peduncles 1–4.5 cm. long, white pubescent; bracts ovate, 3–8 mm. long, 2–4 mm. wide, acuminate, densely pubescent and ciliate. Calyx 1.5–2 mm. long, densely white pubescent. Corolla bright lemon yellow or yellowish green; tube 3–3.5 mm. long, pubescent at apex and with pubescent throat; limb 2.5–4 mm. wide, one lip emarginate, the other 3-lobed, the largest lobes 1.5 mm. long and wide, all lobes micropuberulent within. Fruit not seen.

TANZANIA.   Ufipa District: Mbisi, 6 Oct. 1950, *Bullock* 3418! & 12 Aug. 1950, *Richards* 13092!; Iringa District: 64 km. S. of Iringa on Mbeya road, 14 Nov. 1958, *Napper* 880!; Njombe District: Matamba, 6 Jan. 1957, *Richards* 7514!
DISTR.   T 4, 7; N. Malawi
HAB.   Fireswept grassland particularly on eroded hillsides; 1800–2250 m.

SYN.   *L. baumii* Gürke var. *baumii* forma sensu R. Fernandes in Bol. Soc. Brot., sér 2, 59: 247, 249–251 (1987)

NOTE.   *Batty* 830 (112 km. Iringa–Mbeya, 6 Nov. 1969) shows longer unburnt shoots to 13 cm. but the inflorescences are mostly basal; the 18 cm. in the description above refers to *Richards* 15722 (Iringa District, Sao Hill, 17 Dec. 1961). *Lynes* D15 had been annotated as *L. africana* var. *villosa* by Moldenke in Oct. 1949 and several have been annotated "*L. baumii* var. *baumii* forma" by R. Fernandes and cited in her paper but I think it is distinct . Mildbraed first suggested the name *praecox* but did not describe it. It is too constant over too wide an area to be only a form of *L. baumii* but it could well be a subspecies. No tall shrubby unburnt forms have been seen from T 7 at all and those from T 4 appear to be *L. baumii* var. *nyassensis*. *L. baumii* var. *nyikensis* is similar but differs in flower colour. Nevertheless there is a possibility that Fernandes's belief that these dwarf floriferous stems have been caused by early rains after fires have burnt off normal shoots of the previous season could be correct. It happens in many hundreds of savanna species. In this case normal plants have been overlooked over a vast area.

8. **L.woodii** *Mold.* in Phytologia 2: 318 (1947) & 13: 176 (1966); R. Fernandes in Bol. Soc. Bot., sér. 2, 59: 261 (1987). Type: Malawi, Blantyre, *Buchanan* in *Medley Wood* 6937 (F (83373), holo., NH, iso.)

Strongly aromatic* subshrub or herb with several often unbranched stems, (7–)15–60 (–100) cm. tall from a tough woody root or sometimes from a thin creeping woody rhizome; remains of long burnt-off stout shoots mostly also present remaining after fires; young stems densely pubescent and with longer semi-adpressed white bristly hairs 0.5–1(–2) mm. long. Leaf-blades opposite, elliptic, rhomboid or usually ± obovate-elliptic, 1–8(–11.5) cm. long, (0.5–)1–3(–3.5) cm. wide, rounded to subacute at the apex, cuneate at the base, shallowly crenulate, with adpressed bristly hairs and shorter pubescence above and with venation impressed, sometimes very rugose and bullate, roughly silvery grey velvety beneath with dense pubescence and numerous ± adpressed white hairs as well, densely glandular-punctate; venation prominent beneath; petiole 2–4 mm. long. Inflorescences solitary in the axils of the upper leaves, to 1.2 cm. long and 0.7–1.2 cm. wide, ovoid, oblong or cylindric, silvery grey pubescent and hairy similar to the stems and leaves, becoming up to 2.2 cm. long, 1 cm. wide in fruit, many-flowered; peduncles similarly hairy, (0.6–)2.5–4(–6.5) cm. long, usually not exceeding the leaves; bracts narrowly triangular-lanceolate to elliptic, 6–7 mm. long, 2–3 mm. wide, tapering, acute to cuspidate, enlarging in fruit to 9.5 mm. long and 2.5–3 mm. wide. Calyx 1.5 mm. long, truncate, slit, densely glandular at the base, densely white pilose above. Corolla white, green or lemon-yellow, usually with an orange-yellow centre, narrowly funnel-shaped, 3–4.5 mm. long, the upper part of tube densely spreading pubescent and glandular; limb 2.2–3 mm. wide, the largest lobe rounded, ± 1 mm. long and wide. Fruits enveloped in the thin papery densely white pubescent calyx, brown, transversely oblate, 2 × 3 mm., apiculate, glabrous, dividing into 2 ± hemispherical nutlets, the inner faces covered with white tissue made up of raised areas and bounded by a narrow raised margin.

UGANDA.   Karamoja District: Lonyili [Longili] Mt., Apr. 1960, *J. Wilson* 913!; Ankole District: E. Ankole near Sanga Rest House, Oct. 1932, *Eggeling* 616!; Masaka District: Masaka–Mbarara road, 16 Oct. 1925, *Maitland* 845!

* Tanner on field notes to several specimens says non-aromatic but this contradicts other collectors.

KENYA. W. Suk District: Kapenguria, 13 May 1932, *Napier* 1928!; Nakuru District: Dundori, 9 July 1958, *Verdcourt* 2204!; Nairobi, Karura Forest edge, 10 Nov. 1930, *Napier* 560!
TANZANIA. Musoma District: Wogakuria Guard Post, 30 Dec. 1964, *Greenway & Myles Turner* 11790!; Kondoa District: 96 km. Babati–Dodoma road, 7 Dec. 1965, *Richards* 20924!
DISTR. U 1, 2, 4; K 2–4; T 1, 2, 5; Rwanda, Burundi, Mozambique, Malawi, Zambia, Zimbabwe
HAB. Scattered tree grassland, open woodland, grassland subject to seasonal burning; 1110–2280 m.

SYN. [*Lantana salviifolia* sensu Bak. in F.T.A. 5: 276 (1900), pro parte, *non* Jacq.]
   [*Lippia wilmsii* sensu H. Pearson in Fl. Cap. 5: 196 (1901), pro parte; Moldenke in Phytologia 2: 469, 483 (1948), 3: 77 (1949), 3: 457, 488 (1951), 13: 171, 174 (1966) & 40: 202, 204 (1978); Troupin, Fl. Rwanda 3: 284, fig. 92.2 (1985), omn. pro parte et adnot. permult. in Herb. Hort. Kew, *non* H. Pearson sensu stricto]
   [*L. africana* sensu Mold. in Phytologia 2: 469 (1948), pro parte, *non* Moldenke sensu stricto]
   L. africana Mold. var. *villosa* Mold. in Phytologia 2: 469 (1948). Type: Kenya, E. Elgon, *Holm* 32 (S, holo.; K, photo.!)
   [*L. wilmsii* H. Pearson var. *scaberrima* sensu Mold. in Phytologia 3: 458 (1951), pro parte, *non* Mold. sensu stricto]
   L. wilmsii H. Pearson var *villosa* (Mold.) Mold. in Phytologia 4: 180 (1953) & 13: 176 (1966), excluding some cited specimens
   L. sp. cf. L. plicata; Vollesen in Opera Bot. 59: 82 (1980)

NOTE. Fernandes (Bol. Soc. Brot., sér 2, 59: 250, 262 (1987)) cited *Maas-Geesteranus* 4786 (Kenya, E. of Kitale, Cherangani, 23 May 1949) as *L. woodii* the identity of which she unravelled in the above paper but was doubtful that *L. africana* var. *villosa* Mold. was the same taxon. From the considerable amount of material at Kew now available which she did not see I am convinced that only one taxon is involved. Most of the material had been named *L. wilmsii* or *L. africana* var. *villosa* and it had been assumed that these two names belonged to the same taxon until Fernandes sorted out the taxonomy of this group of species. *Anderson* 1031 (Lindi District, Nachingwea, 6 Apr. 1955) is probably *L. woodii* but differs in facies. No other material has been seen from **T**8 — it may be a form or hybrid of *L. woodii* (e.g. × *L. kituiensis* but that is not recorded for **T**8) or perhaps a distinct taxon.

9. **L. plicata** *Bak.* in F.T.A. 5 : 281(1900); Meikle in Mem. N.Y. Bot. Gard. 9: 36 (1954); F.F.N.R.: 370 (1962); Haerdi in Acta Trop., Suppl. 8: 152 (1964). Lectotype chosen here: Zambia, Lake Tanganyika, Fwambo, *Carson* 81 (K, lecto.!)*

Woody herb or shrub 0.6–3.6 m. tall, with minty aromatic odour; stems brown or reddish purple, ± square, strigose with bulbous-based hairs, the bases remaining to render older stems minutely tuberculate. Leaves opposite or occasionally in whorls of 3; blades ovate to lanceolate, 1.5–10(–12.5) cm. long, 0.7–5.5 cm. wide, acute at the apex, ± rounded at the base, crenate-serrate or minutely crenate, very closely rugose-bullate, reticulate, exceedingly scabrid above with ± white tubercle-based hairs, densely pubescent and much less scabrid beneath, with numerous small glands, sometimes glabrescent; petioles 2–6(–10) mm. long. Inflorescences many-flowered, 1–4 per axil and also forming rather dense terminal panicles, cone-like, ± globose to oblong, eventually expanding, 1.3–5 cm. long, 1.5–2 cm. wide; peduncles 1–3.5(–7) cm. long; bracts ovate to ovate-lanceolate, (5–)7–8(–10) mm. long, 3–5 mm. wide, acute, ± ribbed and venose, ± yellow-green, thinly to densely pubescent on outer face, ultimately falling and leaving a bare rhachis roughened by bract and pedicel-bases. Calyx-lobes 2, oblong, 1.2 mm. long, densely hairy. Corolla slightly or strongly scented, white with yellow inside the tube, or pink or lilac; tube slender, 4–4.5(–?8) mm. long, widened above, narrow part glabrous, the upper part pubescent (often in lines) and glandular; limb 4.5–5 mm. wide, lower lip ± 3-lobed, the lowest lobe 1.5 mm. long, 2 mm. wide. Style 1.5 mm. long; stigma 0.6 mm. long. Nutlets half-ovoid, 1.2–1.5 mm. long, 1 mm. wide, densely spreading pubescent on outer face and glandular. Fig. 5.

TANZANIA. Mpanda District: Mahali Mts., Sibogo, 1 Aug. 1958, *Newbould & Jefford* 1285!; Ufipa District: Mbisi, 27 Nov. 1949, *Bullock* 1945!; Mpwapwa, 17 Sept. 1937, *Mr. & Mrs. Hornby* 817!; Morogoro, Uluguru Mts., 25 Oct. 1934, *E.M. Bruce* 34!
DISTR. T 2**, 4–8; Zaire, Burundi, Mozambique, Malawi, Zambia and Angola
HAB. Tall *Pennisetum* and *Hyparrhenia* grassland, *Acacia-Combretum* thicket, *Brachystegia, Terminalia, Combretum, Julbernardia*, etc., open woodland, old cultivations and roadsides; 450–1950 m.

---

* Baker gives '128', White '198' for the other *Carson* syntype and it could be either.
** This **T** 2 record is based on a sheet 'N. Kilimanjaro, 6000–7000 ft., *D.H.S. Grant* in F.H. 127, July 1924', but curiously no other material has been seen from this well-collected area.

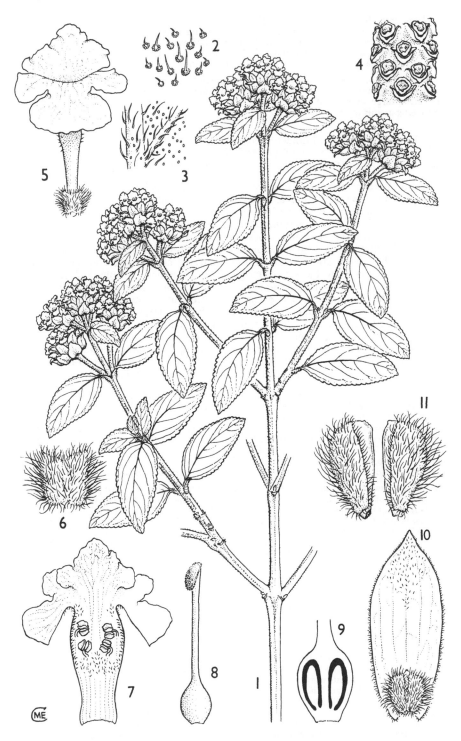

FIG. 5. *LIPPIA PLICATA* — **1**, habit, × ²/₅; **2**, part of upper leaf surface, × 10; **3**, part of lower leaf surface, × 10; **4**, part of peduncle showing scars of bracts and flowers, × 8; **5**, flower, × 6; **6**, calyx, × 14; **7**, corolla, opened out, × 6; **8**, ovary and style, × 14; **9**, longitudinal section of ovary, × 24; **10**, bract and fruit, × 6; **11**, 2 nutlets, × 16. 1–3, 5–9, from *Bullock* 1945; 4, 10, 11, from *Newbould & Harley* 4251. Drawn by Mrs. M.E. Church.

SYN. *L. adoensis* Hochst. var. *multicaulis* Hiern in Cat. Afr. Pl. Welw. 1: 829 (1900). Types: Angola,
    Ambaca, Puri–Cacarambala, *Welwitsch* 5733 & Pungo Andongo, S. of praesidium, *Welwitsch*
    5667, pro parte, 5690, pro parte (LISU, all syn., BM, isosyn., K, 5667 isosyn.!)
    L. strobiliformis Mold. in Phytologia 2: 317 (1947). Type: Tanzania, Mpanda District, Lake
    Tanganyika, Karema, *Storms* 3 (BR, holo.!, K, photo.!)

NOTE. Forms from **T** 4 with the inflorescences under 1 cm. long are not I think worthy of any
varietal recognition. Moldenke commenting on the *Storms* specimen states that it is remarkable for
having a large number of sessile nodules on the long slender roots. Some person should
reinvestigate this in the field.

10. **L. kituiensis** *Vatke* in Linnaea 43:528(1882); Moldenke in Phytologia 3: 292(1951) &
12: 237 (1965), pro parte, & 39: 106 (1974); R. Fernandes in Bol. Soc. Brot., sér 2, 59: 271, t.
1–3 (1987). Type: Kenya, Kitui, *Hildebrandt* 2738 (B, holo.†, M, lecto., K, iso.!, * W, iso.)

Shrubby herb or shrub (even described as a small tree) 1.2–4.5(–6) m. tall; stems
densely hairy with ± rough tubercle-based hairs; bark rough, corky, deeply longitudinally
fissured. Leaf-blades ± stiff, scented, elliptic, ovate, oblong-lanceolate or -elliptic, 1.5–11.5
cm. long, 0.6–3.7(–5) cm. wide, rounded to acute at the apex, cuneate to truncate at the
base, rough or scabrid above with tubercles from which the hairs very soon break off,
rugose and bullate above with venation impressed, softly hairy and glandular beneath
with long white hairs and venation raised; margins closely finely to more coarsely
crenate-serrate save at base and rough with tubercles; petiole 2–7 mm. long.
Inflorescences 1–3 per axil, globose to hemispherical or oblong-ovoid, 0.5–2 cm. long
and wide, mature peduncles 2–6.5(–8) cm. long; lowest bracts ovate, 8–10 mm. long, 4 mm.
wide, acute to acuminate, pubescent or rough with tubercles, others smaller, triangular-
ovate to -lanceolate, 5 mm. long, 2 mm. wide, acute to acuminate, pubescent to woolly and
glandular. Calyx bilobed, 1.2 mm. long, densely pubescent. Corolla cream to yellow-green
with orange-yellow or yellow centre; tube 3–4 mm. long, puberulous and glandular above;
larger lobes semi-circular, 1–1.5 mm. long, 1.5–2 mm. wide. Mericarps half-ellipsoid, 1.5
mm. long, 1 mm. wide, white-granular on inner face, densely pubescent outside.

KENYA. W. Suk District: N. Cherangani Hills, Aug. 1969, *Tweedie* 3736!; Kiambu District: Muguga, 18
    Sept. 1951, *Trapnell* 2134!; Masai District: 48 km. from Narok turn-off on Nairobi–Naivasha road
    towards Narok, 16 June 1956, *Verdcourt* 1501!
TANZANIA. Musoma District: Serengeti, Seronera–Soitayai, km. 24, 28 Mar. 1961, *Greenway* 9915!;
    Arusha District: N. side of Little Meru Mt., Kisimiri, 22 Feb. 1969, *Richards* 24148!; Lushoto District:
    Kwai valley, 25 Apr. 1953, *Drummond & Hemsley* 2245!
DISTR. **K** 2, 4–6, 7 (see note); **T** 1–3, 7; Malawi (but see note)
HAB. Bushland, woodland, rough grassland, often on volcanic soil or lava rocks; 405–2250 m.

SYN. [*Lantana salviifolia* sensu Bak. in F.T.A. 5: 276 (1900), as "*salvifolia*", pro parte, *non* Jacq.]
    [*L. viburnoides* sensu Bak. in F.T.A. 5: 276 (1900), pro parte, *non* (Forssk.) Vahl]
    [*Lippia asperifolia* sensu Bak. in F.T.A. 5: 280 (1900), pro parte, *non* Marthe]
    Lantana scabrifolia Mold. in Phytologia 1: 422 (1940) & Fifth Summ. Verbenac. 1: 237, 241 & 250
    (1971). Type: Kenya, near Nairobi, *Mearns* 267 (NY, holo., US, iso., BM, photo.!)
    [*Lippia ukambensis* sensu auct. pro max. parte, e.g. T.T.C.L.: 640 (1949); K.T.S.: 588 (1961);
    Moldenke in Phytologia 13: 31 (1966), pro parte; Blundell, Wild Fl. E. Afr.: 399, t. 242 (1987)
    non Vatke]
    [*L. burtonii* sensu Mold. in Phytologia 12: 106 (1965), pro parte, quoad *Peter* 51781, *non* Bak. (see
    also Fernandes in Bol. Soc. Brot., sér. 2, 59: 255 (1986))]
    [*L. lupuliformis* sensu Mold. in Phytologia 12: 264 (1965), pro parte, quoad *Peter* 580, 2122 &
    51797, *non* Mold. sensu stricto]
    [*L. schliebenii* sensu Mold. in Phytologia 12: 482 (1966), pro parte, *non* Mold. (1947)]
    [*L. wilmsii* H. Pearson, var. *scaberrima* sensu Mold. in Phytologia 13: 175 (1966), pro parte, *non L.
    africana* var *scaberrima* Mold. (1948)]

NOTE. Fernandes (Bol. Soc. Brot., sér 2, 59: 272 (1986)) mentions a *Whyte* specimen as from
'?Malawi' but the species is not included in her account for F.Z.; the specimen is, however,
annotated as *L. kituiensis* and does appear to be this species. Material from the W. Usambaras, e.g.
Semsei 3865, has elongate lanceolate leaves to 16 × 4.5 cm. which are more finely serrate. *Murray
High School* 19 (Teita Hills, Wusi, 30 Oct. 1965) is similar with elongate leaves scarcely rough above
and the basis of the **K** 7 record. Fernandes has annotated *Davies* 830 (Tanzania, Mbeya, Utengule,
30 Jan. 1933) as *L. kituiensis* but I think it is a form of *L. plicata. L. kituiensis* is it seems practically
non-existent in S., SW. or central Tanzania. The Kikuyu have used it for weaving basket-type grain
stores.

* The confusion of two species on the Kew sheet is explained by R. Fernandes in the reference
cited.

11. **L. carviodora** *Meikle* in K.B. 3: 467 (1949); E.P.A.: 790 (1962); Sebsebe & Puff in Beitr. Biol. Pflanzen 64: 221–230 (1989). Type: Kenya, Northern Frontier Province, Mathews Range, Wamba, *Gilbert Walker* (K, holo.!)

Aromatic much branched dioecious shrub 0.5–1.5 m. tall; branchlets striate, at first pale with peeling bark, later grey or brownish, pubescent but eventually ± glabrous, the youngest parts densely covered with minute pale glands; older stems dark grey, strongly ·ribbed; leaf-blades ovate, elliptic-oblong or almost round, 0.7–3.5(–4.8) cm. long, 0.4–2.2(–3.8) cm. wide, rounded at the apex, truncate, cuneate or subcordate at the base, closely crenate or obtusely serrate, often slightly bullate, adpressed pubescent and sparsely glandular above, densely grey-puberulous and with numerous small glands beneath; petiole 0.4–1(–2.2) cm. long. Inflorescences solitary in the axils, ± 12-flowered spikes, ± 5 mm. long, ± 7 mm. wide. Male: peduncles 1–3 mm. long, white-tomentose; bracts green, linear-lanceolate to elliptic, 3–4 mm. long, 1–2 mm. wide, pubescent, acuminate; calyx membranous, 1 mm. long and wide, white-pubescent, divided into 2 oblong or emarginate lobes; corolla white, sometimes with a yellow centre; tube 3–3.5(–4) mm. long, pubescent outside at least above; lobes 1.2–1.8 mm. long; ovary rudimentary; style up to 2 mm. long. Female: peduncles 4–6(–10) mm. long; bracts broadly ovate to semi-circular, 4–7 mm. long, 5–10 mm. wide, densely glandular, accrescent in fruit to 0.8–1.2 cm. long, 1.1–1.5 cm. wide, pubescent, reticulately veined, acute, cordate at the base, forming hop-like infructescences 2 cm. long, 1.5 cm. wide; calyx lobed as in ♂, enlarging and covering the fruit, membranous, 1–2 mm. long, 1–1.5 mm. wide; corolla ± as in ♂ with the tube 1.7–2.6 mm. long; stamens rudimentary; ovary subglobose, ± 1 mm. long, 2-locular with style 2–3 mm. long. Fruit compressed-globose, ± 3 mm. wide included in the tomentose calyx and crowned by the withered corolla, dividing into 2 brownish pyrenes.

KENYA. Northern Frontier Province: Lolokwi [Ololokwe] Mt., 13 Jan. 1972, *Bally & Smith* 14772!; Turkana District: Oropoi, Feb. 1965, *Newbould* 6978!; Kilifi District: Kibarani, 20 Apr. 1949, *Jeffery* H21/49!
DISTR. **K** 1, 2, ?3, 7 (see note); Somalia
HAB. *Acacia* and mixed *Acacia spp.*, *Commiphora* bushland, usually in rocky places, often forming pure stands; (?0–)700–1200 m.

SYN. *Lantana microphylla* Franch., Sert. Somal.: 49 (1882); Hutch. & Bruce in K.B. 1941: 176 (1941), *non Lippia microphylla* Cham. Type: Somalia [region des Çomalis], *Révoil* (P, holo.!)
   [*L. petitiana* sensu P.E.Glover, Check-list Somaliland Trees, Shrubs & Herbs 268 (1947), *non* A. Rich.]
   Lippia carviodora Meikle var. *minor* Meikle in K.B. 3: 468 (1949); E.P.A.: 791 (1962). Type: Somalia (N.), Sheikh Pass and Maritime plain, *D. Thomson* 72 (K, holo.!)

NOTE. Used for making a tea and also grazed. It is highly likely that the Kibarani plant cited above was cultivated and certain that *Jeffrey* H12/49 from Kilifi itself was.

7. LANTANA

L., Sp. Pl.: 626 (1753) & Gen. Pl., ed. 5: 275 (1754); R. Fernandes in Bol. Soc. Brot., sér. 2, 61: 125–214 (1989)

Erect, scandent or occasionally prostrate herbs or shrubs, with usually tomentose, scabrid-hairy or sparsely to densely prickly stems. Leaves opposite or in whorls of 3(–4), mostly aromatic, usually toothed, often ± bullate or rugose, usually glandular-punctate. Flowers sessile in often colourful, usually axillary, pedunculate heads or spike-like inflorescences, each flower subtended by an ovate to lanceolate usually acuminate bract. Calyx small, membranous, truncate or sinuate-dentate. Corolla red, yellow, purple, blue, mauve or white, sometimes bicoloured with throat yellow to orange, often changing colour after fertilisation, salver-shaped with narrowly cylindrical tube; limb obscurely 2-lipped, 4–5-lobed, the lobes obtuse or even emarginate. Stamens 4, didynamous, inserted at middle of tube, included. Ovary 2-locular with 1 basal erect ovule per locule; style short, with ± thick oblique or sublateral stigma. Axis of fruiting spike with scars ± scattered, not close as in *Lippia*. Drupes with ± fleshy mesocarp, rarely rather dry; endocarp hard, 2-locular or splitting into 2 1-locular pyrenes. Seeds without albumen.

A genus of about 150 species (270 taxa fide Moldenke), mostly in tropical and subtropical America, with a few in the tropics of the Old World. The true number of taxa is probably much less. Several are widely cultivated and there are numerous cultivars. In some areas a few have become numbered amongst the world's worst and most aggressive weeds. The species are notoriously difficult and variable and hybridization is apparently very widespread. Attempts to extend results from one area to a wider context usually fail and the use of different classifications in different areas is likely to be an illogical but practical solution. I have failed to produce a satisfactory account; many possibly hybrid specimens will not key out properly. Where I differ from R. Fernandes, who recognises more taxa, I have given extended notes. Moldenke has caused great confusion by subsequently misidentifying taxa he previously described from single specimens at intervals; the extensive material at K, BM, etc., was not studied as a whole. I have not cited all these numerous misidentifications but they have been exhaustively investigated by R. Fernandes in the paper cited to which reference should be made.

L. montevidensis (Sprengel) Briq. (L. sellowiana Link & Otto) has been grown as a bedding plant in Nairobi and is slender with decumbent or trailing branches and white to lilac flowers in heads with peduncles usually much exceeding the leaves; the corolla-limb is ± 6.5 mm. wide. I have seen only two specimens (Nairobi, Kenyatta National Hospital, 25 Mar. 1974, *Mwangi* 22! & Museum Garden, near Ainsworth Bridge, 18 Apr. 1972, *Mathenge* 837!) and it is not mentioned by Jex-Blake, Gard. E. Afr., ed. 4 (1957). *Chambo* in *E.A.H.* 14850 (Tanga District, Mlingote Estate, 25 Sept. 1964) is similar to *L. montevidensis* and *L. ukambensis* but said to have yellow flowers. There is no data as to whether wild or cultivated. It may be an introduced plant but I have failed to name it. R.A. Howard (Arnoldia 29: 73–109 (1969)) lists the very numerous cultivar names which have been used in the genus.

1. Stems nearly always covered with recurved prickles but sometimes very sparse or even absent; bracts of the inflorescences linear to linear-lanceolate, up to 2 mm. wide; spikes not elongating after flowering; all leaves opposite . . . . 1. *L. camara*
   Stems without prickles; bracts lanceolate to ovate, usually wider; spikes mostly elongating after flowering; leaves opposite or in whorls of 3 (note, a poorly known white-flowered variant or hybrid of *L. trifolia* from **T** 7, Mufindi/Dabaga area, has narrow bracts ± 2 mm. wide) 2
2. Flowers white or with yellowish centres; leaves mostly opposite, less often in whorls of 3 3
   Flowers lilac, pink or mauve, etc.; leaves mostly in whorls of 3 6
3. Inflorescences very shortly pedunculate or sessile, the peduncles mostly shorter than heads, the whole not exceeding the leaves, usually much shorter; leaves small, mostly under 4 cm. long 4
   Inflorescences with longer peduncles often exceeding the leaves or at least longer than the heads; leaves larger, up to 14 cm. long 5
4. Leaves less pubescent, densely glandular beneath, with surface not at all obscured by pubescence, usually less plicate and more coarsely crenate-dentate; bracts more acuminate, sparsely white bristly pilose with long spreading ciliae (**K** 4, 7; **T** 5) . . . . 3. *L. humuliformis*
   Leaves more pubescent, the under-surface obscured, often velvety, the glands not so evident, plicate, finely crenulate; bracts subacute, not acuminate, more finely densely pubescent (**K** ?4; **T** 1, 2) . . . . 2. *L. viburnoides* subsp. *masaica*
5. Bracts narrow, mostly lanceolate and narrowly attenuate at the apex; leaves often in whorls of 3 . . . . . . 5. *L. trifolia* (white variants and hybrids)

   Bracts wider, ovate or ovate-lanceolate, rarely rounded, subacute or obtuse to distinctly attenuate; leaves typically opposite . . . . 2. *L. viburnoides* aggregate
6. Inflorescences mostly shortly pedunculate at maturity, the peduncles mostly not exceeding the heads or young inflorescences ± sessile, typically with basal large bracts, broad and as long as the young inflorescence 4. *L. ukambensis*
   Inflorescences long-pedunculate 7

7. Bracts narrow, mostly lanceolate and narrowly attenuate at
     the apex; leaves mostly in whorls of 3   . . . . .   5. *L. trifolia*
     Bracts broader and more ovate; leaves sometimes opposite   pink-flowered forms
                                                                      of *L. viburnoides* and
                                                                      L. trifolia hybrids

1. **L. camara** L., Sp. Pl.: 627 (1753); Schauer in DC., Prodr. 11: 598 (1847); Bak. in F.T.A.
5: 275 (1900); T.T.C.L.: 639 (1949); U.O.P.Z.: 326 (1949); Jex-Blake, Gard. E. Afr., ed. 4: 117,
239 (1957); K.T.S.: 586 (1961); F.F.N.R.: 368 (1962); Meikle in F.W.T.A., ed. 2, 2: 435 (1963);
Ivens, E. Afr. Weeds: 65 (1967); Verdc. & Trump, Common Poisonous Pl. E. Afr.: 172
(1969); Jafri & Ghafour in Fl. W. Pakistan 77: 9 (1974); Townsend, Fl. Iraq 4: 656 (1980);
Fosberg & Renvoize, Fl. Aldabra: 223, fig. 35/5 (1980); Blundell, Wild Fl. Kenya: 109,
31/197 (1982); Spies & Stirton in Bothalia 14: 101–111 (1982); Moldenke in Rev. Fl.
Ceylon 4: 220 (1983) (extensive bibliography, synonymy and uses); Spies in S. Afr. Journ.
Bot. 3: 231–250 (1984) (cytotaxonomy); Nash & Nee, Fl. Veracruz 41: 70 (1984) (extensive
synonymy); Sinha & Sharma in F.R. 95: 621–633 (1984) (extensive biological references);
Sanders in Syst. Bot. 12(1): 44–60, fig. 9 (1987); Blundell, Wild Fl. E. Afr.: 398, t. 704 (1987).
Type: provenance unknown, *Linnean Herbarium* No. 783/4 (LINN, lecto.*)

Mostly a spreading, rather unpleasantly aromatic shrub 0.35–5 m. tall but can be
scandent to 12 m. or even in cultivation over a long period become a tree to 8 m. with a
bole 10–12 cm. diameter; stems ± square, almost completely unarmed to slightly or
copiously and viciously covered with recurved prickles. Leaves opposite, ovate to ovate-
oblong, 2–12 cm. long, 2–7 cm. wide, mostly fairly small, acute at apex, rounded to broadly
or narrowly cuneate or rarely subcordate at base, scabrid with coarse tubercule-based
hairs and rugose with impressed venation above, ± pubescent or glabrescent beneath but
scabrid on raised venation, closely crenate-serrate; petioles 0.7–2 cm. long. Flowers in
axillary ± flat heads 2–3(–5) cm. diameter, the axis 0.5–1 cm. long; peduncles 1.5–10 cm.
long; bracts linear-lanceolate to lanceolate, 4–8(–13) mm. long, 1–1.5 mm. wide. Calyx
very thin, ± 3 mm. long. Corolla very variously coloured, red, purple, pink with yellow
centre, salmon, orange, pure yellow, white, etc., sometimes the inner flowers a different
colour from the outer, e.g. yellow and red; tube 0.8–1.3 cm. long, densely puberulous
outside; limb 4–9 mm. wide, the lobes oblong to rounded, up to 5 mm. wide. Drupes fleshy,
purple or black, 3–5(–7) mm. diameter (dry); pyrene essentially ovoid, 6 mm. long, 3.5 mm.
wide, laterally somewhat like the head of a bird with a narrower and a wider portion, the
latter more rugose and crenate where the two join.

UGANDA. W. Nile District: Omugo Rest Camp, 8 Aug. 1953, *Chancellor* 137!; Mengo District: Mabira
     Forest, 7 Nov. 1938, *Loveridge* 8! & 9! & Buganda, Kyempisi, Kiwango, *Tanner* 6031!
KENYA. Nairobi, Ainsworth Swamp by Nairobi R., 3 Sept. 1956, *Verdcourt* 1573!; N. Kavirondo District:
     Webuye Falls [Broderick Falls], May 1958, *Tweedie* 1537!; Teita District: Tsavo National Park East,
     Galana R., km. 1.6 from Sobo Rocks, 9 Jan. 1967, *Greenway & Kanuri* 13009!
TANZANIA. Mwanza District: Butimba, 15 Mar. 1952, *Tanner* 568!; Tanga District: Amani Nursery, 31
     Oct. 1969, *Ngoundai* 428!; Rufiji District: Mafia I., Ngombeni, 11 July 1932, *Schlieben* 2561!; Zanzibar
     I., Kitope, 10 Apr. 1950, *R.O. Williams* 11!
DISTR. U 1, 4; K 3–5, 7; T 1–3, 6, 7; Z; probably originally native of the W. Indies, now naturalized
     throughout the tropics and subtropics of both hemispheres
HAB. Forming dense thickets, by roadsides, in forest clearings, old cultivations, etc.; 0–2040 m.

SYN. *L. aculeata* L., Sp. Pl.: 627 (1753). Type: specimen grown in Uppsala Botanic Garden, *Linnean
     Herbarium* No. 783/6 (LINN, lecto.)
     L. antidotalis Schumach. & Thonn., Beskr. Guin. Pl.: 276 (1827); Bak. in F.T.A. 5: 276 (1900).
     Type: Ghana [Guinea], *Thonning* 125 (C, syn., P-JU, isosyn.)
     L. camara L. var. *aculeata* (L.) Mold. in Torreya 34: 9 (1934) & in Rev. Fl. Ceylon 4: 225 (1983)

NOTE. The infraspecific classification of this species is beyond the scope of this Flora, but the
references given will lead enquirers to the enormous literature devoted to this plant, which is now
a dangerous weed in very many parts of the tropics.
     It was formerly cultivated as an ornamental, but this is now recognised as a very dangerous
procedure. In some parts of the world, e.g. arid Middle Eastern areas, its cultivation causes no
problems and it is a useful ornamental, but in many tropical areas its propensity for forming
widespread uncontrollable thickets has led to serious ecological problems. Even here, however,
the problems have sometimes been misunderstood. Frequently the thickets have prevented soil

---

* Given by Moldenke, loc. cit. (1983). Since this specimen bears '3 camara' written in Linnaeus's
hand it appears the most suitable of several syntypes to choose.

erosion and acted as shade and a nursery for forest tree seedlings, leading to re-establishment of the forest and eventual disappearance of the *Lantana*. In different areas and with different variants of the species the problems are different. It is certain the plant should never be cultivated in any tropical areas.

Moldenke keeps var. *aculeata* (L.) Mold. separate but there appears to me to be every intermediate. The common form in East Africa seems to be a pink one with the centre flowers starting yellow and becoming orange then pink. *Mathenge* 479 (Nairobi, University College Campus, 27 Nov. 1969) has yellow flowers and shorter bracts with fine indumentum somewhat different from most specimens seen but I have not attempted to name it in detail. Batty reports that white-flowered specimens occur in **T** 6 but undoubtedly populations in East Africa are far more uniform than in South Africa where the plant has become an unmitigated disaster and has been studied in detail. The earliest specimen seen from East Africa is one from E. Usambaras, Bulwa, *Peter* 18304, collected in Nov. 1916; *Napier* 429, collected in Nairobi in Sept. 1930, states "brought from the coast" and is the earliest I have seen from up-country.

R. Fernandes has described a distinctive forma *glandulosa* in Bol. Soc. Brot., sér. 2, 61: 132 (1989), based on Angola, Pungo Andongo, Barrancos da Pedra Pungo, *Welwitsch* 5676 (COI, holo., BM, K!, LISU, iso.). *L. tiliifolia* Cham. forma *glandulosa* (Schauer) R. Fernandes has been grown in W. Zambia; very similar to *L. camara*, it differs in the spreading indumentum of the bracts.

Peter 39378 (Dar es Salaam, Geresane, 2 Apr. 1926) has been annotated by Moldenke as *L. tiliifolia*. I would agree with Nash and Nee in including it in *L. camara* sensu lato.

2. **L. viburnoides** (*Forssk.*) *Vahl*, Symb. Bot. 1: 45 (1790); Bak. in F.T.A. 5: 276 (1900), pro parte excl. specim. cit.; Blatter in Rec. Bot. Surv. India 8: 363 (1921); Schwartz in Mitt. Inst. Bot. Hamb. 10: 215 (1939); K.T.S.: 587 (1961); Collenette, Fl. Saudi Arabia: 495 (1985); R. Fernandes in Bol. Soc. Brot., sér. 2, 61: 196, t. 15–18 (1989). Type: Yemen, between Bolghore and Moxhaja, *Forsskål* [323] (C, lecto.!)*

Shrub or subshrub 0.3–2.5 m. tall, occasionally scrambling; branches with adpressed short upwardly directed hairs, at length ± glabrous and roughened with distinctly raised petiole-bases. Leaves almost always opposite, but occasionally in whorls of 3, ovate to ovate-lanceolate, elliptic-oblong or elliptic-lanceolate, 1.5–2.5(–12) cm. long, 0.7–4(–7) cm. wide, acute to ± obtuse at the apex, rounded then attenuate-cuneate into the petiole at the base, regularly coarsely to finely somewhat rounded crenate, not to distinctly discolorous, pubescent with ± tubercle-based hairs above and somewhat scabrid, pubescent to velvety or tomentose beneath; petiole up to 1.5 cm. long. Peduncles solitary, rarely paired, (1.5–)2.5–13.5 cm. long, shorter than to usually longer than the subtending leaf, shortly pubescent; heads 0.7–1.5 cm. long, 1.1–2.5 cm. wide, mostly wider than long but elongating in fruit to 1.5–3 cm. or occasionally much longer, to 5 cm. in some forms; lower bracts rounded to broadly ovate or rarely oblong-ovate, 0.6–1.4(–1.5) cm. long, (3–)5–9 mm. wide, ± acuminate to less often long-acuminate or obtuse, equalling or exceeding the upper corollas at flowering but shorter than fruiting spikes, pubescent and glandular to densely velvety tomentose, sometimes strongly ribbed in fruit. Calyx 2 mm. long, undulate, thinly pubescent. Corolla predominantly white or sometimes pink; tube 5.5 mm. long; limb ± 5 mm. wide; lower lip 3.5 mm. wide. Drupes lilac-purple, rhomboid-subglobose, 2.5–3.5 mm. long.

1. Peduncles short, 0.4–1(–2.5) cm. long; leaves 1.5–3 cm.
   long, 1–1.6 cm. wide; inflorescences 0.8–1(–1.7) cm.
   long . . . .   subsp. **masaica**
   Peduncles, leaves and inflorescences longer (subsp.
   viburnoides) 2
2. Leaves very discolorous, densely thick white velvety
   tomentose beneath, the indumentum totally obscuring
   the surface, very rugose above, mostly 4–4.5 cm. long, 2
   cm. wide; inflorescences sometimes paired in axils,
   mostly equalling or not exceeding mature leaves   var. **kisi**
   Leaves less discolorous, not so densely white tomentose
   beneath, often larger; inflorescences mostly exceeding
   the leaves . . . .   var. **viburnoides**

* R. Fernandes has chosen this specimen (of three in Forsskål's herbarium labelled *viburnoides*) but it should be pointed out that only "In Monte Barah" is mentioned in the original reference. One could argue that this should be the holotype.

subsp. **viburnoides**

Leaves mostly larger, exceeding 4 cm. long. Inflorescences usually larger and equalling or exceeding the leaves, with quite long peduncles.

var. **viburnoides**

Leaves ovate to elliptic-lanceolate, up to 12 cm. long, 7 cm. wide, not markedly discolorous nor densely thick white velvety tomentose beneath. Peduncles up to 13.5 cm. long mostly exceeding the leaves.

UGANDA. Karamoja District: Lokitonyale, Sept. 1963, *Tweedie* 2728! & Lodoketemit, 12 July 1958, *Kerfoot* 304! & Amudat, 20 May 1940, *A.S. Thomas* 3408!
KENYA. Northern Frontier Province: Marsabit National Reserve, 14 Jan. 1972, *Bally & Smith* 14793!; Naivasha District: Mt. Margaret, 27 Jan. 1963, *Verdcourt* 3569!; Masai District: 16 km. from Narok on Nairobi road, top of Siyabei Gorge, 15 July 1962, *Glover & Samuel* 3142!
TANZANIA. Shinyanga [New Shinyanga], Jan. 1933, *Napier Bax* 345!; Ufipa District: above Msanzi village, 13 Dec. 1958, *Richards* 10339!; Morogoro District: 12.8 km. NE. of Kingolwira Station, 13 Mar. 1954, *Welch* 216!; Zanzibar I., Chwaka, 24 Feb. 1961, *Faulkner* 2761!
DISTR. U 1; K 1–7; T 1–8; Z; P; Arabia, Egypt, Ethiopia, Somalia, Zambia, Zimbabwe, Malawi, Mozambique and Angola
HAB. Bushland including *Acacia–Commiphora*, *Acacia–Tarchonanthus*, etc., forest clearings and edges, thicket, *Brachystegia* woodland; 0–1890 m.

SYN. *Charachera tetragona* Forssk., Fl. Aegypt.-Arab.: CXV, 115, 378 (1775). Type: Yemen, El Hadiyah, *Forsskål* [315] (C, holo.!)
    C. viburnoides Forssk., Fl. Aegypt.-Arab.: CXV, 116, 379 (1775)
    [*Lantana salviifolia* sensu Bak. in F.T.A. 5: 277 (1900), as "*salvifolia*", pro parte; T.T.C.L.: 639 (1949), pro parte; U.O.P.Z.: 326 (1949), as "*salvifolia*", pro parte, *non* Jacq.]
    L. tetragona (Forssk.) Schweinf., Arab. Pflanzennamen: 145 (1912)
    [*Lippia schliebenii* sensu Mold. in Phytologia 12: 482 (1966), pro parte, *non* Moldenke sensu stricto]
    Lantana viburnoides (Forssk.) Vahl subsp. *richardii* R. Fernandes in Bol. Soc. Brot., sér. 2, 61: 200, t. 19 (1989). Type: Ethiopia, Assaye, *Quartin-Dillon & Petit* (P, holo.!)
    L. viburnoides (Forssk.) Vahl subsp. *richardii* R. Fernandes var. *richardii*; R. Fernandes in Bol. Soc. Brot., sér. 2, 61: 201, t. 19, 20 (1989)

NOTE. R. Fernandes has separated the mainland populations from the Arabian populations as subsp. *richardii*, but there is much variation in Arabia and almost identical specimens from both areas make the separation difficult. She discusses the problem at length in the reference given. She also maintains (tom. cit.: 153, t. 7, 8) *L. petitiana* A. Rich. (Tent. Fl. Abyss. 2: 169 (1850); type: Ethiopia, Wodgerat, *Petit* (P, lecto.!*, K, isosyn.!) distinct, but Sebsebe (pers. comm.) considers it a form of *L. viburnoides*.
    A few otherwise typical specimens with opposite leaves and wide bracts have pink flowers both in Arabia and in Africa. Coastal specimens are often coarser with wider, more attenuate, even long cuspidate bracts. A few specimens, e.g. *Verdcourt, Baring & Williams* 2195 (Kenya, Masai District, Ol Orgesailie, 31 May 1958) and *Richards* 24915 (Tanzania, Masai District, Kitumbeine Mt., 1 Mar. 1969) have the inflorescences distinctly more elongate in fruit, up to 4.5–5 cm. and then longer than the peduncles. *Kerfoot* 304 & 426 (Uganda, Moroto District, Lodoketemit Catchment, July 1958) are said to have orange-buff and bright orange flowers respectively. A duplicate without colour notes was named *viburnoides* by R. Fernandes and I agree. Perhaps the notes refer only to the throat of the corolla.

    var. **kisi** (*A. Rich.*) *Verdc.*, comb. et stat. nov. Type: Ethiopia, Wodgerat, *Petit* (P, syn.!)**

Leaves ovate to elliptic-ovate, mostly 4–4.5 cm. long, 2 cm. wide, very discolorous, densely thick white velvety tomentose beneath, the indumentum totally obscuring the surface, very rugose above, rather closely crenate-serrate. Inflorescences sometimes paired in axils, mostly equalling or not exceeding mature leaves; bracts elliptic, acute but not long-acuminate, ± densely woolly-velvety or velvety pubescent on outer surface.

KENYA. Teita District: Sagala [Sagalla] Hill, 4 Feb. 1953, *Bally* 8720! & Mbololo Hill, *Gardner* in F.D. 2993!
TANZANIA. Masai District; Ngorongoro, near Old Boma, 28 Dec. 1960, *Senga* 10!; Mbulu District: S. slopes of Mt. Hanang, Hamit [Hemit, Himit], 14 Feb. 1946, *Greenway* 7739! & same area, 26 Oct. 1968, *Carmichael* 1560!; Dodoma District: Turu, Itigi-Katarake, km. 629, 31 Dec. 1925, *Peter* 33801!

---

* R. Fernandes (Bol. Soc. Brot., sér. 2, 61, t. 7 (1989)) figures one sheet bearing a type label as the holotype and another t. 8 as the isotype. I regard them as syntypes since Richard did not add the type label and both sheets are from his herbarium. The one indicated as holotype can be considered the lectotype. There are 4 other syntypes at P, all bearing the locality Wodgerat, which she did not see.
** R. Fernandes has indicated one sheet as the holotype, but it would be better called a lectotype since several sheets are involved.

DISTR. **K** 7; **T** 2, 3, ?4, 5, 8; Ethiopia (see note)
HAB. Secondary scrub, bushland with *Themeda, Exotheca* grassland; (900–)1050–2190 m.
SYN. *L. kisi* A. Rich., Tent. Fl. Abyss. 2: 169 (1850); R. Fernandes in Bol. Soc. Brot., sér. 2, 61: 141, t. 4
    (1989)
    [*L. salviifolia* sensu Bak. in F.T.A. 5: 276 (1900), as *"salvifolia"*, *non* Jacq.]
NOTE. Sebsebe had considered retaining this as a species chiefly on account of the paired
inflorescences, but this is inconstant even in the syntypes. R. Fernandes keeps it distinct on
account of leaf-shape and indumentum, but is chiefly concerned with distinguishing it from *L.
rugosa* Thunb. Meikle named *Bally* 8720 *L. kisi* by comparison with a Petit sheet at Kew, authentic
but probably not a true syntype. I also can find no or little difference between these Teita hilltop
specimens and the syntypes of *L. kisi*. The inflorescence-bracts are slightly more attenuate and
produced in the types. Moldenke has annotated *Greenway* 7739 as *L. rugosa* Thunb. and probably
published this somewhere. Some specimens from Ufipa District, e.g. *Sanane* 1397 (Mbizi Forest, 14
Nov. 1970, 2220 m.) are probably best placed here. The **T** 8 record is based on *Evans* 9 (Lindi
District, Nachingwea, 10 Oct. 1951) which has thickly velvety rounded leaves, cylindrical
inflorescences with small uniform 1–3-ribbed bracts 3 × 1.5–2 mm.; unfortunately the flower
colour is not noted. *Gillett* 12900 (Kenya, Northern Frontier Province, Moyale, 23 Apr. 1952), with
broadly ovate velvety leaves up to 3.5 × 3 cm. and almost round velvety lower inflorescence bracts, is
probably a distinct related variety. *Delamere* s.n. (Kenya, ?Northern Frontier Province, Cantalla)
and *Ichawa* 755 (Kenya, ?Northern Frontier Province, just east of Maralal, Karisia [Kalisia] Hill) are
similar but no flower colour is given; Sebsebe has equated the last with *Gilbert & Sebsebe* 8813 from
just over the border in Ethiopia with purple flowers. Two other somewhat similar specimens, but
herbaceous, 22.5–30 cm. tall, one with white flowers and one with pink flowers, have been
collected by *Carter & Stannard* 358 & 359 from the same place (Kenya, Laikipia District,43 km. N. of
Rumuruti, 3 km. N. of Kisima Farm). There are several specimens with short-peduncled small
inflorescences with mauve flowers and small acute but not long attenuate bracts which resemble
subsp. *masaica*; they may represent a separate taxon, e.g. *Kokwaro & Mathenge* 2853 (Laikipia
District, roadside between Rumuruti and Maralal, 20 Aug. 1971). *Hingley* 85 (Nakuru National Park,
N. shore of Lake Nakuru, Baharini, 2 Dec. 1972) is similar but comes close to small bracted forms of
L. ukambensis. Study of these populations in areas close to the capital should be easy.
    The contrast between *L. viburnoides* var. *kisi* and *L. viburnoides* subsp. *masaica* in the 'Winter
Highlands' part of **T** 2 is very marked and in this area they behave as two species and I had at first
considered the Mt. Hanang specimens should be described as a new species.

subsp. **masaica** *Verdc.*, subsp. nov. a subsp. *viburnoide* pedunculis brevioribus 0.4–1(–2.5) cm.
longis plerumque brevibus, foliis minoribus 1.5–3 cm. longis 1–1.6 cm. latis subtus dense velutine
pubescentibus, inflorescentiis parvis 0.8–1(–1.7) cm. longis, bracteis 5 mm. longis 4 mm. latis acutis
vel subacutis haud longe acuminatis, corollae tubo 2.5–3 mm. longo differt; etiam *L. humuliformis*
Verdc. valde affinis sed foliis subtus dense pubescentibus glandulis vix discernandis margine
subtilius crenatis, bracteis inflorescentiae minus acuminatis distinguitur. Type: Tanzania, Masai
District, Engare Naibor, *Richards & Arasululu* 26656 (K, holo.!)

Small shrub or woody herb 0.9–1.5 m. tall. Leaves usually small, 1.5–3 cm. long, 1–1.6 cm. wide,
rarely up to 4 × 2.5 cm., quite densely pubescent beneath, very closely crenate. Inflorescences small,
mostly 0.8–1 cm. long, 1 cm. wide lengthening to 1.7 cm. in fruit; peduncles mostly uniformly very
short, 0.4–1 cm. long, less often up to 1.7 cm. or exceptionally 2.5 cm.; bracts broadly rounded-elliptic
or ovate, acute, subacute or ± obtuse but not long acuminate, white pubescent and ciliate. Corolla-
tube mostly 2.5–3 mm. long.

KENYA. Machakos District: Nairobi–Mombasa road, near Kima turning, May 1971, *Tweedie* 3982! (see
    note) & Kima, 26 Jan. 1986, *Birnie* 515!; Masai District: Ilbisil (Bissel), 19 Dec. 1972, *Mumiukha* 86!
TANZANIA. Masai District: 16 km. E. of Loliondo, Klein's Camp*, 11 Nov. 1953, *Tanner* 1796!; Mbulu
    District: Tarangire National Park, road from Camp to Babati, 27 Nov. 1969, *Richards* 24815!; Arusha
    District: near Ngare Nanyuki, Mkuru road, 21 Dec. 1969, *Richards* 24981!
DISTR. **K** 4, 6, 7; **T** 1 (intermediate), 2, 3; not known elsewhere, but see note
HAB. Coarse grassland, grassland with scattered *Acacia*, etc., bushland, *Acacia, Commiphora*
    bushland; 990–1800 m.

NOTE. Typical specimens of this are strikingly distinctive, but many specimens are intermediate
with more typical subsp. *viburnoides*. *Richards* 25124 has leaves up to 4 × 2.5 cm., but very short
peduncles. *Tweedie* 3982 is perhaps best referred to subsp. *masaica*, having the short peduncles but
the leaves and bracts closely resemble those of subsp. *viburnoides*. *Verdcourt* 2514 (Kenya, Masai
District, 8–16 km. S. of Kajiado) is an intermediate with rather short peduncles. *Livingstone &
Kendall* 59 & 60 (Tanzania, Iringa District, 35.2 km. E. of Iringa on Morogoro road, 15 Jan. 1961)
come close to both subsp. *masaica* and *L. humuliformis*. *Peter* 41139 (Tanzania, Lushoto District,
Buiko–Mkomazi, 9 June 1926) appears to belong here and is the basis of the T3 record.

* Not, presumably, the better-known Klein's Camp in Musoma District.

3. **L. humuliformis** *Verdc.*, sp. nov. affinis *L. viburnoidis* (Forssk.) Vahl inflorescentiis plerumque subsessilibus vel pedunculis 0.3–2.2 cm. longis secus caulem dispersis, foliis plerumque minoribus subtus glandulosioribus, habitu intricato differt. Typus: Kenya, Meru National Park, 3 km. S. of Rojewero R., *Gillett* 18297 (K, holo.!, EA, iso.!)

Shrub 0.6–2(–3) m. tall, with slender branched stems; young shoots square, ridged, pubescent with spreading or upwardly directed white hairs and with scattered shorter hairs and glands. Leaves opposite, ovate-elliptic, (0.6–)1.5–5.5 cm. long, (0.3–)0.5–2.5 cm. wide, ± rounded to subacute at the apex, cuneate at the base, rather coarsely crenate-serrate, with scattered short and long pubescence above and gland-dots but not scabrid, usually rather sparsely to fairly densely pubescent beneath and with the very dense orange-red or greenish yellow glands not obscured; petiole 2–8 mm. long. Inflorescences strobiliform, hop-like, ± globose, 0.5–1.5 cm. long and wide, axillary at leafy and leafless nodes, sometimes almost sessile but peduncle can be 0.3–2.2 cm. long; bracts ± leafy, ovate-triangular to lanceolate, 0.5–1.1 cm. long, 3–7 mm. wide, acute to long attenuate-acuminate at the apex, rounded to cordate at the base, with numerous rather long white marginal cilia, adpressed pubescent and glandular, the apices comose at top of inflorescence. Calyx scarcely 1 mm. long, obscurely 2-lobed, white-pubescent. Corolla white, greenish white or cream; tube narrowly funnel-shaped, 2.5–3 mm. long, pubescent and glandular above; limb ± 3 mm. wide, the upper lip 1.5–2 mm. long and wide and lobes of lower lip 1–1.5 mm. long and wide, pulverulent within. Fruit ± globose, drupaceous, 2–2.5 mm. diameter at first, invested in remains of the membranous calyx, wrinkled in dry state.

KENYA. Machakos District: Nairobi–Mombasa road, Kiboko–Kibwezi, May 1971, *Tweedie* 4000!; Machakos/Teita Districts: Tsavo, 26 Dec. 1945, *Bally* 4729!; Kitui District: Garissa road, Nzui [Nzue] R., *Bally* 1925!
TANZANIA. Dodoma District: Segala/Izava, 7 Jan. 1971, *Myoya & Rajabu* 12!; Mpwapwa District: Kongwa Ranch, 19 Feb. 1966, *Leippert* 6309! & Mpwapwa, 20 Dec. 1933, *Mr. & Mrs. Hornby* 574!; District uncertain: Ageba, 31 Dec. 1965, *Newman* 80!
DISTR. **K** 4, 7; **T** 5; not known elsewhere
HAB. *Acacia, Commiphora* scrub and bushland on red soil; 480–1300 m.

SYN. [*Lippia lupuliformis* sensu Mold. in Phytologia 12: 264 (1965), pro parte quoad *Hornby* 574, *Dummer* 5046, etc., *non* Moldenke sensu stricto]
     Lantana sp. A sensu R. Fernandes in Bol. Soc. Brot., sér. 2, 61: 210 (1989)

NOTE. This had been variously named *Lantana viburnoides, Lippia carviodora, Lippia sp.* and *Lantana sp. nov.*; it is very distinct in having the ± shortly pedunculate strobilate inflorescences scattered in the axils over some distance of the stems and in the very attenuate bracts. It is deceptively similar to *Lippia carviodora* but quite different in fruit. Intermediates with *L. viburnoides* occur in N. Tanzania and *L. viburnoides* subsp. *masaica* is very close, but has less attenuate bracts and smaller leaves and inflorescences. *Drummond & Hemsley* 4091 (Kwale District, Mackinnon Road, 1 Sept. 1953) and *Scott Elliot* 6350 (Machakos District, "Ukambani" probably near Nzaui, Nov. 1893) have very small very plicate leaves and had been called *L. microphylla* Franch. but seem to be dry-season forms of *L. humuliformis*; here, however, the distinction with *L. viburnoides* subsp. *masaica* breaks down.

4. **L. ukambensis** (*Vatke*) *Verdc.*, comb. nov. Type: Kenya, Kitui, *Hildebrandt* 2739 (B, holo.†, K, iso.!, fragm. of holo.!)

Subshrubby herb or rarely ± shrubby, aromatic, with several stems from a woody base, (0.14–)0.3–1.5 m. tall; stems becoming brownish, often ± unbranched, densely adpressed pubescent or hispid and with shorter hairs, or the indumentum more spreading, particularly near to base. Leaves paired or in whorls of 3, ovate to ovate-lanceolate, 1.7–8(–10) cm. long, 0.5–5 cm. wide, ± acute at the apex or actual tip rounded, broadly cuneate at the base, slightly discolorous, rugose and ± rough or even scabrid above with adpressed hairs but not velvety, roughly velvety beneath with dense matted yellow-grey-green hairs and glandular, closely rounded-serrate, the venation usually prominent beneath, subsessile or petiole 1–5(–8) mm. long. Inflorescences axillary, solitary or paired (i.e. up to 6 per node), ± cone-like and shortly pedunculate, mostly not reaching beyond base of leaf but elongating and becoming oblong in fruit, 0.3–3.5(–5) cm. long, 1–1.5 cm. wide; peduncles 0–1.3(–1.8) (or even up to 5–6 cm. in some variants (?hybrids)); bracts very conspicuous, rounded-ovate, obovate, lanceolate or triangular-lanceolate, 0.8–1.3(–2) cm. long, 0.5–0.7(–1.3) cm. wide, larger at base of inflorescence, acute to cuspidate, the cusp often purplish, adpressed pubescent glandular and ciliate, often clearly 5-nerved or ribbed. Calyx thin, 2.5 mm. long, ± truncate. Corolla deep purple, crimson or magenta,

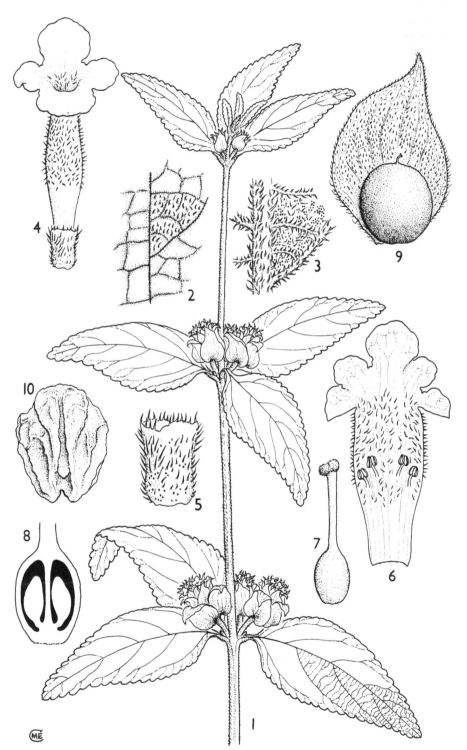

Fig. 6. *LANTANA UKAMBENSIS* — **1**, habit, × ⅔; **2**, part of upper leaf surface, × 8; **3**, part of lower leaf surface, × 8; **4**, flower, × 8; **5**, calyx, × 16; **6**, corolla, opened out, × 8; **7**, ovary and style, × 16; **8**, longitudinal section of ovary, × 30; **9**, fruit with bract, × 4; **10**, pyrene, × 8. 1–8, from *Greenway et al.* 14112; 9, 10, from *Tanner* 4188. Drawn by Mrs. M.E. Church.

sometimes with white or yellow centre, scarcely exceeding the bracts above and shorter in lower part of inflorescence; tube slender, (2-)3-4(-5) mm. long, densely spreading pubescent; limb 1.5-3 mm. wide, the lobes rounded, ± 1 mm. long, not so pubescent. Fruits blue or purple, subglobose, 2-4 mm. long and wide, often ± shiny, invested at first in the thin membranous calyx. Fig. 6.

UGANDA. Karamoja District: Bokora, Iriri, May 1957, *J. Wilson* 342!; Toro Game Reserve, 8 km. SW. of Sindikwa triangulation marker, 14 Dec. 1962, *Buechner* 34!; Teso District: Serere, June 1926, *Maitland*!
KENYA. NE. Elgon, Oct. 1952, *Tweedie* 1078!; Uasin Gishu District: Ol Dane Sapuk, 16 May 1951, *G.R. Williams Sangai* 182!; S. Nyeri District: Fort Hall-Nyeri road, 95 km. NE. of Nairobi, 12 Jan. 1972, *Bally & Smith* 14764!
TANZANIA. Musoma District: Majita, Nyambono, 22 Apr. 1959, *Tanner* 4188!; Mbulu District: Lake Manyara National Park, above Endabash, 14 June 1965, *Greenway & Kanuri* 11850!; Ufipa District: Sumbawanga, near Chapota, Ipeta [Empeta] Swamp, 8 Mar. 1957, *Richards* 8571!
DISTR. U 1-4; K 1-5; T 1-7; Guinea Bissau to Cameroon, Central African Republic, Zaire, Burundi, Rwanda, Sudan, Ethiopia, Mozambique, Malawi, Zambia and Zimbabwe (see note)
HAB. *Hyparrhenia* grassland, *Combretum* and other wooded grassland, open woodland including *Brachystegia*, old cultivations, sometimes amongst granitic rocks; 75-2070 m.

SYN. ?*Lippia schimperi* Walp., Repert. 4: 53 (1845), quoad descr. et *Schimper* 257, *nom. illegit.** (see note)
　　L. ukambensis Vatke in Linnaea 43: 528 (1882)
　　[*Lantana viburnoides* sensu Bak. in F.T.A. 5: 276 (1900), pro parte, *non* (Forssk.) Vahl (see note)]
　　[*L. salviifolia* sensu Bak. in F.T.A. 5: 276 (1900), pro parte, quoad *Scott Elliot* 6341, *non* Jacq.]
　　L. salviifolia Jacq. var. *ternata* Chiov. in Racc. Bot. Miss. Consol. Kenya: 97 (1935). Type: Kenya, NE. Meru, *Balbo* 74 (TOM, syn.!)
　　L. viburnoides (Forssk.) Vahl var. *velutina* Mold. in Phytologia 3: 120 (1949). Type: Sudan, Equatoria, Azza Forest, *Myers* 6529 (K, holo.!)
　　L. milne-redheadii Mold. in Phytologia 3: 268 (1950); F.F.N.R.: 370 (1962), adnot. Type: Zambia, Mwinilunga, W. of Matonchi Farm, *Milne-Redhead* 3542 (K, holo.! & iso.!, BR, iso.)
　　L. rhodesiensis Mold. in Phytologia 3: 269 (1950); F.P.S. 3: 196 (1956); K.T.S.: 587 (1961); F.F.N.R.: 370 (1962); Meikle in F.W.T.A., ed. 2, 2: 435 (1963), pro parte; R. Fernandes in Bol. Soc. Brot., sér. 2, 61: 156, t. 9-11 (1989). Type: Zambia, Mumbwa, *Macaulay* 735 (K, holo.!)**

NOTE. R. Fernandes (Bol. Soc. Brot., sér. 2, 59: 264-270, t. 1, 2(1986)) has exhaustively explained the confusion attending the Kew sheet labelled *Hildebrandt* 2738 and *Lippia kituiensis;* it bears two different plants, one a *Lippia*, the other a *Lantana*. This confusion resulted in a common East African *Lippia* being wrongly called *L. ukambensis* for 80 years when it is in fact *L. kituiensis*. The fragment from the holotype must have been requested by someone at Kew (after publication of F.T.A. Verbenaceae) but was never correctly interpreted until Fernandes pointed out it was a *Lantana*. There are now several sheets from near Kitui clearly conspecific with *L. ukambensis* and I believe it is the correct name for the species more usually known as *L. rhodesiensis* although the types are by no means identical. Fernandes maintains two species, *L. angolensis* Mold. (type: Angola, Mossamedes, Humpata, *Fritzsche* 109 (S, holo.)) discussed at length by R. Fernandes in Bol. Soc. Brot., sér. 2, 61: 128-131, t. 1, 2 (1989), and *L. rhodesiensis*, distinguishing the former by its usually paired leaves, more shrubby habit and smaller lower flowering bracts (5.5-)7-10(-13.5) × (3.5-)4.5-6.5 mm., which are less conspicuously nerved, but admits there are numerous intermediates. Many East African specimens come close to *L. angolensis* including the type of *L. ukambensis*. I have preferred to maintain a broad taxon and although I have not included *L. angolensis* in the synonymy I believe it is difficult to maintain as a separate species. A difficult situation might be resolved by having three subspecies based on *ukambensis*, *rhodesiensis* and *angolensis*. Meikle in F.W.T.A., ed. 2, 2: 435 (1963) records *L. rhodesiensis* from Guinea Bissau to S. Nigeria and Cameroon but Fernandes points out that the W. African material has usually relatively broader leaves, more shrubby habit and smaller bracts and suggests it could be a distinct subspecies. Similarly she excludes *L. mearnsii* Mold. var. *latibracteolata* Mold. in Phytologia 2: 313 (1947); F.P.N.A. 2: 138 (1947) (type: Zaire, Kabare, *Bequaert* 5490 (BR, holo., K, photo.!)) which was considered a synonym of *L. rhodesiensis* by Meikle. It differs from typical and W. African material by the retrorse hairs on the stem, obtuse leaves, narrower spikes, smaller bracts and smaller drupes. There are, however, many specimens in W. Africa which seem to me very little different from material from E. Africa. The whole situation is confused by the almost certain existence of hybrids between this complex and both *L. trifolia* and *L. viburnoides*.
　　Fernandes has annotated *Schimper* 257 as *L. viburnoides* subsp. *richardii* var. *schimperi* Mold. (type: Ethiopia, Dewari, 5 Oct. 1863, *Schimper* s.n. (5 Oct. 1863) (S, holo.)), and published this view in Bol. Soc. Brot., sér. 2, 61: 207, t. 21 (1989), whereas Sebsebe and I both believe it is the species treated above. Walpers indicates that the flowers are 'roseae'. Fernandes has similarly annotated a specimen *Davies* 670 (Tanzania, Mbeya District, Mbosi, 20 Nov. 1932) but this appears to be abnormal, the apparently pedunculate inflorescences actually having small leaves immediately

* Walpers cites *Lantana abyssinica* Dietr. & Otto in synonymy.
** Some other specimens also labelled 735 are not this species but perhaps (*fide* R. Fernandes) hybrids between *L. rhodesiensis* and *L. trifolia*.

below the heads and the stem-leaves irregularly placed; unfortunately no flower colour is given, but I think it is probably *L. ukambensis*.
Johnston 91 cited by Bak. in F.T.A. 5: 276 (1900) and T.T.C.L.: 639 (1949) as *L. viburnoides* is probably a hybrid of *L. ukambensis*.

5. **L. trifolia** *L.*, Sp. Pl.: 626 (1753); Sims in Bot. Mag. 35, t. 1449 (1812); Schauer in DC., Prodr. 11: 606 (1847); Briq. in E. & P. Pf. 4, 3a: 151 (1895); Urban in F.R. Beih. 5: 51 (1920) & Symb. Antill. 8: 593 (1921); Moldenke in Lilloa 4: 287 (1939); Meikle in Mem. N.Y. Bot. Gard. 9: 35´(1954); Moldenke in Fl. Madag., Fam. 174: 11 (1956); K.T.S.: 587 (1961); E.P.A.: 789 (1962); F.F.N.R.: 368 (1962); Meikle in F.W.T.A., ed. 2, 2: 435 (1963); Backer & van den Brink jr., Fl. Java 2: 597 (1965); Adams, Fl. Pl. Jamaica: 628 (1972); Moldenke in Ann. Missouri Bot. Gard. 60: 48 (1973); Blundell, Wild Fl. E. Afr.: 398, t. 616 (1987); Fernandes in Bol. Soc. Brot., sér. 2, 61: 180, t. 15, 16 (1989). Type: West Indies, copy of Plumier plate seen by Linnaeus in Holland which later became t. 70 of Plumier, Pl. Americ. (fasc. III), ed. Burmann, 1756 (lecto.)

Shrub or subshrubby herb 0.9–3 m. tall with often purplish adpressed or subadpressed pubescent stems or sometimes the indumentum of short to long ± spreading bristly hairs. Leaves aromatic, opposite or more usually in whorls of 3(–4), ovate-lanceolate to lanceolate or ovate, ovate-oblong or elliptic, (1.5–)5–14 cm. long, (0.5–)2–6 cm. wide, mostly 2.5–3 times as long as wide, narrowed to an acute or rarely subobtuse apex, ± cuneate at the base, finely crenate or crenate-serrate, glandular, adpressed pubescent to scabrid with tubercle-based hairs above, densely to sparsely puberulous, shortly tomentose, velvety or rarely spreading pilose beneath; venation ± impressed or not impressed above; petiole 2–5 mm. long. Peduncles 1–2 (i.e. up to 6 per node), 1.5–13 cm. long, mostly shorter than the subtending leaf, elongating somewhat in fruit, the heads of flowers hemispherical, 0.8–1 cm. long, 1.3–1.7 cm. wide, becoming subcylindric in fruit and 1.7–4.5(–5) cm. long; lower bracts lanceolate to narrowly ovate-elliptic or ovate, (0.4–)0.8–1.4 cm. long, (0.6–)3–6.5 mm. wide, long-acuminate or acute, the upper bracts smaller. Calyx 1.2–2 mm. long, pubescent. Corolla white, mauve-white, pinkish white, or usually pink or purple, usually with yellow throat; tube 5–7 mm. long, with limb 4–9 mm. wide; lower lip (4–)5–5.5 mm. long and wide. Drupes mauve to purple, subglobose, 3–3.5 mm. long, 2.5–3 mm. wide; axis of fruiting spike up to 5 cm. long.

UGANDA. Karamoja District: Moroto, July 1930, *Liebenberg* 258!; Bunyoro District: Budongo Forest, 24 Nov. 1938, *M.V. Loveridge* 94*!; Masaka District: Kalungi Swamp, 12 Nov. 1934, *G. Taylor* 1641!
KENYA. Northern Frontier Province: Maralal–Baragoi, 20 km. N. of Maralal, 10 Nov. 1978, *Hepper & Jaeger* 6698!; Elgon, *T.H.E. Jackson* 360!; Embu District: Karue, 17 Oct. 1932, *M.D. Graham* 2284!
TANZANIA. Ngara District: Bugufi, Mabawe, 11 Jan. 1961, *Tanner* 5686*!; Moshi District: Gararagua, Rongai Ranges, 19 Apr. 1957, *Greenway* 9178!; Ufipa District: Sumbawanga, Malonje Farm, 5 Mar. 1957, *Richards* 8443!; Morogoro District: above Bunduki, Mgeta R. valley, 12 Mar. 1953, *Drummond & Hemsley* 1507!
DISTR. U 1–4; K 1–7; T 1–4, 6–8; widespread in Africa but probably introduced from America, Central and S. America, West Indies, also introduced into various parts of Asia and some Pacific islands
HAB. Grassland, bushland, bracken associations, *Brachystegia* woodland, forest/grass edges, ?forest, also abandoned cultivations; 0–2400 m.
SYN. *Camara trifolia* (L.) Kuntze, Rev. Gen. Pl. 2: 504 (1891)
    [*Lantana salvifolia* sensu Bak. in F.T.A. 5: 277 (1900); Hiern, Cat. Afr. Pl. Welw. 1: 827 (1900), pro parte; T.T.C.L.: 639 (1949), *non* Jacq. (1798)]
    L. mearnsii Mold. in Phytologia 1: 421 (1940) & Fifth Summ. Verbenac.: 212 etc. (1971). Type: Kenya, near Fort Hall, *Mearns* (NY, holo.)
    Lippia schliebenii Mold. in Phytologia 2: 316 (1947) & 12: 482 (1966); Fernandes in Bol. Soc. Brot., sér. 2, 59: 263 (1986). Type: Tanzania, Lindi District, Rondo [Mwera] Plateau, Bakari, *Schlieben* 5596 (BR, holo.!, K, photo.!)
    Lantana trifolia L. forma *hirsuta* Mold. in Phytologia 3: 113 (1949) & in Ann. Missouri Bot. Gard. 60: 49 (1973) & in Rev. Fl. Ceylon 4: 213 (1983); Fernandes in Bol. Soc. Brot., sér. 2, 61: 191 (1989). Type: Colombia, Valle del Cavea, between Cabuyae and La Solorza, *Cuatrecasas* 14438 (NY, holo.)
    L. trifolia L. forma *oppositifolia* Mold. in Phytologia 4: 179 (1953) & in Ann. Missouri Bot. Gard. 60: 49 (1973). Type: Java, Bogor [Batavia], *Backer* 21771 (BO, holo.)
    L. trifolia L. forma *albiflora* Mold. in Phytologia 6: 327 (1958) & in Rev. Fl. Ceylon 4: 215 (1983). Type: Ecuador, El Oro, Zaruma, *Asplund* 15821 (S, holo.)

---

* Cited specimens marked with an asterisk are forma *hirsuta* Mold.

L. mearnsii Mold. var. *punctata* Mold. in Phytologia 12: 428 (1965). Type: Malawi, Thyolo [Cholo] Mt., *Brass* 17719 (NY, holo., K!, US, iso.)

[*L. mearnsii* Mold. var. *congolensis* sensu Mold., Fifth Summ. Verbenac.: 230, 252 (1971), pro parte, *non* Mold. (1947)]

NOTE. *L. trifolia* is extremely polymorphic over its wide range and R. Fernandes, following Moldenke, has recognised several forms; forma *trifolia*, forma *congolensis* (Mold.) R. Fernandes and forma *hirsuta* Mold. — both first and third occur in East Africa but there are so many specimens which are intermediate that I have not formally recognised it; forma *trifolia* has the hairs on branches and peduncles slender, short and adpressed to somewhat spreading, whereas forma *hirsuta* has longer stronger more bristly spreading hairs. In the Flora Zambesiaca area Fernandes found that the corollas were predominantly white but in East Africa and throughout much of the species range they are pink to deep mauve. Albinos occur in populations otherwise red (e.g. Kenya, Elgon, 5 Dec. 1930, *Lugard* 103a (white) & 103 (mauve)). *Tanner* 6008 (Tanzania, Biharamulo District, Nyakahura, no date) despite its paired leaves and white flowers is clearly *L. trifolia* with typical narrow bracts. In some areas, e.g. Iringa District, Dabaga–Mufindi area, white-flowered forms seem to predominate. *Lynes* Dabaga 70 (Dabaga, 21 Feb. 1932) has paired leaves, white flowers and distinctly narrow lanceolate bracts 4–8 × 0.6–2.2 mm.; it was thought when it was first received at Kew to be an introduced species but there are other similar specimens known from the area, all of which seem to be from consistently white populations and with bracts too narrow to be hybrids with *L. viburnoides*, e.g. *Davies* M1 (Mufindi), *Shabani* 984 (Mufindi, Lake Ngwazi), *Lynes* D.g. 154 (Njombe), *Issa* 31 (Mufindi, Lake Ngwazi); *St. Clair-Thompson* 567 (Mbeya, Mporoto Sawmill) is said to have yellow flowers but is similar. Fernandes has determined *Shabani* 984 and *Davies* M1 as *L. trifolia* and suggested *St. Clair-Thompson* 567 might be a hybrid between *L. camara* and some other species. Collectors state this is very common in the area so I suspect it is a distinct variant of *L. trifolia*.

Hybrids

Despite a natural reluctance to use putative hybridisation as an excuse for failing to work out the taxonomy it seems clear that hybridisation is common in *Lantana*, but without experimental evidence it is not clear whether or not some of the so-called hybrids are not in effect distinct taxa.

L. trifolia × L. ukambensis

Specimens with leaves in whorls of 3, large basal bracts, purple flowers and rather short peduncles, found in **K** 3, 4, 6, **T** 3, 6, 8, appear to be this hybrid, e.g. Tanzania, Tanga District, near Mnyusi station, Kisarake, 15 July 1972, *Semsei* 4263 and Kilosa District, Mikumi Nat. Park, 30 Apr. 1968, *Renvoize & Abdallah* 1824. Some Peter specimens from Tanzania annotated *L. mearnsii* var. *latibracteata* Mold. by Moldenke seem to belong here.

L. trifolia × L. viburnoides

Specimens with leaves opposite or in whorls of 3, usually lilac-pink to purple flowers, long peduncles and bracts broader than in *L. trifolia* have been recorded from **K** 3, 4, 7; **T** 3–8, e.g., Kenya, Masai District, 19.2 km. from Narok on road to Olokurto, Orengitok, 17 May 1961, *Glover et al.* 1214.

L. viburnoides × L. ukambensis

Several sheets with leaves in whorls of 3 or paired, broad bracts and mauve flowers have been referred to this hybrid, e.g. *Migeod* 668 (Tanzania, Lindi District, Tendaguru, 26 Apr. 1930) and *Anderson* 756 (Tanzania, Lindi District, Nachingwea, 16 Apr. 1952), the latter hairy, having densely pubescent lower bracts 11 × 6 mm. and a 10 mm. long corolla-tube and 7 mm. wide limb. *Tweedie* 2728 (Uganda, Karamoja District, Lokitanyala, Sept. 1963), with 4 cm. long inflorescences exceeding the leaves, 4.5 cm. long peduncles, large ovate cuspidate ciliate and densely pubescent bracts 13 × 8 mm. and white flowers, is a more convincing hybrid but may be just a form of *L. viburnoides*. It is impossible to decide the status of single distinctive specimens; they could easily be distinct taxa. Only experimental work will throw light on these problems.

8. DURANTA

L., Sp. Pl.: 637 (1753) & Gen. Pl., ed. 5: 284 (1754)

Erect, trailing or subscandent unarmed or spiny shrubs or small trees. Leaves small, opposite or verticillate, simple, entire to serrate. Flowers few to many in terminal and axillary racemes or panicles; bracts small. Calyx 5-toothed, persistent. Corolla irregular with straight or curved tube; limb 2-lipped, with 5 unequal lobes. Stamens 4–5, slightly didynamous, inserted above the middle of the tube, included; in the case of 4 stamens a small staminode is sometimes present also. Ovary globose, 8–10-locular, each locule with a single erect anatropous ovule; style equalling or shorter than the corolla-tube; stigma oblique, ± capitate. Fruit fleshy, with 4–5 2-seeded pyrenes, tightly enclosed by the persistent calyx. Seeds erect, without albumen; testa membranous.

A tropical American genus variously estimated at 20–35 species but probably fewer; one is widely cultivated throughout the tropics and subtropics and now often naturalised.

D. erecta L., Sp. Pl.: 637 (1753); Hiern, Cat. Afr. Pl. Welw. 1: 831 (1900); Urban, Symb. Antill. 4: 536 (1911) & 8: 599 (1921); Caro in Revista Argentina Agron. 23: 6, fig. 1 (1956); Bromley in K.B. 39: 803 (1984). Type*: Plumier plate subsequently published as t. 79 in Plumier, Plantarum Americanum, ed. J. Burman 4: 70 (1756), Groningen Univ. Library, (lecto.)

Spreading, drooping, trailing or subscandent often thorny shrub or small tree, 1.8–6 m. tall; bark pale; young shoots adpressed pubescent. Leaves opposite or verticillate; blades ovate, obovate, elliptic or ovate-lanceolate, 1.5–8 cm. long, 1–4.5 cm. wide, obtuse to acute at the apex, cuneate at the base, the margins entire, serrate or crenate above the middle, glabrous or finely adpressed pubescent; petioles 0.5–1.2 cm. long, usually adpressed pubescent. Racemes ± lax, 3–16 cm. long, often in very lax panicles; pedicels 1–5 mm. long. Calyx 3–4(–6) mm. long, pubescent inside and out. Corolla lilac, lavender-blue or white; tube 7 mm. long; limb spreading, 0.7–1.3 cm. wide. Style shorter than corolla-tube; stigma capitate. Fruit orange-yellow and shining, globose, 0.5–1.1 cm. diameter, the persistent calyx shrunk over it and produced as a curved beak; pyrenes 3.5–5.5 mm. long, 2.5–3.5 mm. wide. Seeds cuneiform, white, shining, 2.5–3 mm. long, 1–1.5 mm. wide. Fig. 7.

UGANDA. Moroto township, May 1956, *J. Wilson* 240!; Teso District: Serere, Mar. 1932, *Chandler* 603!; Mengo District: Kiwoko–Tweyanze, *Langdale-Brown* 2079!
KENYA. Nairobi District: Karura Forest, 14 Oct. 1967, *Mwangangi & Abdalla* 269! & Nairobi Arboretum (cult.), 27 July 1952, *G.R. Williams* 488!; Kericho District: Kibosek, 30 Mar. 1961, *Glover et al.* 211!
TANZANIA. Musoma District: Grumeti R., 10 Feb. 1968, *Greenway et al.* 13166!; Lushoto District: Tanga, 12 Aug. 1932, *Geilinger*!; E. Province, probably Ulugurus, June 1952, *Paol* 12!
DISTR. U 1, 3, 4; K 4–6; T 1, 3, 6; Z (fide U.O.P.Z.); Argentine to Mexico and Florida; now widely cultivated and naturalized throughout the tropics and subtropics of the world and even in Europe, Cyprus, etc.
HAB. Scrub and riverine thicket, often on termite-mounds, *Euclea*, *Olea* forest and clearings; 1080–1950 m.
SYN. *D. repens* L., Sp. Pl.: 637 (1753); F.P.N.A. 2: 140 (1947); U.O.P.Z.: 238, fig. (1949); T.T.C.L.: 639 (1949); Jex-Blake, Gard. E. Afr., ed. 4: 111, 240, 356 (1957); K.T.S.: 586 (1961); F.F.N.R.: 368 (1962); U.K.W.F.: 616 (1974); Townsend, Fl. Iraq 4: 657 (1980); Troupin, Fl. Rwanda 3: 278, fig. 90/1 (1985). Type: "Castorea repens spinosa" Plumier, Nov. Pl. Amer. Gen.: 30, t. 17 (1703)
D. plumieri Jacq., Enum. Syst. Pl.: 26 (1760) & Select. Stirp. Amer. Hist.: 186, t. 176/76 (1763); L., Sp. Pl., ed. 2, 1: 888 (1762); Bak. in F.T.A. 5: 287 (1900), *nom. illegit.* (cites the types of both Linnean species in synonymy)
NOTE. Hiern united *D. repens* and *D. erecta*, both of equal date and now considered forms of one species and used the latter name which must stand under the rules although the former is universally known and is unlikely to be given up by horticulturalists, etc. Known under many names, golden dewdrop and pigeon berry are the two always used in East Africa where it is one of the best-known garden shrubs; it also makes a good fast-growing hedge. The white form sometimes seen is presumably var. *alba* (Masters) Caro.

* Plate 79 itself, chosen by Caro and Sandwith, cannot possibly be the type of a name published in 1753. Linnaeus did, however, see the original Plumier drawing much earlier when in Holland.

FIG. 7. *DURANTA ERECTA* — 1, flowering branchlet, × ⅔; 2, portion of branchlet, showing spines, × ⅔; 3, flower, × 3; 4, corolla, opened out, × 4; 5, ovary, × 8; 6, transverse section of ovary, × 14; 7, part of fruiting branchlet, × ⅔; 8, fruit, × 3; 9, fruit, with calyx removed, × 3; 10, pyrene, inner surface, × 4. 1, from *J. Wilson* 240; 2, from *G.R. Williams* 534; 3–6, from *G.R. Williams* 488; 7–10, from *Greenway et al.* 13166. Drawn by Mrs. M.E. Church.

9. VITEX

L., Sp. Pl.: 635 (1753) & Gen. Pl., ed. 5: 285 (1754); Pieper in E.J. 62, Beibl. 141 ('142'): 1–91 (1928)*; Moldenke in Phytologia 5–6 (1955–1958)** & in Rev. Fl. Ceylon 4: 348 (1983) (full generic synonymy)

Trees, shrubs or occasionally lianas, the stems mostly 4-angled, glabrous to densely hairy. Leaves opposite or in whorls of 3, (1–)3–7-foliolate, mostly petiolate; leaflets entire, toothed or lobed, often glandular. Cymes mostly dichasial, short and dense to open and spreading, sessile or pedunculate, the bracts often well developed; cymes sometimes aggregated into thyrsoid or lax panicles. Calyx campanulate to tubular, (3–)5(–6)-lobed or toothed or nearly truncate. Corolla white and/or coloured, mostly shades of blue, lilac or mauve, ± zygomorphic; tube shortly cylindrical to long-tubular or ± funnel-shaped; limb spreading, ± 2-lipped, the upper lip 2-fid, the lower 3-fid. Stamens 4, didynamous, inserted in the tube, mostly exserted. Ovary at first ± 2-locular, later usually 4-locular, with single ovule in each locule; style terminal, filiform, shortly bifid. Drupes fleshy with hard 4-celled endocarp; fruiting calyx usually accrescent, plate-like or cupular. Seeds without endosperm.

About 250 species in the tropics of both Old and New Worlds and a few in temperate areas. Moldenke gives 380 taxa including varieties, etc.

Several of the indigenous species are also cultivated, some being excellent timber trees. Apart from these, mentioned in the text, *V. agnus-castus* L. is listed by Jex-Blake (Gard. E. Afr., ed. 4: 129 (1957)) but I have seen only one incomplete specimen from East Africa (Tanzania, 17 km. N. of Dar es Salaam, Africana Hotel, 4 Dec. 1988, *Donald* 41). Jex-Blake describes it as a very lovely deciduous shrub or spreading tree with branched spikes of lavender-blue flowers. A specimen collected in Mombasa (*MacNaughton* 171 in *F.D.* 2836) labelled "an exotic tree — 9 m. — flowers small and faint purple" has been annotated as *V. trifolia* L. by Moldenke; the corolla-tube is very short. *V. agnus-castus*, *V. pseudo-negundo* (Bornm.) Hand.-Mazz., *V. trifolia* and *V. negundo* L. form a difficult group and several varieties of these species (some with deeply divided leaflets) are often (wrongly I think) maintained at specific rank. I am following Moldenke and Singhakumara in treating the wild plant usually called *V. negundo* in East Africa as *V. trifolia* var. *bicolor* (see p. 52). *MacNaughton* 171 is probably best considered a form of *V. trifolia* var. *trifolia*.

Species 1–6 belong to *Vitex* subgenus *Vitex* section *Vitex* (*Terminales* Briq.) and the rest to section *Axillares* Briq.

1. Leaves unifoliolate, sometimes a few 2–3-foliolate 2
   Leaves 3–7-foliolate, only occasionally an odd leaf
   unifoliolate 3
2. Native savanna shrub or tree with leaflets 3–25 × 1.8–11.5
   cm., glabrous to velvety, not so discolorous; corolla-
   tube 5–6 mm. long; fruit 1.4–2.5 cm long (**U** 1, 3)   12. *V. madiensis*
                                                                   (in part)

   Introduced tree with leaflets 2.5–7 × 1–4 cm., very discolorous,
   white or silvery beneath; corolla-tube 8–9 mm. long;
   fruit 5–6 mm. long; cultivated . . . .   1. *V. trifolia*
                                                              cultivated variants
3. Inflorescences terminal, usually together with additional
   axillary ones from upper axils 4
   Inflorescences all axillary or if appearing terminal then
   very short, 1–2 cm. long 10

* Several of Pieper's species have been considered to date from his 1929 Fedde Repert. paper but are in fact adequately validated in the keys given in this 1928 revision; in the 1928 revision several syntypes are often cited but in the later paper a specific specimen is selected as the type and I have accepted these cases as lectotypifications.
** This series of materials towards a revision of *Vitex* is a vast source of information mostly compiled but much taken from a worldwide survey of material. Details of the papers are 5: 142–174, 186–224, 257–280 & 293–336 (1955); 5: 343–393, 403–464 (1956); 5: 465–507, 6: 13–64 & 70–128 (1957); 6: 129–192 & 197–231 (1958). I have considered that the constant citing of these references and of frequent misidentifications in synonymy is not worth the time and effort involved and serves no useful purpose for the average user of a local flora.

4. Leaflets densely finely white to buff tomentose beneath so
   as to obscure the whole surface 5
   Leaflets glabrous to pubescent and glandular beneath but
   surface clearly evident 6
5. Lower lip quite glabrous or with a few hairs at basal
   corners; leaflets narrowly lanceolate (cultivated) . .   *V. agnus-castus*
   Lower lip with semicircular line of dense hairs at base near
   throat; leaflets usually more broadly lanceolate (wild
   or naturalised) . . . . 1. *V. trifolia*
6. Inflorescences with lateral branches very short or
   suppressed so that panicles are spike-like; leaflets
   3-foliolate; fruits not densely covered with yellow
   glands 7
   Inflorescences with lateral branches longer, forming a
   more extensive panicle of spike-like elements; leaflets
   3–5-foliolate; fruits densely-covered with yellow glands
   or not 9
7. Inflorescences long and unbranched; leaflets scabrid
   above; fruiting calyx with lobes 5 mm. long; fruits 2.3
   cm. long, densely minutely glandular . . . . .   4. *V. sp. A*
   Without above characters combined 8
8. Inflorescences with lateral branches short but distinctly
   developed and bearing several cymes; fruits yellow or
   buff-brown (widespread) . . . .   5. *V. strickeri*
   Inflorescences with lateral branches so reduced that
   pedicels appear direct from main axis; fruits dull
   brick-red (T 5) . . . .   6. *V. ugogoensis*
9. Leaves 3-foliolate, the leaflets sessile; petioles sometimes
   widely winged; fruits not covered with yellow
   glands . . . .   2. *V. zanzibarensis*
   Leaves 3–5-foliolate, the petiolules up to 2.5 cm. long;
   petioles never winged; fruits thickly coated with
   yellow glands . . . .   3. *V. buchananii*
10. Pyrophytic herb . . . .   12. *V. madiensis*
                                                        subsp.*milanjiensis*
                                                        (var. *epidictyodes*)

    Shrubs to tall forest trees 11
11. Inflorescences very short, axillary or sometimes appearing
    4⁵ terminal, only 1–2(–5) cm. long on peduncles 1–1.2
    cm. long; leaflets 5, sessile or petiolules very
    short . . . .   7. *V. schliebenii*
    Inflorescences usually much longer or if not then
    peduncles longer 12
12. Leaflets glabrous save sometimes midrib puberulous
    beneath (or rarely hairy outside East Africa); ovary
    densely hairy above; calyx and pedicels tomentose
    with short hairs; widespread savanna tree with fruits to
    3 × 2 cm. . . . .   13. *V. doniana*
    Leaflets hairy at least beneath; if ovary hairy above then
    calyx and pedicels densely hairy with longer hairs 13*
13. Ovary (and tips of young fruits) densely hairy 14
    Ovary glabrous but often with glands 16
14. Leaflets glabrous above (save for var. *amaniensis***), always
    narrowly acuminate; predominantly forest trees . .   16. *V. ferruginea*
    Leaflets hairy above (soon glabrescent in some
    specimens), rounded to acuminate; predominantly
    savanna trees 15

---

* *V. sp. B* (species 17) will key near here; known only in fruit which is characteristic, very elongate-oblong unlike the other species
** Technically this will key to *V. mombassae* but has more narrowly oblong-elliptic leaflets and is a rain-forest tree to 15 m.; the inflorescences, however, are exceedingly similar.

15. Leaflets always sessile; bracts linear; corolla-tube 3–4 mm.
long; stamens not or scarcely exserted, at least not
beyond the length of shortest corolla-lobes . . .   14. *V. payos*
Leaflets sometimes sessile but more usually some or all
petiolulate; bracts narrowly oblong-oblanceolate;
corolla-tube 7–9 mm. long; stamens and style well
exserted (5–9 mm.) . . . .   15. *V. mombassae*
16. Inflorescences lax and few-flowered, the actual pedicels
short but lateral elements of each ultimate dichasium
1–3-flowered, the stalk well overtopping central flower;
peduncles and secondary peduncles long and slender;
fruits 6–7 mm. diameter . . . .   8. *V. mossambicensis*
Inflorescences denser; fruits larger 17
17. Rain-forest trees; foliage often dries dark 18
Savanna trees; foliage often dries pale 19
18. Fruit 0.9–1.1 cm. long, 6.5 mm. wide (**T** 3, 6, 7) . . . .   9. *V. amaniensis*
Fruit (1.1–)2.2–2.4 cm. long, (0.8–)1.3–1.5 cm. wide (**K** 4)   10. *V. keniensis*
19. Leaves always 5-foliolate, the nerves rather closely placed;
inflorescences extensive, up to 11(–15) cm. wide; bracts
0.3–2.2 cm. long, broader, showing glabrous insides;
corolla-tube ± 3 mm. long; fruits 1–1.3 cm. long, 0.8–1
cm. wide . . . .   11. *V. fischeri*
Leaves 1–5-foliolate, the nerves not so closely placed;
inflorescences not so wide; bracts 3–6 mm. long, very
narrow, not showing glabrous insides; corolla-tube 5–6
mm. long; fruits 1.4–2.5 cm. long, 1–1.9 cm. wide . .   12. *V. madiensis*

1. **V. trifolia** *L.*, Sp. Pl.: 638 (1753), as "*trifoliis*"; C.B. Cl. in Fl. Brit. India 4: 583 (1885); Trimen, Handb. Fl. Ceylon 3: 356 (1895); Alston in Handb. Fl. Ceylon 6: 232 (1931); Moldenke in Rev. Fl. Ceylon 4: 378 (1983); Singhakumara, The Biology of *Vitex* in Sri Lanka, D.Phil. thesis, Oxford (unpublished). Lectotype chosen by Singhakumara: India, without exact locality and collector unknown, *Herb. Linnaeus* 811.7 (LINN, lecto.!)*

Shrub or small tree to 6.5 m. or some varieties procumbent or creeping and rooting at the nodes, mostly much branched below with slender stems; bark brown, smooth; young branches densely pubescent with greyish hairs. Leaves opposite, 1–5-foliolate; leaflets oblong-elliptic or obovate to oblanceolate or sometimes almost round, 1.5–7 cm. long, 0.8–4 cm. wide, obtuse, emarginate or rounded to acute or acuminate at the apex, cuneate to attenuate or rarely rounded at the base, entire or in one variety 2–3-sect, puberulous or glabrescent and ± gland-dotted above, tomentose to densely velvety pubescent beneath; petioles 0.6–3.3 cm. long; petiolules 0.1–2 cm. long. Inflorescences terminal and axillary in upper leaf-axils, much branched panicles of many-flowered cymes 3–23 cm. long, 2–4 cm. wide; peduncles 1–5 cm. long together with the rhachis grey-pubescent; cymes short, 0.4–1.5 cm. long; pedicels ± 1 mm. long; main bracts leaf-like; bracteoles linear, 1–3 mm. long. Calyx 4–5 mm. long, 2.5–3.5 mm. wide with acute or blunt teeth, densely white tomentose outside. Corolla blue, lavender or purple, sometimes pale, puberulous outside; tube 1–1.3 cm. long, limb 2-lipped, the lower up to 6 mm. long. Fruits yellow or reddish turning blue or black, globose or ovoid, about 5 mm. long and wide, the calycine cup 5 mm. long and wide, pubescent outside.

var. **bicolor** (*Willd.*) *Mold.*, Known Geogr. Distrib. Verbenaceae, ed. 2: 79 (1942) & Rev. Fl. Ceylon 4: 386 (1983); Singhakumara, The Biology of *Vitex* in Sri Lanka, D.Phil. thesis (unpublished). Type: India, without location, *Herb. Willdenow* 11709 (B, holo., microfiche!)

Leaflets often 5, the 3 in middle with distinct petiolules, entire. Cymes of panicle distinctly stalked.

KENYA. Kwale District: Diani, Shaitani Forest, 17 Nov. 1978, *Brenan et al.* 14522!; Mombasa, 19 May 1952, *Templar* 23!; Kilifi, 10 Oct. 1945, *Jeffery* 348!
TANZANIA. Uzaramo District: Dar es Salaam, Feb. 1874, *Hildebrandt* 1254!; Zanzibar I., Chukwani, 8 July 1950, *R.O. Williams* 41! & Mkokotoni, 26 Oct. 1959, *Faulkner* 2389! & Pemba I., Panza I., 13 Feb. 1929, *Greenway* 1399!

* The sheet bears two specimens, the upper annotated 413 which clearly refers to Fl. Zeyl.: 413 (1748).

DISTR. **K** 7; **T** 6; **Z**; **P**; Mozambique, Madagascar, Natal, Comoro Is. to India, Sri Lanka, Malaya, Indochina to Hainan, Philippines, Indonesia to New Guinea, Fiji, New Caledonia, Hawaii, Solomon Is., Samoa and Australia; also very widely cultivated

HAB. Open vegetation near shore, strand formations and behind mangroves; ± sea-level–20 m.

SYN. *V. bicolor* Willd., Enum. Hort. Berol. 2: 660 (1809)
    V. trifolia L. var. *parviflora* Benth., Fl. Austral. 5: 67 (1870), quoad syntypum Moreton Bay, *Fitzalan* (K!) (see note)
    [*V. negundo* sensu Vatke in Linnaea 43: 533 (1882); Bak. in F.T.A. 5: 328 (1900); Pieper in E.J. 62, Beibl. 141: 53, 77 (1928); U.O.P.Z.: 485 (1949); T.T.C.L.: 642 (1949); K.T.S.: 597 (1961), *non* L. sensu stricto]
    V. negundo L. var. *bicolor* (Willd.) Lam, Verben. Malay Arch.: 191 (1919); Townsend in Fl. Iraq. 4: 663 (1980)

NOTE. Lam (1919) treats this as a variety of *V. negundo* and Townsend (1980) follows this; others treat it as a hybrid between the two. Lam lists the differences between *V. negundo* and *V. trifolia* as follows: 3–5-foliolate (1–3-foliolate), leaflets ovate-lanceolate to narrow lanceolate (ovate or broadly ovate, sometimes ± obovate), acuminate (subacute to rounded), sometimes crenate (entire), entirely glabrous above (midrib pubescent or ? sometimes pubescent all over), central leaflet distinctly petiolulate the others ditto or sessile (all leaflets sessile), panicle 4–24 × 3–13 cm. (4–13 × 2–5), calyx 5-ribbed, 1.5–2 mm. long (not ribbed, 3–4 mm. long), corolla-tube 3–4.5 mm. and lip 5–8 mm. (tube 8.5 mm. and lip 1.4 cm.). There is certainly a grey area between the two, many specimens being difficult to place, but I would hesitate to suggest they should be combined. Corolla size certainly suggests that all East African material is nearer *V. negundo* and this is the name which has been used in all local literature. There is no doubt, however, that the general facies and particularly the inflorescence structures point to *V. trifolia*. East African material is certainly a good match of the type of *V. bicolor* Willd. Singhakumara points out that the inflorescences of *V. negundo* are broad thyrses with long spreading branches, the whole 15–40 cm. long and 10–30 cm. wide and the individual cymes are dense; the thyrses of *V. trifolia* on the other hand are narrower, about 20 cm. long and without major branches; the cymes are fewer-flowered and not dense. He also demonstrates that *V. trifolia* and particularly the var. *bicolor* are coastal plants whereas *V. negundo* is not and extends up to 1830 m. in Nepal. *V. negundo* has several varieties with serrate leaves. *V. trifolia* var. *trifolia* extends over much the same range as var. *bicolor* but Moldenke suggested that the correct varietal epithet might be Bentham's var. *parviflora* and one syntype does agree, but when Domin raised (Bibl. Bot. 22 (89): 1114, 1117, fig. 182 (1928)) Bentham's variety to specific rank as *V. benthamiana* he cited only the *Landsborough* syntype from the Gulf of Carpentaria, and specifically said the other was something different. The Moreton Bay specimen is certainly the same as the East African plant but the other is rather different so I have kept to Willdenow's epithet. I believe it is wild in East Africa but it has been suggested otherwise (e.g. K.T.S. 'often cultivated and has in places gone wild on the East African Coast'). Only one cultivated specimen has been seen, *Kahurananga & Kilu* 2982 (Nairobi City Park, 24 July 1976) with trifoliolate leaves below the inflorescence and simple ones on lateral shoots; it has variegated leaflet margins and is what is called cv. *variegata* — it is probably also var. *subtrisecta* (Kuntze) Mold.

2. **V. zanzibarensis** *Vatke* in Linnaea 43: 533 (1882); Gürke in P.O.A. C: 339 (1895), as "*sansibarensis*"; Bak. in F.T.A. 5: 318 (1900); Pieper in E.J. 62 Beibl. 141: 53 (1928); T.T.C.L.: 642 (1949); Vollesen in Opera Bot. 59: 83 (1980). Type: Tanzania, Bagamoyo, *Hildebrandt* 1303* (B, holo.†, BM, K, iso.!)

Habit very variable; shrub 4–5 m. or tree 5–16(–24**) m. tall and 1.3 m. diameter at breast height or scandent shrub to genuine liane; bark yellowish or brown, gnarled, flaking and longitudinally fissured or striated; underbark bright yellow with red streaks; stems grey, ± glabrescent. Leaves with pungent odour, 3–5-foliolate; leaflets sessile, oblong-lanceolate to lanceolate, 3–13 cm. long, 1–6.5 cm. wide, long-acuminate at the apex, cuneate at the base, entire, ± glabrous but gland-dotted above and beneath; petiole winged or unwinged, 1.5–9 cm. long, 0.2–1 cm. wide. Panicles terminal and axillary combined and lower axillary ones also sometimes present, extensive, ± 15 cm. long, 6–15 cm. wide; peduncles 3–7 cm. long; pedicels obsolete or very short, axes densely adpressed pubescent; bracts linear, 1 cm. long; bracteoles narrowly triangular, 1 mm. long. Calyx 1–1.5 mm. long, ribbed, adpressed pubescent and glandular, shallowly toothed. Corolla white, bluish purple or mauve, adpressed pubescent outside; tube 3–4 mm. long, densely pubescent and glandular; limb 7 mm. wide, the small lobes 2.5 mm. long and wide and larger lip 3–5 mm. long, 3–3.5 mm. wide. Fruit orange to black, globose, 8–10 mm. diameter, juicy, the juice dark blue; calycine cup plate-like, 4–5.5 mm. wide.

---

* The BM specimen is numbered 1302 and this has been pencilled in their copy of the original Linnaea description and the 1303 deleted but the Kew sheet is numbered 1303 and there is no evidence that the 2 is not just an error made on a duplicate label.
** Semsei's '80 ft.' may well be a considerably exaggerated estimate.

KENYA. Kwale District: Kaya Kinondo, 16 July 1987, *S.A. Robertson & Luke* 4910! & Gongoni [Gogoni]
Forest, Gazi, Sept. 1943, *Conservator of Forests* H 65/43!
TANZANIA. Uzaramo District: Kola, 29 Jan. 1938, *Vaughan* 2744! & Pugu Forest Reserve, 9 Mar. 1964,
Semsei 3678!; Kilwa District: Selous Game Reserve, Malemba Thicket, 25 Feb. 1976, *Vollesen* in
M.R.C. 3281! & W. of Lungonya R., 22 Jan. 1968, *Rodgers* 164!
DISTR. **K** 7; **T** 3, 6, 8; probably in Mozambique
HAB. Deciduous coastal thicket and dry forest, particularly at edges, woodland; 0–10 & 360–600 m.

SYN. *V. bunguensis* Mold. in Phytologia 35: 419 (1977). Type: Tanzania, Rufiji District, 105 km. Dar es
Salaam–Kilwa road, Bungu, *Minjao & Raya* in *D.S.M.* 1908 (NY, holo., DSM, iso.)

NOTE. When a large tree the bole is rather crooked and the wood hard, dark reddish brown.
This has been confused with *V. altissima* L.f. but, as Vollesen adnot. points out, in that species the
inflorescences and flowers are very different.

3. **V. buchananii** *Gürke* in P.O.A. C: 339 (1895); Bak. in F.T.A. 5: 319 (1900); Pieper in
E.J. 62, Beibl. 141: 54 (1928); T.T.C.L.: 642 (1949); F.F.N.R.: 371 (1962); Haerdi in Acta
Trop., Suppl. 8: 153 (1964); Coates Palgrave, Trees S. Afr.: 807 (1977); Vollesen in Opera
Bot. 59: 83 (1980). Type: Malawi, *Buchanan* 782 (B, holo.†, K, iso.!)

Shrub 2–4.5 m. or branched tree 3–15 m. tall or even described as a herb to 1.2 m.; bark
smooth, yellow-green; young shoots often square, with dense pubescence of short curled
and adpressed and rather longer spreading hairs. Leaves aromatic, 3–5-foliolate; leaflets
ovate to elliptic or obovate-elliptic, 3.5–18 cm. long, 2–9 cm. wide, acuminate at the apex,
rounded to cuneate at the base, entire or toothed, ± discolorous, subscabrid above with
short hairs or glabrescent but with small yellow glands, sparsely to densely pubescent to
subtomentose beneath particularly on the venation and densely dotted with small yellow
glands; petiole 4–12.5 cm. long; petiolules 0–2.5 cm. long. Inflorescences much branched,
paniculate, 10–30 cm. long, densely pubescent with short and long hairs and glandular;
bracts linear, 2–3 mm. long; pedicels 0–1.5 mm. long, enlarging to 2–3 mm. in fruit. Calyx
2 mm. long, ribbed, denticulate or shallowly toothed, densely pubescent. Corolla white or
greenish yellow; tube 3–4.5 mm. long, the upper part glandular and tomentose outside;
upper lip bilobed with lobes 1–1.5 mm. long, lower lip 3-lobed, 2.5 mm. long. Style 2–5
mm. long. Fruit very characteristic, globose, 5–7 mm. diameter, so densely glandular as to
appear yellow or yellow-brown but white under the coating; calycine cup 7–9 mm. wide,
pubescent and gland-dotted.

TANZANIA. Lushoto District: Mlinga Mt., 18 Feb. 1937, *Greenway* 4907!; Uluguru Mts., Morogoro, 9
Jan. 1933, *Schlieben* 3221!; Rungwe District: Mitalula Coffee Station, 16 Feb. 1968, *S.A. Robertson*
962!
DISTR. **T** 3, 6–8; Malawi, Zambia, Mozambique
HAB. Thicket, secondary bushland, deciduous woodland of *Ricinodendron*, etc., also *Brachystegia*,
Uapaca woodland, grassland and derived plantations; 360–1500 m.

SYN. *V. volkensii* Gürke in P.O.A. C: 339 (1895); Bak. in F.T.A. 5: 318 (1900); T.T.C.L.: 642 (1949):
Moldenke in Phytologia 8: 94 (1961). Type: Tanzania, Lushoto District, Derema, *Volkens* 132
(B, holo.†, BM, K, iso.!)
V. quadrangula Gürke in E.J. 28: 463 (1900); Bak. in F.T.A. 5: 520 (1900). Type: Tanzania, S.
Uluguru Mts., *Goetze* 157 (B, holo.†, K, iso.!)
V. radula Pieper in E.J. 62, Beibl. 141: 42, 55 (1928) (in key) & in F.R. 26: 161 (1929); T.T.C.L.: 642
(1949). Lectotype chosen by Pieper (1929): Tanzania, Rungwe District, ? Kyimbila, *Stolz* 1398
4(B, lecto.†, EA, K, isolecto.!)
V. buchananii Gürke var. *quadrangula* (Gürke) Pieper in E.J. 62, Beibl. 141: 54 (1928); T.T.C.L.:
642 (1949)

4. **V. sp. A**

Shrub with older stems hollow, ± pale and ± glabrous, ± ridged; young shoots darker,
purplish brown with pale lenticels, with prominent ridges on the 4 angles, shortly
pubescent. Leaves 3-foliolate; leaflets elliptic, (4–)7.5–14.5 cm. long, (2–)4.3–7.5 cm. wide,
acuminate at the apex, cuneate at the base, the central leaflet more narrowly so, the
lateral leaflets with midrib noticeably excentric, quite scabrid with bulbous based hairs
above, with 2–3 obscure crenations towards apex on each side, slightly paler beneath and
with the rather closely reticulate raised venation with dense horizontally spreading hairs
and dense small yellow glands in the areoles; petiole 5.5–9 cm. long, shortly pubescent;
petiolules 2–5 mm. long. Inflorescence unbranched, terminal, 25 cm. long, similar to
stems in colour and pubescence; fruiting stalks ± 7 mm. long, jointed, the true pedicels ± 3
mm. long; bracts linear, 3 mm. long. Flowers unknown. Fruits yellow, broadly ovoid, 1.3 cm.

long, 1 cm. wide, obscurely 4-grooved, densely minutely punctate with yellow glands; calycine cup 1 cm. wide, 5 mm. deep, with reflexed oblong lobes, 5 mm. long, 1.5 mm. wide which break off; all parts densely shortly spreading pubescent and with yellow glands.

TANZANIA. Morogoro District: Kimboza Forest Reserve, July 1983, *Rodgers, Hall & Mwasumbi* 2581! DISTR. **T** 6; not known elsewhere
HAB. Lowland evergreen forest on limestone; 300–350 m.

NOTE. It is related to *V. buchananii* but quite different, being characterised by its unbranched inflorescence, scabrid leaflets, larger fruits and distinct calyx-lobes.

5. **V. strickeri** *Vatke & Hildebrandt* in Linnaea 43: 532 (1882); Gürke in P.O.A. C: 340 (1895) & in E.J. 28: 463 (1900); Bak. in F.T.A. 5: 318 (1900); T.C.E. Fries in N.B.G.B. 8: 702 (1924); Pieper in E.J. 62, Beibl. 141 ('142'): 55 (1928); T.T.C.L.: 642 (1949), pro parte; Moldenke in Phytologia 6: 143 (1958)*; K.T.S. 597 (1961); Vollesen in Opera Bot. 59: 83 (1980). Type: Tanzania, Dar es Salaam, *Hildebrandt* 1250 (B, holo.†, BM, K, iso.!)

Aromatic erect or scrambling shrub or small tree 1.5–4.5(–5.4) m. tall; bark pale and ridged; young stems ± square, densely yellow or pale brown pubescent, ± glabrous when older. Leaves 3-foliolate; leaflets ovate or elliptic, 1.8–10.5 cm. long, 1.3–5.5 cm. wide (the median longest), acute to long-acuminate at the apex, very rarely rounded, cuneate to rounded at the base, the laterals more rounded or subtruncate, entire, crenate or ± toothed, sometimes very slightly bullate, scabrid and ± rugose above, sparsely to densely pubescent beneath particularly on nerves and with dense glands or ± glabrous (particularly in **K** 7); venation raised beneath; petioles 1.3–5(–7) cm. long; petiolules 1–5(–18) mm. long, the lateral leaflets with them short or subsessile. Inflorescences narrow, 2–12.5 cm. long, pubescent, the lateral cymes mostly short so that panicle appears almost subspicate; young inflorescence with bracts slightly comose at apex, soon falling; main bracts to each cyme purplish, 3 mm. long, 1–1.5 mm. wide. Calyx shortly tubular, 1.5–3 mm. long, shortly toothed to distinctly lobed, pubescent and glandular. Corolla white, dull yellow or very pale pink; tube 2–2.5 mm. long, glabrous to pubescent-tomentose outside; lobes ovate to ± round, 1.5 mm. long, 1.2–1.5 mm. wide, glandular and tomentose outside at tips, the upper lip distinctly bilobed or not. Fruit buff-brown, ovoid to globose, 6 mm. long, glossy, glabrous and not glandular, the wall very thin and easily splitting; endocarp very hard; fruiting calyx ± 7 mm. wide, irregularly lobed.

KENYA. Kiambu District: near Nairobi, Karura Forest, 22 Nov. 1966, *Perdue & Kibuwa* 8058!; Teita District: Bura, *Gardner* in F.D. 2989!; Kilifi District: Arabuko, Mar. 1930, *R.M. Graham* 814!
TANZANIA. Mwanza, about 21 km. along railway, 21 Mar. 1937, *B.D. Burtt* 5538!; Tanga District: 8 km. SE. of Ngomeni, 31 July 1953, *Drummond & Hemsley* 3579!; Bagamoyo District: Kikoka Forest Reserve, Apr. 1964, *Semsei* 3751!
DISTR. **K** 4, 6, 7; **T** 1–3, 6, 8; **Z**; not known elsewhere
HAB. Bushland, thickets or rocky outcrops, woodland, forest edges, sometimes riverine; 0–1710 m.

SYN. *V. lamiana* Pieper in E.J. 62, Beibl. 141: 55 (1928) (in key) & in F.R. 26: 161 (1929); T.T.C.L.: 642 (1949); Moldenke in Phytologia 5: 430 (1956); K.T.S.: 597 (1961). Lectotype chosen by Pieper (1929): Tanzania, Kilimanjaro, *Endlich* 777 (B, lecto.†)**

NOTE. There is considerable variation in the indumentum of the foliage.

6. **V. ugogoensis** *Verdc.* in K.B. 45: 699, fig. 1 (1990). Type: Tanzania, 11 km. on Dodoma–Kongwa road, *Leippert* 5513 (K, holo.!)

Shrub or tree 1.5–5 m. tall; stems pubescent when very young, soon glabrous, often darker brown or grey than in *V. strickeri* and sometimes more terete. Leaflets 3, elliptic, ovate, obovate or ± round, 1–9 cm. long, 0.6–4 cm. wide, acute or acuminate at the apex, cuneate at the base, discolorous, crenate towards the apex, pubescent and glandular on both surfaces, very slightly scabrid and ± bullate above; petiole 0.4–2 cm. long; petiolules obsolete or central one up to 8 mm. long. Inflorescences 2.5–5 cm. long, extending to 7 cm. in fruit, reduced so that side-branches appear as fascicles with ± 5 flowers, the pedicels arising ± directly from main axis, the secondary axes ± suppressed; pedicels 1–1.5 mm. long but fruit-stalks 3–6.5 mm. long; bracts oblong to linear, 3–5 mm. long, 0.3–1 mm.

* His reference to *Hildebrandt* 1303 from 'Bagamoro' as the type is an error; that number is *V. zanzibarensis*
** The syntypes *Scott Elliott* 6417 & *Fries* 1988 are at Kew and confirm this synonymy.

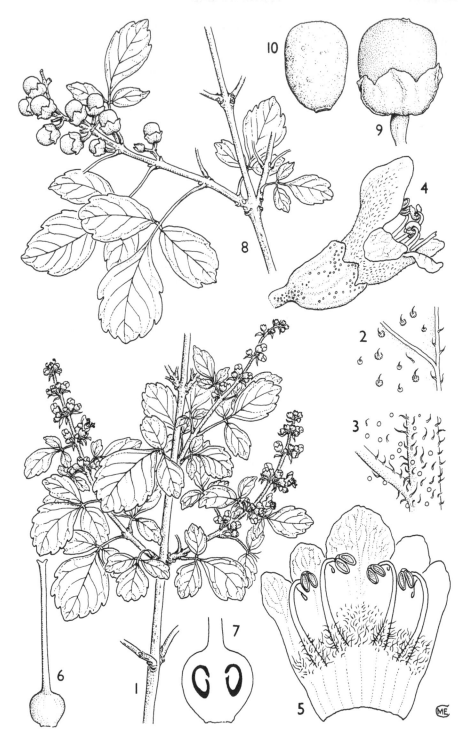

FIG. 8. *VITEX UGOGENSIS* — **1**, flowering shoot, × ²/₅; **2**, part of upper leaf surface, × 20; **3**, part of lower leaf surface, × 20; **4**, flower, × 6; **5**, corolla, opened out, × 6; **6**, ovary and style, × 6; **7**, longitudinal section of ovary, × 12; **8**, fruiting shoot, × ²/₅; **9**, fruit, × 3; **10**, seed, × 3. 1–3, from *Leippert* 5513; 4–7, from *Richards* 20932; 8–10, from *B.D. Burtt* 3923. Drawn by Mrs. M.E. Church.

wide. Calyx 2 mm. long, glandular, the lobes rounded or very broadly triangular, 0.5 mm. long. Corolla white or pinkish yellow; tube 4 mm. long, tomentose outside, largest lobes narrowly oblong, 3 mm. long, 2.2 mm. wide, not bifid, others 1.5–2 mm. long. Fruits ± red or brick-red to brown, ± globose, usually somewhat elongated, to 9 mm. long, 8 mm. wide, very glossy; calycine cup 0.8–1 cm. wide, venose, sparsely gland-dotted. Fig. 8.

TANZANIA. Kondoa District: Farkwa, 21 Dec. 1965, *Newman* 63!; Dodoma District: between Mukodowa and Dodoma, 11 Dec. 1965, *Richards* 20932!; Mpwapwa, Matomondo, Kisalu, near PWD camp 4, 17 May 1976, *Magogo & Ruffo* 706! & Mpwapwa, 21 Apr. 1932, *B.D. Burtt* 3923!
DISTR. T5; not known elsewhere
HAB. Deciduous thicket, bushland and *Brachystegia* woodland on grey sandy soil; 1140–1500 m.
SYN. [*V. strickeri* sensu T.T.C.L.: 642 (1949), quoad *Hornby* 1263, *non* Vatke & Hildebrandt]
NOTE. Hornby and Burtt independently refer to the fruits as like brown boot buttons in colour and shape. I had at first considered treating this as a subspecies of *V. strickeri* but it seems well defined.

7. **V. schliebenii** *Mold.* in Phytologia 7: 85 (1959); Vollesen in Opera Bot. 59: 83 (1980). Type: Tanzania, Lindi District, Kimuera*, *Schlieben* 6008 (B, holo., K, iso.!)

Shrub 2.4–3 m. or tree 8–10 m. tall; bark grey-brown, smooth; innovations grey-tomentose; branches slender, 4-angled, densely spreading pubescent with brown hairs, at length ± glabrous. Leaves (3–)5-foliolate; leaflets elliptic to elliptic-lanceolate, 1.2–17 cm. long, 1.5–6 cm. wide, acuminate or less often rounded at the apex, acute to attenuate at the base, slightly scabrid above, densely pubescent with brown hairs beneath, longer on main venation, glandular; petioles 2.5–9 cm. long; petiolules ± 0–2.3 mm. long. Inflorescences short, axillary near the apices of the branchlets, 2–3(–5) cm. long, 2 cm. wide, several-flowered; peduncles 1–1.2 cm. long, brown pubescent; pedicels filiform, 1.5–6 mm. long. Calyx campanulate, 1.5–2 mm. long, shallowly toothed. Corolla white, the lobe-tips cream or pale yellow, hypocrateriform; tube 2–2.5(–4 fide Moldenke) mm. long; lobes oblong-ovate 2 mm. long, 1.5 mm. wide. Fruits violet or purple, 5 mm. diameter.

KENYA. Tana River District: Kurawa, 10 Oct. 1961, *Polhill & Paulo* 649! & Garsen–Malindi road, 1.5 km. towards Malindi from turn-off to Oda, 22 July 1974, *R.B. & A.J. Faden* 74/1191!; Lamu District: Kiunga, 31 July 1961, *Gillespie* 91!
TANZANIA. Kilwa District: Kingupira, 26 Feb. 1976, *Vollesen* in *M.R.C.* 3285 & 5 Mar. 1975, *Vollesen* in *M.R.C.* 1894!; Masasi District: near Masasi, Chironga Hill, 8 Mar. 1991, *Bidgood et al.* 1847!; Lindi District: 100 km. W. of Lindi, Kimuera, Mbamba, 17 Feb. 1935, *Schlieben* 6008!
DISTR. K 7; T 8; S. Malawi, Mozambique
HAB. *Euphorbia, Commiphora, Strychnos* bushland, *Acacia* scrub and thicket, often riverine or black clay; 5–550 m.

8. **V. mossambicensis** *Gürke* in P.O.A. C: 340 (1895); Bak. in F.T.A. 5: 329 (1900); Pieper in E.J. 62, Beibl. 141: 57 (1928); Vollesen in Opera Bot. 59: 83 (1980). Type: Mozambique, *Carvalho* (B, holo.†)

Shrub or small tree 4–6(–18 fide *Gillman* 1136) m. tall; young shoots densely pubescent with pale ferruginous hairs; old stems pale fawn, glabrous, roughened with nodular petiole-bases. Leaves 5-foliolate; leaflets oblong-elliptic, 1.5–9.5 cm. long, (0.7–)3–3.8 cm. wide, acuminate, the tip acute or obtuse, rounded to cuneate at the base, entire, glabrous above, pubescent along midrib beneath and in axils of nerves or sparsely to fairly densely pubescent all over the lower surface; petiole (1.3–)2.5–5.5 cm. long; petiolules up to 2 cm. long. Flowers fragrant, few in very slender dichasia 2–6 cm. long; peduncle 2–12 cm. long; true pedicels very short save that of central flower up to 2 mm. but other flowers when branches reduced to 1 flower appear to have stalks up to 9 mm. long. Calyx conic, 2 mm. long, truncate or denticulate, glabrous. Corolla pale blue or white with blue lower lip; tube 4 mm. long; limb ± 4 mm. long. Fruit globose, 6–7 mm. diameter, glabrous; calycine plate flat, 6 mm. diameter, glabrescent or puberulous.

TANZANIA. Kilwa District: Malemba Thicket, 24 Jan. 1977, *Vollesen* in *M.R.C.* 4378!; Lindi District: Mtange, 23 Mar. 1943, *Gillman* 1384! & Rondo Plateau, Mchinjiri, Feb. 1952, *Semsei* 652!
DISTR. T 8; Mozambique
HAB. Coastal scrub and thicket; 400–800 m.

---

* Kew isotype label states Nambiranji [Nambilanje].

SYN. *V. oligantha* Bak. in F.T.A. 5: 327 (1900). Types: Tanzania, Kilwa, *Kirk* 108 & s.n. (K, syn.!)
    V. mossambicensis Gürke var. *oligantha* (Bak.) Pieper in E.J. 62, Beibl. 141: 57 (1928); T.T.C.L.: 643
    (1949)

9. **V. amaniensis** *Pieper* in E.J. 62, Beibl. 141: 46, 60 (1928) (in key) & in F.R. 26: 164
(1929); T.T.C.L.: 643 (1949). Lectotype chosen by Pieper (1929): Tanzania, Lushoto
District, Amani, *Warnecke* 221 (B, lecto.†)

Tree 12–20 m. tall, with smooth white bark or pale grey-brown ± corky bark on twigs;
branchlets crisply pilose. Leaves drying dark, 5-foliolate; leaflets elliptic to obovate-
elliptic, 6.5–18 cm. long, 3.2–9(–11) cm. wide, acuminate, cuneate at the base, glabrous
above, pubescent beneath only on the nerves which are rather close and parallel,
glandular-punctate, entire to remotely serrulate with upwardly directed teeth; petioles
6–12 cm. long; petiolules 0.5–2 cm. long. Inflorescences axillary, 8–25 cm. long, ±
few-flowered dichasial cymes, the axes ferruginous tomentose or shortly crisped-pilose;
bracteoles linear, 0.5–1 cm. long; peduncles 3–11.5(–13) cm. long; pedicels of central
flowers up to 5 mm. long but others much shorter. Calyx campanulate, 3–4 mm. long,
somewhat 2-lipped, densely grey crisped-hairy, irregularly toothed, the teeth ± 1 mm.
long. Corolla lilac-white or pale blue with deeper blue lips and mauve lines; tube 3.5 mm.
long, ferruginous tomentose outside; larger lip 3-fid, the main lobe 2.75 × 3 mm., the
others much smaller; smaller lip with 3 ovate-triangular lobes 2.2 × 1–2 mm. Ovary
glandular at apex. Fruit black, ellipsoid-oblong, 0.9–1.1 cm. long, 6.5 mm. wide, glabrous;
calycine cup 0.7–1 cm. wide, pale brown pubescent.

TANZANIA. Lushoto District: Amani, by R. Dodwe on path to Nderema, 19 Feb. 1906, *Braun in Herb.
    Amani* 1503!; Morogoro District: Uluguru Mts., Tanana, 5 Feb. 1936, *E.M. Bruce* 774! & Bunduki
    Forest Reserve, Mar. 1953, *Paulo* 53! & NW. Uluguru Mts., 16 Jan. 1933, *Schlieben* 3255!
DISTR. **T** 3, 6, 7; not known elsewhere
HAB. Rain-forest, forest edges and relicts; 900–1400 m.

SYN. [*V. lokundjensis* sensu Moldenke in Phytologia 5: 444 (1956), pro parte, *non* Pieper]

NOTE. *Mayuga* 99 (Kilosa District, Chonwe Mts., Dec. 1968) I thought might be young foliage of this
    species, both entire and toothed leaflets being present, but the fruit is said to be '1" diam. edible'
    possibly due to confusion with a larger fruited species since the information on the label does not
    refer directly to this specimen.

10. **V. keniensis** *Turrill* in K.B. 1915: 47 (1915); Pieper in E.J. 62, Beibl. 141: 60 (1928);
Wimbush, Cat. Kenya Timbers, ed. 2: 73 (1952); K.T.S.: 595, fig. 109/b, photo. 79, 80
(1961). Type: NE. & E. Mt. Kenya, without exact locality, probably Meru, *D.K.S. Grant in
F.D.* 846 (K, syn.!, EA, isosyn.!)

Tree 12–30 m. tall, up to 1.8(–2.3) m. in diameter; bole 12–18 m. tall; bark very thin,
rough and slightly fissured; slash creamy yellow turning dirty green; stems, petioles and
leaf-venation beneath with longer more shaggy indumentum than in *V. fischeri*. Leaves
5-foliolate; leaflets obovate, 5.5–17 cm. long, 3.2–8.5 cm. wide, broadly rounded to
obtusely acuminate at the apex, cuneate to rounded at the base, coriaceous, usually drying
darker than in *V. fischeri*, sparsely puberulous above, paler beneath and completely
covered with soft ochraceous tomentum and glands but becoming sparser; petiole
13.5–17 cm. long; petiolules ± absent or intermediate one 0.6–4 cm. long. Cymes
ochraceous tomentose, somewhat more lax than in *V. fischeri* forming axillary panicles up
to 12 cm. long, 24 cm. wide; peduncles 6.3–13 cm. long; secondary branches to 4.5 cm.
long; pedicels 1–4 mm. long, densely tomentose; bracts 0.5–1 cm. long, 1–4 mm. wide.
Calyx densely ochraceous tomentose, rather more distinctly toothed than in *V. fischeri*;
tube 2 mm. long, lengthening to 5–6 mm. in fruit; teeth broadly triangular, 0.5 mm. long,
lengthening to 3 mm. Corolla white or white tinged blue with largest lobe mauve; tube
curved, 5–6 mm. long, ochraceous tomentose above outside; limb 8–10 mm. wide, the
largest lobe round or ovate, 3–5 mm. long, 3.5 mm. wide, entire, the others 2.5–3 mm. long,
2 mm. wide, slightly emarginate, densely ochraceous tomentose outside. Ovary globose,
with simple hairs and sparse glands; style filiform, 6 mm. long; stigma-lobes linear.
Drupes black with white spots, obovoid, (1.1–)2.2–2.4 cm. long, (0.8–)1.3–1.5 cm. wide,
shiny, glabrous, rounded at the apex, ± contracted to the base; mesocarp fleshy; endocarp
very woody, the putamen ± 2 cm. long, 1.1–1.2 cm. wide, 6–7 mm. diameter at the base,
obtusely 4-angled; calycine cup 1.1 cm. wide.

KENYA. S. Nyeri District: Mt. Kenya, near Rogati R., Kagochi, 28 Feb. 1960, *Kerfoot* 1504! & NE. of Sagana, near Baricho, 14 Feb. 1964, *Brunt* 1453!; Meru District: Nyambeni Hills, junction of S. Circular road to Maua and R. Thangatha at Stone Bridge, 10 Sept. 1960, *Polhill & Verdcourt* 275!; Teita Hills, Ngangao Forest, *Beentje* in *Taita Hills Exped.* 466*

DISTR. K 4, 7; also cultivated in Uganda and Tanzania (see note)

HAB. Evergreen forest, sometimes riverine; 1290–2100 m.

SYN. *V. balbi* Chiov. in Racc. Bot. Miss. Consol. Kenya: 99 (1935); K.T.S.: 593 (1961). Types: Kenya, Mt. Kenya, Nyeri, *Balbo* 55 & Meru, *Balbo* 78 & 861 (TOM, syn.)

NOTE. Usually synonymised with *V. fischeri*, there is no doubt that it is a distinct taxon and I think worthy of specific rank. It has been cultivated in Uganda (Bunyoro District, Budongo Forest Reserve, Waibira Block, Compt. 19, 24 Jan. 1963, *Styles* 329!), Kenya (Nairobi Arboretum, 18 Feb. 1952, *G. Williams Sangai* 339!) and Tanzania (Arusha District, slopes of Ngurdoto crater, Usa Saw Mill, Jan. 1967, *Procter* 3478! & Lushoto Arboretum, 16 Jan. 1963, *Semsei* 3603! & Magamba, 2 Apr. 1966, *Semsei* 3994!). The timber is pale greyish brown, coarse textured with marked growth zones and often an attractive wavy grain; the heartwood of trees over 60 cm. diameter is often dark and very decorative. The wood works easily and is used for cabinet work, panelling, furniture and coffin boards, etc. A single large leaf (*Verdcourt & Polhill* 2938B, Kenya, Meru District, Nyambeni Tea Estate, 8 Oct. 1960, 2040 m.), glabrescent with large petiolate coarsely serrate leaflets to 26.5 × 11.5 cm., might possibly be juvenile foliage. *Robertson & Luke* 6322 (Kwale District, Gongoni Forest Reserve, NE. side, 2 June 1960, 30 m. in lowland moist forest) seems nearest to this species but is (unless perhaps cultivated) very far from its usual range. The material is only in fruit and flowering material is needed to come to a decision.

11. **V. fischeri** *Gürke* in E.J. 18: 171 (1893) & in P.O.A. C: 339 (1895); Bak. in F.T.A. 5: 330 (1900); Pieper in E.J. 62, Beibl. 141: 60 (1928); T.T.C.L.: 643 (1949); I.T.U., ed. 2: 445 (1952); F.P.S. 3: 200 (1956); K.T.S.: 595, fig. 109/a,c,d (1961); F.F.N.R.: 371 (1962). Types: Tanzania, Mwanza District, near Lake Victoria, Kayenzi [Kagehi], *Fischer* 476 & Uzinza [Usindji], *Stuhlmann* 3576 & Mwanza, *Stuhlmann* 4137, 4184 & Biharamulo District, Kimwani [Kimoani] Plateau, *Stuhlmann* 3394 (all B syn. †)

Savannah shrub or small to fairly large tree (1.8–)4.5–9 or sometimes to 15(–18) m. tall with rounded crown; bark grey to dark brown, shallowly striate and fissured, slash white to light brown with white sapwood; young branches and petioles densely clothed with velvety orange-brown hairs or yellow-tomentose. Leaves scented, 5-foliolate; leaflets oblong, ovate-elliptic to elliptic or slightly obovate-elliptic, 5–19 cm. long, 3–10 cm. wide, shortly to long-acuminate at the apex, the tip acute, cuneate to rounded at the base, entire, somewhat scabrid-puberulous above but later ± glabrous, velvety woolly tomentose beneath and with sparse to dense glands, the surface usually totally obscured but sometimes indumentum much sparser with glands and venation evident; lateral nerves rather numerous, up to ± 15 pairs; petioles 6.5–16.5 cm. long; petiolules 1–2(3–8) cm. long. Cymes axillary, extensive, many-flowered, up to 11(–15) cm. wide; peduncle 6–14 cm. long; pedicels 0–1 mm. long or that of central flower of cyme up to 12 mm.; all axes woolly-tomentose; bracts and bracteoles narrow, 0.3–2.2 cm. long, dark and ± glabrous inside, velvety outside. Calyx often tinged red, 3–4 mm. long, truncate or shallowly toothed, the teeth up to 1 mm. long. Corolla white with blue or mauve upper lip and lower cream or both blue; throat yellow; tube ± 3 mm. long, the lip 3.5 mm. long and lower lobes 2.5 mm. long and wide. Fruit black or purple-green to black with green spots, oblong-globose, 1–1.3 mm. long, 0.8–1 cm. wide, edible; calycine cup ± 1 cm. wide.

UGANDA. Teso District: SW. of Katakwi, Mt. Abela, 10 May 1970, *Katende* 266!; Busoga District: Lake Kyoga, Malima [Mulima] Gombolola, Jan. 1931, *Brasnett* 58!; Mengo District: 24 km. Kampala–Entebbe road, Mar. 1931, *Hansford* 1994!

KENYA. Trans-Nzoia District: Kitale, May 1963, *Tweedie* 2599!; N. Kavirondo District: Kakamega Forest, near forester's house, 15 Oct. 1953, *Drummond & Hemsley* 4752!; Kisumu-Londiani District: Songhor, 15 July 1935, *McDonald* in A.D. 2728!

TANZANIA. Mwanza District: Capri Point Forest Reserve, Nov. 1952, *Procter* 117!; Shinyanga, Mwantine [Mantini] Hills, Mar. 1935, *B.D. Burtt* 5076!; Kigoma District: Kabogo Mts., May 1963, *Kyoto Univ. Exped.* 471!

DISTR. U 1, 3, 4; K 3, 5, 6; T 1, 4; Zaire, Sudan, Zambia, Angola (see note)

HAB. Wooded grassland and thicket particularly on granite rocks, also rich bushland on termite mounds and *Bridelia, Maesopsis, Albizia* forest; 980–2080 m.

SYN. *V. andongensis* Bak. in F.T.A. 5: 329 (1900); Pieper in E.J. 62, Beibl. 141: 60 (1928). Type: Angola, Pungo Andongo, *Welwitsch* 5696 (K, holo., BM, iso.!)

* *Fide* H. Beentje; I have not seen this sheet.

V. bequaertii De Wild. in F.R. 13: 142 (1914); Pieper in E.J. 62, Beibl. 141: 60 (1928). Types: Zaire, Lubumbashi [Elisabethville], *Bequaert* 319 & *Homblé* 202 (BR, syn.)
[*V. keniensis* sensu T.T.C.L.: 643 (1949), *non* Turrill]
V.payos (Lour.) Merr. var. *stipitata* Mold. in Phytologia 8: 72 (1961). Type: Uganda, Mengo District, Kampala, *E.H. Wilson* 194 (UC, holo.!)

NOTE.   Has been cultivated in Nairobi Arboretum (1 July 1926, *F.D.* 1353!), said to be from Nairobi area but given a Nandi name; a specimen was also collected there in 1919 (*Battiscombe* 960!) grown from seed collected by J.L. Moon in Kakamega Forest. The reference to its occurrence in Kwale District (K.T.S.: 595 (1961)) must be an error. A number of sterile sheets particularly from the Kigoma area have been difficult or impossible to name; the indumentum of coppice or juvenile material is frequently atypical — one would expect it to be thicker but it appears to be the opposite. Only study of ranges of trees in the field will throw light on this problem. The specimens have frequently been named *V. ferruginea* but I believe they are states of *V. fischeri*. Examples are *Sharman* 34 (Uganda, Bukedi [Budama], Apr. 1966, leaflets hairy above but stems, etc., lacking the woolly brown tomentum), *Suzuki* 108, 121, *Azuma* 470, 476, *Clutton-Brock* 433, *Kano* 253 (Kigoma District, various chimpanzee study areas in Kabogo Mts. and Gombe Stream Reserve). *Azuma* 476 has very characteristic inflorescence galls like balls of brownish white cotton wool 2–3 cm. in diameter. *Pirozynski* 303 (Buha District, Gombe Stream Reserve, Mkenke valley, 26 Jan. 1964) with leaflets 32 × 17 cm. virtually glabrous beneath save for ± dense glands and petiolules ± 7 mm. long might be from sapling or coppice shoots. *V. mombassae* also occurs in the Gombe area and has differently shaped leaves and much larger fruits — *Clutton-Brock* 17B and 173 appear to belong here.

12. **V. madiensis** *Oliv.* in Trans. Linn. Soc., Bot., 29: 134, t. 131 (1875); Gürke in P.O.A. C: 359 (1895); Bak. in F.T.A. 5: 322 (1900); Pieper in E.J. 62, Beibl. 141: 61 (1928); Aubrév., Fl. For. Soud.-Guin.: 504, 507, t. 115/1,2 (1950); I.T.U., ed. 2: 445 (1952); F.P.S. 3: 200 (1956); Huber in F.W.T.A., ed. 2: 447 (1963); Geerling, Guide de Terrain Lign. Sahél. et Soud.-Guin.: 331, t. 92/4 (1982). Type: Uganda, W. Nile District, Madi Woods, Dec. 1862, *Grant* 649 (K, holo.!)

Shrub or small tree, 1.2–7.5 m. tall, with rough bark, or pyrophytic herb to small shrub, 0.3–1.5 m. tall, forming patches 1 m. wide from a massive subterranean woody rootstock; stems chestnut or purplish, striate, shortly densely velvety pale ferruginous pubescent or quite woolly or sometimes finely pubescent, soon or tardily glabrescent. Leaves fragrant if crushed, often in whorls of 3, 1–5(–6)-foliolate, often drying yellow-green, becoming subcoriaceous; leaflets round, elliptic or obovate to oblanceolate, 2–25 cm. long, 1.8–11.5 cm. wide, usually shortly acuminate at the apex, rounded, cordate or cuneate at the base, very coarsely shallowly crenate, ± glabrous above and rather shining save for venation with pubescence as on the stems, densely velvety beneath at first or slightly rough above with very short scattered almost imperceptible hairs and densely puberulous on the nerves beneath or leaves entirely glabrous; very closely reticulate beneath with rather coarse veinlets which thicken with maturity, smooth or sometimes very finely reticulate above when dry; young leaflets have a very different-looking finer or even quite unraised venation; petioles 5–15 cm. long; petiolules 0.1–3 cm. long. Inflorescences scented, few–many-flowered, 3–7 cm. long, densely pale ferruginous pubescent often with quite long hairs or ± glabrous; bracts linear, 3–6 mm. long; peduncle 3.5–12.5 cm. long; pedicels 1–3(–4) mm. long. Calyx densely pubescent and glandular; tube 2–4 mm. long; teeth triangular, 1–2 mm. long. Corolla pink or white with violet or blue lips, densely velvety outside or tube glabrescent; tube 4–6 mm. long; biggest lobes 2–3 mm. long, 1.5–5 mm. wide, shorter lobes 1–1.5 mm. long, 1 mm. wide. Ovary globose with few glands. Fruit, often white-spotted, oblong-ellipsoid, globose or obovoid, (0.8–)1.4–2.5 cm. long, 1–1.9 cm. wide, shiny, very woody; fruiting pedicels 2.5–3 mm. long; calycine cup 1–1.3 cm. wide, distinctly toothed, slightly hairy.

subsp. **madiensis**

Shrub or small tree 1.2–7.5 m. tall; leaves mostly 1–3(–5)-foliolate, with usually broader leaflets with venation often plane above.

UGANDA.   W. Nile District: Terego, May 1938, *Hazel* 450! & E. Madi, Pakelle [Pakelli], May 1932, *Hancock* 761! & N. of Arua, 16 Mar. 1945, *Greenway & Eggeling* 7214; Teso District: Serere, *Eggeling* 747!
DISTR.   U 1, 3; Gambia, Senegal and Mali to N. Nigeria, Cameroon, Gabon, Central African Republic, Zaire, Sudan, Zambia, Mozambique and Angola
HAB.   Bushland, *Lophira, Butyrospermum, Terminalia, Combretum* woodland; 600–1350 m.

SYN.   *V. no. 1* sensu Thomson, App. Speke's Journal: 644 (1863)
      V. no. 2 sensu Tomson, App. Speke's Journal: 644 (1863)

V. simplicifolia Oliv. in Trans. Linn. Soc., Bot., 29: 133, t. 130 (1875); Gürke in P.O.A. C: 339 (1895); Bak. in F.T.A. 5: 320 (1900); Pieper in E.J. 62, Beibl. 141; 65 (1928); I.T.U., ed. 2: 442 (1952), adnot.; F.P.S. 3: 200 (1956); Huber in F.W.T.A., ed. 2, 2: 447 (1963); Geerling, Guide de Terrain Lign. Sahél. Soud.-Guin.: 331, t. 91/4 (1982). Type: Sudan, Madi, *Grant* 701/5 (K, holo.!)

V. camporum Büttn. in Verh. Bot. Ver. Brandenb. 32: 35 (1890); Bak. in F.T.A. 5: 323 (1900); F.W.T.A. 2: 276 (1931). Types: Zaire, Lower Congo, Tondoa (Underhill), *Büttner* 428 (B, syn.†) Angola, San Salvador, *Büttner* 427 (B, syn. †) & Golungo Alto, *Welwitsch* 5728 (B, syn. †)

V. schweinfurthii Gürke in E.J. 18: 170 (1893). Types: Sudan, Jur [Ghattas Zeriba], sphalm. "Bongo", *Schweinfurth* 2030 (B, syn. †) & Mittu, Muolo, *Schweinfurth* 2848 (B, syn. †)

V. vogelii Bak. in F.T.A. 5: 319 (1900); Aubrév., Fl. For. Soud.-Guin.: 506, t. 114/5 (1950). Type: Nigeria, S. Bornu, Musgu, *Vogel* 97 (K, holo.!)

V. schweinfurthii Bak. in F.T.A. 5: 322 (1900). Types: Sudan, Lao, Dinka Territory, *Schweinfurth* 1303, Jur Ghattas [Gir], *Schweinfurth* 1519 & Jur [Seriba Ghattas], *Schweinfurth* 1953 (all K, syn.!)

V. barbata Bak. in F.T.A. 5: 323 (1900); Pieper in E.J. 62, Beibl. 141: 61 (1928); F.W.T.A. 2: 276 (1931); Aubrév., Fl. For. Soud. Guin.: 504, t. 115/4 (1950). Types: Senegal, *Heudelot* 30, Senegambia, *Heudelot,* & Sierra Leone, Talla Hills, half-way up Gonkwi, *Scott-Elliot* 4881 (all K, syn.!)

V. diversifolia Bak. in F.T.A. 5: 323 (1900); F.W.T.A. 2: 276 (1931); Aubrév., Fl. For. Soud. Guin.: 506, t. 114/1–2 (1950). Types: Nigeria, Lagos, Yoruba District, *Barter* 1096 (K, syn.!) & Nupe *Barter* 1644 (both K, syn.!)

V. madiensis Oliv. var. *schweinfurthii* (Gürke) Pieper in E.J. 62, Beibl. 141: 63 (1928)

V. simplicifolia Oliv. var. *vogelii* (Bak.) Pieper in E.J. 62, Beibl. 141: 65 (1928)

V. pobequinii Aubrév., Fl. For. Soud.-Guin.: 506, t. 115/3 (1950), *nom. invalid.*

NOTE.   A very variable plant with foliage and inflorescence varying from almost glabrous to velvety, leaves 1–5-foliolate and venation coarse to very fine. Even the type of *V. simplicifolia* has one leaf with a small subsidiary leaflet. Despite the very different appearance of extreme specimens even varieties are not really worthy of recognition. Nevertheless both Huber and Girling keep *V. simplicifolia* separate. If this is maintained both occur in Uganda. Recently certain routine determiners have kept up *V. vogelii* Bak., usually (e.g. F.W.T.A.) sunk into *V. simplicifolia,* and it must be admitted many West African specimens are distinctive and do not match well with the Sudanese type of *V. simplicifolia.*

subsp. **milanjiensis** (*Britten*) F. White, F.F.N.R.: 372, 455 (1962). Types: Malawi Mulanje [Milanji], at 1800 m., *Whyte* 138 & Zambia, *Whyte* (both BM, syn.!)

Tree, shrub or subshrubby pyrophytic herb 0.3–2 m. tall; leaves mostly 5-foliolate with usually narrowly obovate, elliptic or oblanceolate leaflets; venation usually closely reticulate and raised above.

var. **milanjiensis**

Tree or shrub up to about 8 m.

TANZANIA.   Ngara District: Bushubi, *Tanner* 6014!; Kigoma District: Mtunda Hill Forest Reserve, Nov. 1954, *Procter* 282!; Tabora District: Kakoma, Oct. 1935, *C.H.N. Jackson* 115!

DISTR.   T 1, 4, ?7; Zaire, Burundi, Malawi, Zambia, Zimbabwe and Angola (Cabinda)

HAB.   *Combretum, Terminalia* woodland, *Brachystegia* woodland and flood pan/*Brachystegia* interzones; ± 980–1250 m.

SYN.   *V. milanjiensis* Britten in Trans. Linn. Soc., ser 2, Bot. 4: 36 (1894); Bak. in F.T.A. 5: 330 (1900) *V. madiensis* Oliv. var. *milanjiensis* (Britten) Pieper in E.J. 62, Beibl. 141: 63 (1928); Coates Palgrave, Trees S. Afr.: 809 (1977)

NOTE.   T.T.C.L.: 643 (1949) mentions that *V. madiensis* had been recorded from Tanzania but casts doubt on the record and also mentions var. *milanjiensis* with doubt. Britten himself suggests his taxon might be a variety of *V. madiensis.*

var. **epidictyodes** (*Pieper*) Verdc., comb. nov. Type: Tanzania, Rungwe District, Tukuyu [Langenburg], *Stolz* 556 (B, lecto. †, K!, NY, PRE, S, isolecto.)

Pyrophytic subshrubby herb or small shrub* 0.3–2 m. tall forming patches ± 1 m. wide from a massive subterranean woody rootstock, sometimes somewhat succulent.

TANZANIA.   Buha District: Kasulu–Kibondo, km. 112, 15 Nov. 1962, *Verdcourt* 3327!; Mpanda District: Mahali Mts., Kasangazi, 29 Sept. 1958, *Newbould & Jefford* 2762!; Rungwe District: Bomalakitana, 25 Jan. 1910, *Stolz* 556! & Mulinda, 30 July 1912, *Stolz* 1478!

DISTR.   T 4, 7; Zaire (especially Shaba), Burundi, Malawi, Angola (Cabinda)

HAB.   Tall *Hyparrhenia* grassland, wooded grassland, woodland including dense *Brachystegia, Pterocarpus* woodland and *Combretum, Terminalia* woodland; 1060–1800 m.

_____

* Pieper calls it a tree; the label of the Kew isotype is difficult to read but 1½ m. is clear.

SYN. *V. hockii* De Wild. in F.R. 13: 143 (1914). Type: Zaire, Shaba, Manika, *Hock* s.n. (BR, holo.)
V. epidictyodes Pieper in E.J. 62, Beibl. 141: 48, 61 (1928) (in key) & in F.R. 26: 164 (1929); T.T.C.L.: 643 (1949); Moldenke in Phytologia 5: 332 (1955)

NOTE. It is highly probable that the suffruticose habit is maintained by continuous burning and that the variety has no taxonomic significance; extremes are so distinctive, however, that it is felt a name should be available. Whether a plant would grow into a tree if protected could easily be ascertained by experiment — I rather suspect there is some genetic basis.

13. **V. doniana** *Sweet*, Hort. Brit.: 323 (1827); Bak. in F.T.A. 5: 323 (1900); Pieper in E.J. 62, Beibl. 141: 64 (1928); T.T.C.L.: 642 (1949); I.T.U., ed. 2: 443, fig. 93 (1952); Brenan in Mem. N.Y. Bot. Gard. 9: 38 (1954); Moldenke in Phytologia 5: 322 (1955) (very extensive bibliography and information); F.P.S. 3: 200, fig. 52 (1956); K.T.S.: 593, fig. 108 (1961); F.F.N.R.: 371 (1962); Huber in F.W.T.A., ed. 2, 2: 446, fig. 308 (1963); Coates Palgrave, Trees S. Afr.: 808 (1977); Vollesen in Opera Bot. 59: 83 (1980); Geerling, Guide de Terrain Lign. Sahél. Soud.-Guin.: 328, t. 92/3 (1982); Von Maydell, Arbres et Arbustes du Sahel: 375, 374 & 504–505 (plates) (1983); Troupin, Fl. Rwanda 3: 291, fig. 94.4 (1985). Type: Sierra Leone, Freetown, *Don* (BM, holo.!)

Deciduous much-branched tree 4.5–12(–24*) m. tall, with dark green rounded crown; bole up to 7.2 m. long and 2.6 m. girth; bark pale brown to greyish white with long narrow vertical fissures and stringy ridges or occasionally smooth; slash yellowish white darkening brown; young branchlets glabrous or puberulous to velvety tomentose. Leaves 5(–7)-foliolate; leaflets obovate, obovate-oblong to -elliptic or nearly obtriangular, 4–24.5 cm. long, 2.5–10.5 cm. wide, rather glaucous beneath, rounded, emarginate, very slightly apiculate or cuspidate at the apex, cuneate at the base, coriaceous, entire, glabrous or midrib puberulous beneath; petiole 5–20 cm. long; petiolules 0.3–2.5 cm. long. Cymes axillary, much branched, dense to fairly lax, ± 20-flowered, 2.5–7 cm. long, 3–16 cm. wide, velvety brown pubescent; peduncles 2–7.5 cm. long; pedicels 0–2 mm. long; bracts linear, to 6 mm. long; bracteoles narrowly triangular, 2 mm. long. Calyx 3–5 mm. long, shallowly toothed, rusty brown hairy, accrescent. Corolla white to rose with lower lip violet or purple or all mauve, at least sometimes with yellow honey guide, densely brown velvety tomentose outside; limb with 4 ovate lobes 3 mm. long and one larger, obovate and crinkly, 4.5 mm. long. Fruit black, oblong-ellipsoid, (1.8–)3 cm. long, 1.5–2 cm. wide, glabrous, sitting in a shallow cup 1–1.5 cm. wide.

UGANDA. W. Nile District: Terego, Mar. 1938, *Hazel* 449!; Bunyoro District: Budongo Forest, Busingiro rest house, 11 Mar. 1971, *Synnott* 536!; Teso District: Serere, Mar. 1932, *Chandler* 530!
KENYA. Elgon, May 1932, *Chater Jack* 282!; S. Kavirondo District: Bukuria, Sept. 1933, *Napier* 2899 in C.M. 5329!; Kwale District: Vanga, Kirao, Jan. 1930, *R.M. Graham* 756 in F.D. 2226!
TANZANIA. Bukoba District: Ruzinga, 19 Sept. 1936, *Gillman* 174!; Kondoa District: Bukulu, 14 Jan. 1962, *Polhill & Paulo* 1170!; Morogoro District: Kibembawe [Kibambawe], Sept. 1930, *Haarer* 1851!; Zanzibar I., Massazini, 22 Dec. 1959, *Faulkner* 2441!
DISTR. U 1–4; K 2 (Murua Nysigar), 3–7; T 1–8; Z; P; widespread throughout tropical Africa from Senegal to Cameroon, Mozambique, Malawi, Zambia, Zimbabwe and Angola, extending to the Comoro Is., also cultivated at Nairobi and various other forest stations
HAB. Essentially a savanna species, *Combretum* woodland, elephant grassland but extends into forest; mostly in rather wet areas; 0–1950 m.

SYN. *V. umbrosa* Sabine in Trans. Hort. Soc. London 5: 455 (1824), non Sw. (1788), nom. illegit.
V. cuneata Thonn. in Schumach. & Thonn., Beskr. Guin. Pl.: 289 (1827); Gürke in P.O.A. C: 339 (1895); Bak. in F.T.A. 5: 328 (1900); Pieper in E.J. 62, Beibl. 141: 71 (1928); U.O.P.Z.: 484, fig. (1949); Aubrév., Fl. For. Soud.-Guin.: 504, t. 113/1, 2 (1950) & Fl. For. Côte d'Ivoire, ed. 2, 3: 230, t. 335 1,2 (1959). Type: Ghana [Guinea], *Thonning* 244 (C, syn., LE, isosyn.)
V. cienkowskii Kotschy & Peyr., Pl. Tinn.: 27, t. 12 (1867); Bak. in F.T.A. 5: 328 (1900). Types: Sudan, confluence of White Nile and Bahr el Ghazal near Meschra-Req, *Heuglin* 43 (W, syn.) & above Fassoglu towards Kassan, *Cienkowski* 64 (W, syn.)
V. paludosa Vatke in Linnaea 43: 534 (1882). Type: Zanzibar I., Kidoti, *Hildebrandt* 1123 (B, syn., BM, K, isosyn.!) & without exact locality, *Kirk* (B, syn., K, isosyn.!)

NOTE. This species, one of the commonest savanna trees in many areas and widely known as 'mfuro' in many languages, is one of the most useful members of the family in East Africa. The white, yellowish white or pale brown wood is easily worked and useful for boxes, crates, shelves, low grade furniture, roof timbers and canoes. It does not possess great strength but within limits is a fine timber, one of the most useful for local use. The edible fruit is sweet and mealy and used as a sugar substitute.

---

* *Styles* 79 'Budongo Forest' gives a height of 24 m. and *Bryce* 218 Kilombero Valley, 21.6 m.

14. **V. payos** (*Lour.*) *Merr.* in Trans. Amer. Phil. Soc. 24: 334 (1935); T.T.C.L.: 644 (1949); Coates Palgrave, Trees Centr. Afr.: 430–3, plates (1957); K.T.S.: 597 (1961); F.F.N.R.: 372 (1962); Palmer & Pitman, Trees S. Afr. 3: 1962 (1972); Coates Palgrave, Trees S. Afr.: 811 (1977); Vollesen in Opera Bot. 59: 83 (1980); Verdc. in Taxon 38:155 (1989), *nom. spec. conserv. propos.* (see note). Conserved type: Tanzania, Tanga, *Volkens* 1 (BM, neo.!)

Shrub or small tree 1.8–10.5 m. tall, stiffly branched and with a rounded crown; bark brown or grey-brown, deeply fissured; stems densely pale ferruginous woolly hairy becoming glabrous, some thick and corky with prominent large petiolar scars. Leaflets (3–)5-foliolate; leaflets elliptic to obovate, 3–19 cm. long, 1.5–9.5 cm. wide, broadly rounded or obtuse at the apex or acute or even very shortly acuminate in young apical leaves, rarely emarginate, cuneate at the base, very discolorous particularly when young, entire, ± roughly pubescent above, densely floccose velvety beneath, sessile but in large leaves the basal narrowing may simulate a winged petiolule; venation ± impressed above, even ± bullate, raised and reticulate beneath under the indumentum; petiole 6–15 cm. long, the apex woolly ferruginous. Cymes few-many-flowered, dense to rather lax, axillary, sometimes almost globose, 6–15 cm. long; axes all densely pubescent; peduncles 2.5–10 cm. long; pedicels ± 1 mm. long, that of central flower up to 2 mm.; bracts linear, 0.8–2 cm. long, 0.5–1 mm. wide, projecting in very young inflorescences. Calyx obconical, densely ferruginous pubescent; tube 2.5–3 mm. long; teeth triangular-ovate, 1–2 mm. long, 1.5 mm. wide, very accrescent. Corolla fragrant, white, blue or mauve; tube 3–4 mm. long, hairy outside; main lip lavender, oblong, ovate or ± round, 3–5 × 4–5 mm., also with 4 white lobes 1.5 × 1.5 mm. Stamens not or scarcely exserted at least not beyond length of shortest corolla-lobes. Ovary hairy. Fruits black, oblong or ± subglobose, 2–2.5 cm. long, 1.4–2.4 cm. wide, ± shiny, glabrous, edible. Calycine cup dark, 1–2 cm. high, 1.5–2.5 cm. wide, broadly crenate; fruiting pedicel 2–5 mm. long.

var. **payos**

Leaflets densely velvety floccose beneath.

KENYA. Machakos District: Nzaui Hill, 16 Feb. 1969, *Kokwaro* 1885! & E. of Mtito Andei, Dec. 1966, *Greenway* 12630!; Kwale District: Shimba Hills, Pengo Hill area, 27 Mar. 1968, *Magogo & Glover* 504!
TANZANIA. Handeni District, Kangata, near Mligasi R., Nov. 1949, *Semsei* in *F.H.* 2903!; Morogoro District: 13 km. NE. of Kingolwira Station, 6 Dec. 1957, *Welch* 451!; Kilwa District: 50 km. from Liwale on road to Njinjo, Mtondo Road Camp, 26 Oct. 1978, *Magogo & Rose Innes* 420!
DISTR. K 4, 5, 7; T 2–8; Mozambique, Malawi and Zimbabwe
HAB. Wooded grassland with *Combretum, Piliostigma*, etc., *Combretum*, and *Brachystegia, Acacia* woodland, rock outcrops, also almost unwooded grassland; 30–1600 m.

SYN. *Allasia payos* Lour., Fl. Cochinch.: 85 (1790)
    Vitex allasia Planch. in Ann. Sci. Nat., sér 4, 2: 262 (1854). Type as for *V. payos, nom. illegit.*
    V. hildebrandtii Vatke in Linnaea 43: 534 (1882); Gürke in P.O.A. C: 339 (1895); Bak. in F.T.A. 5: 326 (1900); Pieper in E.J. 62, Beibl. 141: 65 (1928). Type: Tanzania, Dar es Salaam, *Hildebrandt* 1249 (B, holo.†)
    [*V. mombassae* sensu Bak. in F.T.A. 5: 326 (1900), quoad "*Johnston* Kilimanjaro", *non Hildebrandt*]
    V. shirensis Bak. in F.T.A. 5: 326 (1900), pro parte quoad syn. *Whyte*, Zomba*
    [*V. isotjensis* sensu S. Moore in J.B. 45: 94 (1907), quoad *Eyles* 1201, *non* Gibbs]
    V. eylesii S. Moore in J.B. 45: 154 (1907) & J.L.S. 40: 168 (1911). Type: Zimbabwe, Bulawayo, *Eyles* 1201 (BM, holo.!)

NOTE. The historical type of *Allasia payos* is "in ora Africae Orientale" [possibly near R. Payosi, a tributary of the Zambesi], *Loureiro* (P, holo., microfiche!). Planchon in Ann. Sci. Nat., sér. 4, 2: 262 (1854) had pointed out it was a *Vitex* and Merrill wrongly identified it with *V. hildebrandtii* from descriptions. The specimen is actually conspecific with *V. mombassae* but in order to avoid a totally unacceptable transference of names use has been made of the new powers to conserve the existing usage of names widely and persistently employed in a sense excluding their types.**
    Richards 11816 (Tanzania, Ufipa District, Sumbawanga, Kasanga, 24 Nov. 1959, has the calyx-lobes much more narrowly triangular and in parts of **T** 4 the leaflets are more persistently acuminate.

var. **glabrescens** (*Pieper*) *Mold.*, Known Geogr. Dist. Verben., ed. 2: 79 (1942) & in Phytologia 6: 47 (1957) & in Phytologia 8: 72 (1961). Types: Mozambique, Gazaland, Kurumadzi R., *Swynnerton* 1059

---

* See footnote p. 64.
** Unfortunately the proposal was not accepted but I refuse to adopt the "correct names" since the resulting confusion would necessitate an explanatory phrase every time the name was used. Another attempt will be made.

(K, syn.) & Chirinda Forest, *Swynnerton* 34 (K, syn.); Malawi, *Buchanan* 6970 (K, syn.) & Likoma, *Johnson* 17, Madadoda, *Johnson* 96, *Allen* 151 (K, syn.); Zimbabwe, Umtali, *Engler* 3120 (B, syn.†) and a number of other syntypes from Tanzania.

Leaflets soon with pubescence restricted to venation beneath and very sparse elsewhere, the venation thus very obvious beneath.

TANZANIA.   Shinyanya District: Kisumbi, 4 Jan. 1933, *B.D. Burtt* 4511!; Masai District: 50 km. Kinjungu–Njoge, 27 July 1965, *Leippert* 6058!; Mpwapwa, Jan. 1937, *Lindeman* 277!
DISTR.   T 1–3, 5, 7: scattered throughout much of distribution of the species, Mozambique, Malawi, Zimbabwe
HAB.   *Combretum, Terminalia* woodland; 1170–1600 m.

SYN.   *V. iringensis* Gürke in E.J. 28: 464 (1900); Bak. in F.T.A. 5: 521 (1900); T.T.C.L.: 644 (1949). Type: Tanzania, Iringa District, Lugalu [Rugaro], *Goetze* 548 (B, holo.†)
   V. hildebrandtii Vatke var. *glabrescens* Pieper in E.J. 62, Beibl. 141: 66 (1928)

NOTE.   The status of this variety is doubtful; it is really too erratic and scattered to deserve more than a mention. In fact I suspect that the leaves of all specimens start velvety but that this is lost much more quickly in some areas.

15. **V. mombassae** *Vatke* in Linnaea 43: 533 (1882); Gürke in P.O.A. C: 339 (1895); Bak. in F.T.A. 5: 326 (1900), pro parte; Pieper in E.J. 62, Beibl. 141: 66 (1928); T.T.C.L.: 644 (1949); K.T.S.: 597 (1961); F.F.N.R.: 371 (1962); Coates Palgrave, Trees S. Afr.: 810 (1977); Vollesen in Opera Bot. 59: 83 (1980). Type: Kenya, near Mombasa, *Hildebrandt* 1972 (B, holo.†, K, iso.!)

Much-branched shrub or small tree 1.2–6 m. tall, with white, grey or reddish brown corrugated or fissured bark which becomes stringy, less often smooth. Shoots more slender than in species 13, densely orange ferruginous velvety with more adpressed pubescence but sometimes more shaggy in some areas (e.g. T 1). Leaves 3–5-foliolate; leaflets elliptic, 0.5–13.5 cm. long, 0.5–5.5 cm. wide, shortly acuminate (always in young apical leaves) to broadly rounded or rarely emarginate at the apex, cuneate at the base, with scattered pubescence above, ferruginous velvety beneath with tangled hairs but sometimes glabrescent; venation impressed above, ± bullate and finely reticulate beneath; petiole 2–5.5 cm. long; leaflets sessile or central leaflets with petiolule 6–7(–10 in T 4) mm. long in longest leaves or all leaflets petiolulate even in small leaves in some areas (e.g. T 4). Inflorescences ± 4 cm. long, densely ± adpressed ferruginous; peduncle 2–5 cm. long; bracts narrowly oblong to oblanceolate, 1.1–1.8 cm. long, 3–4 mm. wide, acuminate, glabrous inside. Calyx densely ferruginous hairy; tube 3.5–4 mm. long; lobes 1.5–2 mm. long. Corolla pale blue, pale purple or blue-mauve with throat and the smallest lobes creamy white; tube 7–9 mm. long, pubescent outside; larger lower lip 0.8–1.3 cm. long, 0.7–1 cm. wide, sometimes yellow at base, hairy in middle; small lobes round or triangular, 5 × 3–5 mm. Stamens and style exserted 5–9 mm.; filaments rather stout, micropapillate. Ovary densely pubescent. Fruits dark purple, globose, (1.5(T 4)–)2–3.2(–3.5) cm. diameter, ± slightly pubescent at apex. Calycine cup 1.5–1.7 cm. long, 1 cm. wide, ± venose, distinctly toothed, pubescent. Dried mesocarp a dense mass of fibres. Fig. 9.

KENYA.   Kwale District: Shimba Hills, Mwele Ndogo Forest, 6 Feb. 1953, *Drummond & Hemsley* 1156! & Shimba Hills, Pengo Hill area, 27 Mar. 1968, *Magogo & Glover* 505! & N. side of Gongoni [Gogoni] Forest, 2 Mar. 1977, *R.B. & A.J. Faden* 77/633!
TANZANIA.   Mbulu District: Tarangire National Park, Boundary Hill, 25 Nov. 1969, *Richards* 24767!; Tanga District: Mwanyungu [Mwanyunge], 8 Feb. 1941, *Greenway* 6115!; Lindi District: Rondo Plateau, Nyenea, 11 Dec. 1955, *Milne-Redhead & Taylor* 7628!
DISTR.   K 7; T 1–8; Zaire, Burundi, Mozambique, Malawi, Zambia, Zimbabwe, Angola, Caprivi Strip and South Africa (Transvaal)
HAB.   Scrub and thicket, *Hyphaene, Albizia,* etc., *Sterculia, Pterocarpus,* secondary bushland, *Brachystegia* woodland, often on rocky granite hills; 20–1580 m.

SYN.   *V. flavescens* Rolfe in Bol. Soc. Brot. 11: 87 (1893–94); Bak. in F.T.A. 5: 321 (1900). Type: Angola, between Zambo and Pungo Andongo, Cazella, *Welwitsch* 5731 (K, syn.!, BM, isosyn.!) & Malange, *Marques* 8 (K, syn.!)
   V. mechowii Gürke in E.J. 18: 167 (22 Dec. 1893). Type: Angola, see above, *Welwitsch* 5731 (B, syn.†, BM, K, isosyn.! & Malange, *Mechow* 247 (B, syn.†)
   V. goetzei Gürke in E.J. 28: 464 (1900); Bak. in F.T.A. 5: 521 (1900). Type: Tanzania, Usagara, steppe S. of Rufiji R., *Goetze* 85 (B, holo.†, K, iso.!)
   V. shirensis Bak. in F.T.A. 5: 326 (1900), quoad syn. *Buchanan* 231*

_____

* Not lectotypified, the other syntype is referred to *V. payos*.

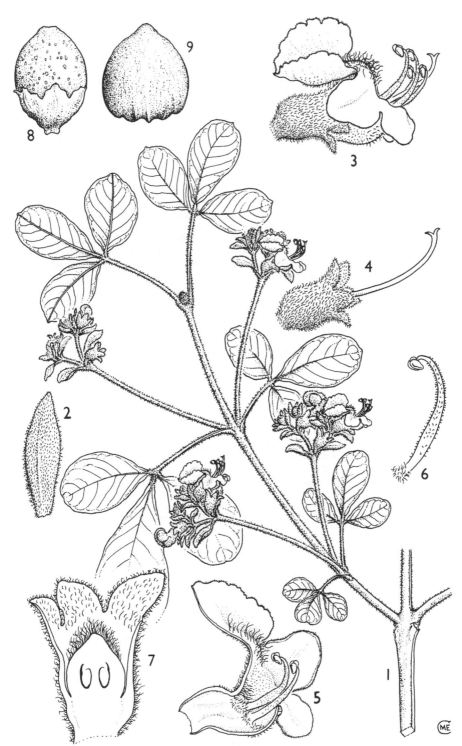

Fig. 9. *VITEX MOMBASSAE* — **1**, habit, × ²/₅; **2**, inflorescence bract, × 2; **3**, flower, × 3; **4**, calyx and style, × 3; **5**, longitudinal section of corolla, × 3; **6**, stamen, × 4; **7**, longitudinal section of ovary, × 6; **8**, fruit, × 1; **9**, endocarp, × 2. 1, from *Mgaza* 141; 2–7, from *Tanner* 447; 8, 9, from *Milne-Redhead & Taylor* 8678. Drawn by Mrs. M.E. Church.

V. mombassae Vatke var. *acuminata* Pieper in E.J. 62, Beibl. 141: 68 (1928). Type: Kenya, without locality, *Battiscombe* 2 (K, holo.!)

NOTE. *V. goetzei* is the form with prominent reticulate venation beneath. Populations from the Shimba Hills mostly have elongate elliptic acuminate leaflets but rounded ones do occur sometimes on the same plant; the type of var. *acuminata* probably came from the Shimba Hills. Plants with 3-foliolate leaves are common in **T** 8 and even rarely the odd 1-foliolate one. Peter reported that the edible fruits were sold in Tabora market in 1913. South African material may be subspecifically distinct.

16. **V. ferruginea** *Schumach. & Thonn.*, Beskr. Guin Pl.: 62 (1827); Bak. in F.T.A. 5: 324, 521 (1900); Pieper in E.J. 62, Beibl. 141: 70 (1928); Huber in F.W.T.A., ed. 2, 2: 447 (1963); Hepper, W. Afr. Herb. Isert & Thonning: 130 (1976). Type: Ghana, near Aquapim, *Thonning* 265 (C, syn., P-JU, isosyn.)

Tree 6–13.5 m. or shrub 1.5–4.5 m. tall, with light grey smooth bark, often deciduous and flowering when leaves very young; slash very thin, dirty white, rapidly turning greenish brown; branchlets rusty pubescent or finely adpressed velvety ferruginous tomentose and with some spreading rusty hairs. Leaves 3–7-foliolate; leaflets oblanceolate-oblong, elliptic, elliptic-lanceolate or oblong, 2.5–16.5 cm. long, 1–6.5 cm. wide, acute to narrowly acuminate or ± obtuse at the apex, cuneate at the base, glabrous above or very slightly hairy on nerves when young or pubescent in var. *amaniensis*, densely finely adpressed velvety rusty pubescent beneath to almost glabrous save for glands, entire or rarely shallowly crenate-dentate; petiole 2.5–12.5 cm. long; longest petiolules 0.2–1.3 cm. long. Cymes axillary, dense, ± 5 cm. long; peduncles 0.5–3 cm. long; bracts lanceolate to elliptic-oblong, 0.5–1.2 cm. long, 2–4 mm. wide, ferruginous pubescent or velvety outside. Calyx ferruginous pubescent outside; tube 2–2.5 mm. long; lobes triangular, 0.7 mm. long. Corolla white and lilac, pale blue or violet, sometimes yellow at throat, ferruginous pubescent outside; tube 6–9 mm. long, hairy at the throat; main lobe of limb 6 × 6–7 mm., the others smaller, 2–5 × 3–4 mm. Ovary densely hairy. Style and stamens exserted ± 4 mm. Fruit probably black when ripe, globose, 1.4–4 cm. in diameter, the endocarps oblong, woody, 1.9 × 1.6 cm., vaguely 4-grooved; fruiting calyx plate-like, 1.4–1.5 cm. wide, spreading pubescent.

subsp. **ferruginea**

Young shoots with ± more spreading indumentum and with some longer hairs; leaflets usually more narrowly acuminate with longer tips.

UGANDA. Bunyoro District: Budongo Forest, Sonso R., Apr. 1933, *Eggeling* 1206!; Mengo District: Kiagwe, Bukasa, Apr. 1932, *Eggeling* 386! & Entebbe Botanic Gardens (original cover), July 1934, *Chandler* 1196!
TANZANIA. Bukoba District: Minziro Forest, 2 Mar. 1957, *Makwilo* 29!
DISTR. **U** 2, 4; **T** 1; Guinea Bissau to S. Nigeria, Zaire and Angola
HAB. Forest including riverine and lakeside; 1140–1200 m.

SYN. *V. rufescens* Gürke in E.J. 18: 169 (1893), *non* A. Juss., *nom. illegit.* Type: Angola, Golungo Alto, Mt. Cungulungulo, *Welwitsch* 5632 (B, holo.†, BM, K, iso.!)
V. guerkeana Hiern in Cat. Afr. Pl. Welw. 1: 835 (1900). Type as for *V. rufescens* Gürke
V. fosteri Wright in K.B. 1908: 437 (1908); Pieper in E.J. 62, Beibl. 141: 68 (1928); F.W.T.A. 2: 276 (1931); Aubrév., Fl. For. Côte d'Ivoire, ed. 2, 3: 233, t. 337/1–5 (1959). Type: Nigeria, Lagos, *Foster* 34 (K, syn.!)
[*V. amboniensis* sensu I.T.U., ed. 2: 442 (1952); F.P.S. 3: 200 (1956), *non* Gürke sensu stricto (also Pieper adnot. on Uganda sheet (K) but not cited)]

NOTE. It is recorded from **U** 1 (Madi, Zoka Forest) in I.T.U., ed. 2: 443 (1952).

subsp. **amboniensis** (*Gürke*) *Verdc.*, comb. nov. Type: Tanzania, Tanga District, Amboni, *Holst* 2578 (B, holo.†)

Young shoots with more adpressed velvety indumentum but occasionally some longer hairs; leaflets acute to more shortly acuminate with shorter tips, sometimes ± obtuse.

var. **amboniensis**

Leaflets glabrous above or nearly so; mostly shrubby or sometimes a tree to 6 m.

KENYA. Mombasa District: Nyali Beach, 29 May 1934, *Napier* 3304!; Kilifi District: 6.4 km. N. of Malindi, 6 Nov. 1961, *Polhill & Paulo* 723!; Lamu District: Kui I., June 1956, *Rawlins*!
TANZANIA. Tanga District: Sawa, 23 Apr. 1956, *Faulkner* 1853! & near Moa, Mtotohovu, 23 Feb. 1902, *Kassner* 50!; Uzaramo District: Pugu Forest Reserve, June 1954, *Semsei* 1765!

DISTR. **K** 7; **T** 3, 6, 8; Somalia, Mozambique, Zambia, Zimbabwe, Caprivi Strip and South Africa (Transvaal)

HAB. Bushland, edges of cultivations, *Adansonia, Sterculia, Lannea*, association, dry lowland forest and coastal *Brachystegia* woodland; 0–570 m.

SYN. *V. tangensis* Gürke in P.O.A. C: 339 (1895); Bak. in F.T.A. 5: 321 (1900); Pieper in E.J. 62, Beibl. 141: 68 (1928); Chiov., Fl. Somala 2: 364, fig. 209 (1932); T.T.C.L.: 644 (1949); K.T.S.: 598 (1961). Type: Tanzania, Tanga, *Volkens* 92 (B, holo.†)
   V. amboniensis Gürke in P.O.A. C: 340 (1895); Bak. in F.T.A. 5: 329 (1900); Pieper in E.J. 62, Beibl. 141: 69 (1928); T.T.C.L.: 643 (1949); K.T.S.: 593 (1961)*; F.F.N.R.: 371 (1962); Coates Palgrave, Trees S. Afr.: 807 (1977)
   V. polyantha Bak. in F.T.A. 5: 321 (1900). Type: Kenya, Mombasa, *Wakefield* (K, holo.!)
   V. laevigata Bak. in F.T.A. 5: 330 (1900). Type: Tanzania, Dar es Salaam, *Kirk* (K, holo.!)
   V. swynnertonii S. Moore in J.L.S. 40: 168 (1911); Pieper in E.J. 62, Beibl. 141: 70 (1928). Type: Mozambique, Gazaland, *Swynnerton* 1054 (BM, holo.)

NOTE. Has been cultivated in Nairobi Arboretum at 1650 m. (*F.D.* 1366! & May 1932, *Dale* in *F.D.* 2776!). Many *Peter* specimens annotated *V. amboniensis* by Moldenke are in fact *V. mombassae. Busse* 2613 (Rondo [Mwera] Plateau) has petiolules to 3 cm. long.

var. **amaniensis** (*Pieper*) *Verdc.*, comb. nov. Type: Tanzania, Lushoto District, Amani, collector not known, *Inst. Amani* 221a (B, holo.†)

Leaflets pubescent above; tree to 15 m.

TANZANIA. Lushoto/Tanga District: Sigi, 28 Jan. 1931, *Greenway* 2880!; Lushoto District: Amani, Maramba, 28 Dec. 1917, *Peter* 22531!

DISTR. **T** 3; not known elsewhere

HAB. Rain-forest; 570–950 m.

SYN. *V. amboniensis* Gürke var. *amaniensis* Pieper in E.J. 62, Beibl. 141: 69 (1928); T.T.C.L.: 644 (1949)

NOTE. See footnote on p. 51; the position of this taxon needs clarification. Several E. Usambara sheets, e.g. *Engler* 3410 and *Zimmermann* in *Herb. Amani* 2929 have the indumentum on the upper surface of leaflets scarcely perceptible. I have not seen *V. amboniensis* var. *schlechteri* Pieper described from Maputo [Lourenço Marques]. The similar variant of *V. mombassae* described by Pieper as var. *acuminata* is not a rain-forest plant.

17. V. sp. B

Tree ± 18 m. tall; young stems brownish, slightly pubescent and with raised pale lenticels; older stems greyish white, somewhat corky, glabrous. Leaves 5-foliolate; leaflets elliptic, 5–10 cm. long, 2–4 cm. wide, acuminate at the apex, cuneate at the base, glabrous above, pubescent beneath, particularly on main raised venation; petioles 3.5–7.5 cm. long; petiolules up to 1.5 cm. long. Fruiting inflorescences only known, pendent; peduncles 8.5 cm. long, shortly pubescent; secondary axes up to 3.5 cm. long; pedicels or apparent pedicels 0–1 cm. long; bracts linear, 1.5 mm. long. Calyx cupular, 7 mm. long and wide, constricted at base, undulate with short obtuse lobes ± 2 mm. long, glabrous and closely reticulate. Fruits oblong–oblong-obovoid, 1.4–1.7 cm. long, 5.5–8.5 mm. long, glabrous.

TANZANIA. Uzaramo District: 25 km. NW. of Dar es Salaam, Pande Hill Forest Reserve, 5 Feb. 1973, *Harris et al.* 6756!

DISTR. **T** 6; not known elsewhere

HAB. Presumably evergreen forest; 120–150 m.

NOTE. Probably close to *V. mossambicensis* but the elongate fruits are unlike anything seen in East African species.

Doubtful Records

V. angolensis *Gürke* in E.J. 18: 167 (1893); Bak. in F.T.A. 5: 325 (1900); T.T.C.L.: 644 (1949)

Brenan reports *B.D. Burtt* 517 (Tanzania, Mbulu District, near Ufiome Mt.) as this species but I have not seen the specimen. Both *V. angolensis* Gürke and *V. gillettii* Gürke are very close to *V. mombassae*, if not conspecific, and the above specimen is probably the latter species.

* The description in K.T.S. obviously includes undeleted information from I.T.U. given under *V. ferruginea*, e.g. tree to 45′, slash thin, etc.

V. carvalhi *Gürke* in P.O.A. C: 339 (1895); Bak. in F.T.A. 5: 326 (1900); Pieper in E.J. 62, Beibl. 141: 57 (1928); K.T.S.: 593 (1961)

This species described from Mozambique is recorded from Kenya, Lamu District, Wituland, on the basis of *Tiede* 23. I have found no duplicates of this, the original of which was burnt at Berlin.

V. grisea *Bak.* in F.T.A. 5: 325, 521 (June 1900); Pieper in E.J. 62, Beibl. 141: 61 (1928); T.T.C.L.: 643 (1949)

SYN. *V. huillensis* Hiern in Cat. Afr. Pl. Welw. 1: 839 (Aug. 1900), *nom. superfl.* based on same type

Pieper records this species from Tanzania, Ufipa District, Msamvia, on the ground of *Münzner* 65 which I have not seen. It is very probably a form of *V. madiensis* subsp. *milanjiensis*.

Juvenile Forms

Moldenke has annotated a number of specimens obtained from juvenile treelets growing in the E. Usambaras as *V. oxycuspis* Bak. (type: Nigeria, Old Calabar R., *Mann* 2243). The material involved is as follows — *Peter* 21764, between Amani and Mt. Bomole, 14 Oct. 1917, 950 m.; *Peter* 17900, mountain near Nguelo, 18 Oct. 1916, 1000 m.; *Peter* 20249, between Amani and Derema, 25 May 1917, 860 m.; *Peter* 18138, between Kwamkoro and Sangerawe, 9 Nov. 1916, 900 m.; *Peter* 39654 from near the coast, Udigo, near Amboni, 21 Apr. 1926 is similar. All these specimens have thin or very thin, very coarsely crenate-serrate leaflets but are not, I think, all conspecific; 21764 and 17900 may be *V. amaniensis* Pieper; 20249 and 18138 are similar but differ in having fine dark dots on the lower surface; 39654 is, I feel certain, *V. ferruginea* Schumach. & Thonn. subsp. *amboniensis* (Gürke) Verdc. Undoubtedly all these sheets are very similar to *V. oxycuspis*, but that usually retains toothed leaves in the flowering state and is reported to have fruits 2.5 cm. long. *Itawa & Izawa* 35, a sterile glabrescent specimen from the Kigoma area, Kasakati, also has strongly toothed leaflets but I have no idea which species it is. Fieldwork would be needed to correlate these juvenile forms and some intelligent observation might clear up the difficulties.

10. PREMNA

L., Mant. Pl. Alt.: 154, 252 (1771), *nom. conserv.*

Trees (sometimes of timber dimensions), shrubs, climbers or rarely small pyrophytic herbs*, glabrous to densely hairy. Leaves opposite or in whorls of 3–4, usually distinctly petiolate, entire or toothed, often dotted with resinous glands. Flowers ☿ or polygamous, mostly small and dull-coloured, usually numerous in corymbs or thyrses of cymes. Calyx small, campanulate or cupular, truncate or 2–5-lobed and often ± 2-lipped, somewhat accrescent in fruit. Corolla-tube shortly cylindrical; limb spreading, with 4 rounded equal or subequal lobes; throat often hairy. Stamens 4, ± didynamous, included or less often ± exserted. Ovary 2-locular or spuriously 4-locular; ovules 4, 1 or 2 per locule, attached laterally. Style subulate, shortly bifid at the apex. Drupes small, globose, with thin fleshy mesocarp and bony endocarp; pyrene 4-locular; calycine cup persistent, often venose. Seeds oblong, with albumen.

About 200 species in Old World tropics and subtropics, Africa, Asia, Australia, Oceania and extending into China and Japan. Moldenke gives 226 species and subspecific taxa; 17 species occur in the Flora area.

NOTE. *Premna macrodonta* Bak. is actually *Bourreria petiolaris* (Lam.) Thulin (Boraginaceae), see Verdcourt in K.B. 43: 666 (1988).

---

* These are occasionally separated as *Pygmaeopremna* Merr.; they do not occur in Africa.

1. Foliage with distinctly stellate or branched golden-brown
   or ferruginous hairs; stems densely golden-brown
   velvety at least when young 2
   Foliage glabrous to densely velvety with simple or
   multicellular hairs, not stellate nor branched 4
2. Trees to 16 m. tall; leaves up to 25 × 14 cm.; inflorescences
   to 7 × 12 cm. . . . .   17. *P. schliebenii*
   Shrubs or small trees 1.8–6 m. tall or scandent; leaves 1.5–10
   × 1–6 cm.; inflorescences smaller 3
3. Calyx with paler hairs; tube 1.8 mm. long, truncate or with
   irregular lobes 0.5 mm. long . . . .   16. *P. chrysoclada*
   Calyx with very dense ferruginous hairs; tube ± 3 mm. long,
   with distinct triangular teeth 1 mm. long (T 8)   15. *P. tanganyikensis*
4. Tall forest trees or sometimes smaller trees (4.5–)12–30(–
   40) m. tall, with large leaves 4.5–20.5 × 3–13 cm.
   opposite or more usually in whorls of 3–4; petioles
   .3–10 cm. long; inflorescences extensive, 12–30 × 11–14 cm. 5
   Small trees, shrubs or climbers without above characters
   combined; petioles 0.2–1.8(–3.2) cm. long 6
5. Leaves mostly glabrous or glabrescent but occasionally
   finely velvety pubescent (U1); fruit 4–5 mm. diameter
   when dry (U 1–4, K 3, 5, T 1, 4) . . . .   1. *P. angolensis*
   Leaves woolly tomentose beneath; fruit ± 8 mm. diameter
   when dry (K 1, 4) . . . .   2. *P. maxima*
6. Leaves small, 1–2.5 × 0.5–1 cm., and inflorescences very
   short, scarcely 1 cm. long, terminal on very short shoots
   borne in axils of nodular petiole-bases (T 8) . . .   18. *P. sp. B*
   Not as above 7
7. Leaves glabrous, glabrescent or sparsely pubescent; calyx-
   lobes poorly developed save in the small-leaved *P.
   resinosa* 8
   Leaves densely pubescent to velvety save sometimes in *P.
   senensis* and *P. oligotricha* which have well-developed
   calyx-lobes at flowering stage 11
8. Mainly littoral (up to 375 m.) shrubs or trees; leaves 2.5–17 ×
   2–14 cm.; inflorescences 4–7(–15) × 4–14 (–23) cm. 9
   Mostly shrubs or small trees of semi-desert areas but some
   littoral or coastal; leaves 0.5–8 × 0.4–4 cm. and
   inflorescences 1–8.5 × 2–8.5 cm. but both usually
   small 10
9. Shrubs or trees; leaves rounded to subcordate at base, not
   markedly discolorous; inflorescence-axes pubescent   5. *P. serratifolia*
   Climbers or ± shrubby; leaves cuneate at base, slightly to
   very distinctly discolorous; inflorescence-axes
   glabrous or pubescent, usually orange-brown . .   13. *P. discolor*
10. Inflorescences very delicate and few-flowered, often
   appearing lateral, 1–1.5 × 1–3 cm.; leaves entire to
   coarsely toothed (K 7, T 8) . . . .   4. *P. gracillima*
   Inflorescences coarser, more obviously terminal; leaves
   entire or obscurely lobed; widespread semi-desert
   shrub or littoral (subsp. *holstii*) . . . .   3. *P. resinosa*
11. Leaves, stems and particularly calyces densely covered with
   obvious pale yellow glands; leaves ovate, 6.5–12 × 3.5–
   9.5 cm., acuminate; inflorescences 4–12(–15) cm. long,
   slender (T 5) . . . .   9. *P. richardsiae*
   Leaves, etc. without or with less conspicuous glands 12
12. Leaves densely velvety, at least beneath 13
   Leaves usually only pubescent or glabrescent but sometimes
   velvety when young 17
13. Leaves above with dense covering of hairs arising from
   incrassated glands so that surface appears granular   7. *P. hans-joachimii*
   Leaves above without basally incrassated hairs 14

14. Petioles short, 0.4–1 cm. long; leaves discolorous, rounded
    or very shortly acuminate, mostly distinctly toothed in
    upper part, ferruginous velvety beneath; stems
    ferruginous velvety; restricted to eastern areas (**K** ?4, 7;
    T 3, 6, 8) . . . .   6. *P. velutina*
    Petioles longer, (0.5–)0.9–5 cm. long; leaves ovate,
    distinctly long-acuminate, entire or shallowly and
    obscurely but regularly crenate or toothed 15
15. Leaves usually coarsely toothed; inflorescences mostly
    small, 1–6 (usually ± 3) cm. long, usually on short lateral
    shoots; calyx-teeth 1–1.5 mm. long . . . . .   12. *P. senensis*
    Leaves entire or closely minutely toothed; inflorescences
    more extensive, 1.5–10 cm. long and wide 16
16. Calyx-lobes small, 0.5–1 mm. long, the fruiting calyx with ±
    erose margin; leaves usually more velvety pubescent
    above without close gland-dots (**K** 4, 7; **T** 6, 8)   8. *P. hildebrandtii*
    Calyx-lobes rather longer, 1–1.5 mm. long, distinctly
    triangular in fruiting calyx; leaves less pubescent above
    but closely covered with minute gland-dots (**U** 2, 4;
    T 1) . . . .   10. *P. schimperi*
17. Leaves without or much more sparsely gland-dotted above,
    entire to more usually coarsely toothed, drying ± pale
    green (**K** 7; **T** 1, 2, 4–8) . . . .   12. *P. senensis**
    Leaves with dense minute gland-dots on upper surface and
    usually elsewhere, often drying rather blackish 18
18. Leaves 2.5–12 × 1.8–8.5 cm. long, ± entire or closely
    shallowly dentate-crenate, mostly large; inflorescences
    dense corymbose panicles, 1.5–8 × 3–8 cm. . . .   10. *P. schimperi*
    Leaves mostly drying dark, 0.7–5(–7) × 0.5–3(–6) cm., mostly
    rather small, entire to crenate or coarsely toothed;
    inflorescences usually much smaller, 2–5.5 cm. long
    and wide . . . .   11. *P. oligotricha***

1. **P. angolensis** *Gürke* in E.J. 18: 165 (1893); Hiern, Cat. Afr. Pl. Welw. 1: 831 (1900); Bak.
in F.T.A. 5: 289 (1900); I.T.U., ed. 2: 442 (1952); Wimbush, Cat. Kenya Timbers, ed. 2: 65
(1952); K.T.S.: 589 (1961); E.P.A.: 795 (1962); Hepper, F.W.T.A., ed. 2: 438 (1963), ? pro
parte; Berhaut, Fl. Sénégal, ed. 2: 112 (1967); Malaise in Fl. Rwanda 3: 287, fig. 93/2
(1985). Types: Angola, Loango, Chinchocho [Tschintschotscho], Pondebach, *Soyaux* 159
(B, syn.†, K, isosyn.) & Zaire, Cuango [Quango], Majakalla, *Mechow* 527 (B, syn.†)

Tree 4.5–21(–27) m. tall with spreading crown, less often shrubby; bole often crooked,
usually hollow, sometimes fluted; bark pale grey or reddish and grey, finely furrowed;
slash soft pale brown to white, yellowish towards wood with scattered orange-brown
flecks; branches ± horizontal; branchlets sparsely pubescent and glandular then
glabrous, less often ± densely pubescent. Leaves mostly in whorls of 4, less often paired;
blades ovate, oblong or elliptic, 4.5–20.5 cm. long, 3–13(–17) cm. wide, acuminate at the
apex, rounded, cuneate or subcordate at the base, glabrous above and pubescent beneath
on main nerves, later glabrescent or rarely densely pubescent all over (see note),
glandular-punctate; midrib impressed above and the ± parallel venation mostly finely
prominent; all main venation prominent beneath; petiole 3–10 cm. long. Flowers small in
large thyrsoid panicles 12–30 cm. long, the main branches in whorls of 2–5, 3–9 cm. long;
bracts linear to lanceolate, 0.2–1 cm. long; pedicels obsolete or under 1 mm. long. Calyx
1.5–2 mm. long, truncate or obscurely 2-lipped, glabrescent to pubescent. Corolla white,
sometimes tinged green or ? mauve, glabrous outside; tube 2–2.5(–3) mm. long; lobes 1
mm. long, ± semicircular, tomentose-pubescent inside. Anthers brown, just exserted.
Fruits green turning purple, globose, 4–5 mm. diameter; calycine cup shallow, 3.5 mm.
diameter.

* See also 14, *P. sp. A* (**T** 8).
** If no satisfactory result obtained return to couplet 12; some species are difficult to separate in a
key but easily named by comparison with authentic specimens.

UGANDA. Kigezi District: Kirima, Nyamigoye [Nyamugoye], June 1950, *Purseglove* 3460!; Elgon, Ririma [Irima], 24 May 1924, *Snowden* 887!; Mengo District: Kampala, Makerere, May 1935, *Chandler* 1234!
KENYA. Nakuru District: Molo, 3 Feb. 1964, *Brunt* 1446!; N. Kavirondo District: Kakamega Forest, May 1933, *Dale* in *F.D.* 3145! & Kakamega, Apr. 1933, *Dale* in *F.D.* 3146!
TANZANIA. Biharamulo District: Nyakahura, 20 Dec. 1960, *Tanner* 5631!; Mwanza District: Geita, Maisome I., Kigasi, 15 May 1962, *Carmichael* 864!; Buha District: 6.4 km. N. of Kibondo, Feb. 1955, *Procter* 384!
DISTR. U 1 (see note)–4; K 3, 5; T 1, 4; Senegal to Gabon, S. Tomé, ?Ethiopia, S. Sudan, Rwanda, Burundi, Zaire, Angola
HAB. Forest, mainly at edges but also in bushland and grassland; (900–)1170–1650(–?1800) m.
SYN. *P. zenkeri* Gürke in E.J. 33: 292 (1903); F.W.T.A. 2: 272 (1931); Berhaut, Fl. Sénégal: 100, 119 (1954). Type: Cameroon, Yaoundé, *Zenker* 1432 b (B, holo.†, BM, iso.!)
NOTE. A note on *Chandler* 1234 suggests that calliphorine flies may be the pollinators of this species. Material from Sierra Leone named as *P. angolensis* is described as a climber but examination suggests it is an extreme variant needing further field assessment. The yellow-brown sweet-scented wood works well and is good for furniture and excellent for carving. *Alonzie* 3 (Uganda, W. Nile District, Koboko/Maracha, Liru [Liri] Hill group, June 1952) appears to be this species but the stems and leaves are densely softly pubescent. The material (said to be conspecific with *Obina* 103 from W. Nile District, Terego, which I have not seen) may represent a minor variant and had been determined as *P. zenkeri*. *P.zenkeri* seems no more than a pubescent variety and is not retained by Hepper at any rank. It is particularly common in the S. Sudan, much more so than the glabrous variant.

2. **P. maxima** *T.C.E. Fries* in N.B.G.B. 8: 700 (1924); Wimbush, Cat. Kenya Timbers, ed. 2: 65 (1952); K.T.S.: 591, t. 107 (1961). Type: Kenya, N. Mt. Kenya, near Meru, *R.E. & T.C.E. Fries* 1712 (UPS, holo., K, fragm. iso.!)

Large tree 12–30(–40) m. tall; bole up to 60–90 cm. diameter, generally fluted, bent or crooked; bark dark, rough, flaking vertically in small scales; blaze plain white; wood grey-brown; branchlets at first ferruginous hairy, soon glabrous. Leaves opposite or in whorls of 3; blades broadly ovate to elliptic or rounded ovate, 7–12(–18) cm. long, 5–10.5(–15) cm. wide, rounded to acuminate at the apex, rounded at the base or less often subcordate or shortly cuneate, entire, yellowish puberulous above, grey woolly tomentose beneath to ± glabrous on both surfaces; petioles 3–8 cm. long, ferruginous hairy. Inflorescences subcorymbose, 12–16 cm. long, 11–14 cm. wide, yellow tomentose; secondary axes up to 6 cm. long; pedicels ± 1 mm. long; bracts triangular, to 5–9 mm. long, 1–4 mm. wide. Calyx campanulate, 3.5 mm. long, yellow, tomentose, ± 2-lipped, one 2-toothed, the teeth triangular, 1 mm. long, 1.8 mm. wide, the other ± entire. Corolla creamy white, narrowly funnel-shaped; tube 3 mm. long; limb 4-lobed, 2-lipped; lobes rounded, ± 2.5 mm. long, 2.2 mm. wide. Style 1.5 mm. long. Fruits purplish, "pea-sized", ± 8 mm. diameter when dry.

KENYA. Northern Frontier Province: Marsabit, 13 Jan. 1957, *T. Adamson* in *E.A.H.* 11305!; Meru District: NE. Mt. Kenya, Meru Forests, 5 Aug. 1914, *Grant* in *F.D.* 845! & Meru, on old road to Embu, 17 Mar. 1964, *Brunt* 1549!
DISTR. K 1, 4; not known elsewhere
HAB. Upland forest; 1190–1650 m.
NOTE. Produces straight grained timber which is easy to work and carve, also reported to contain an oil rendering it durable in the ground and for use in damp conditions but Dale (K.T.S.) reported this to be unfounded. The tree is in any case too rare for anything but local or very specialised use.

3. **P. resinosa** (*Hochst.*) *Schauer* in DC., Prodr. 11: 637 (1847); A. Rich., Tent. Fl. Abyss. 2: 172 (1850); Bak. in F.T.A. 5: 289 (1900); K.T.S.: 592 (1961); E.P.A.: 796 (1962); Collenette, Fl. Saudi Arabia: 496 (1985). Types: Sudan, Kordofan, Arasch Cool, *Kotschy* 198 (B, syn.†, BM, K, isosyn.!) & at mouth of valley near Milbes*, [*Kotschy* 414] (B, syn., K, isosyn.!)

Shrub, small bushy tree or subscandent, 1–3.6(–4.2) m.; branches slender, at first pale and with very scattered or denser pubescence, later dark grey, glabrous, with longitudinally peeling often whitish epidermis, sometimes nodular with very much reduced lateral branches. Leaf-blades elliptic, rounded-ovate or lanceolate, 0.5–8 cm. long, 0.4–4 cm. wide, obtuse to subacute or bluntly acuminate at the apex, rounded to ±

---

* Hochstetter cites this locality but omits the number; clearly it is a syntype.

cuneate or subcordate at the base, entire to very obscurely or less often distinctly lobed, glabrous to very shortly pubescent, minutely glandular punctate and with scattered yellowish glands; petiole 0.4–1.2 cm. long, pubescent and glandular. Inflorescences ± few-flowered, terminal or appearing lateral when on the very short side shoots, 1–8.5 cm. long and wide; peduncle slender, 0.5–1.7 cm. long, pubescent; pedicels 2–3.5 mm. long, pubescent and glandular; bracts linear, up to 2 mm. long. Calyx glabrous or slightly pubescent and glandular; tube cupular, ± 1.5 mm. long; lobes triangular or rounded, 0.5–1.5 mm. long. Corolla greenish yellow, cream or white; tube 2–2.2 mm. long, glandular outside, hairy at the throat; lobes oblong, 2.2 mm. long, 1.5 mm. wide, puberulous outside towards the apex. Fruit red, ? turning black, globose or obovoid, 6.5 mm. in diameter, glandular at apex, sitting in shallow calycine cup 3.5 mm. wide, sometimes obscurely 4-lobed; pyrene longitudinally reticulately ridged.

subsp. **resinosa**

Leaf-blades smaller, 0.5–2.5(–3.5) cm. long, 0.4–2.2 cm. wide, obtuse to subacute at the apex; petiole 0.3–1 cm. long. Inflorescences usually smaller, 1–1.5(–6) cm. long and wide.

UGANDA. Karamoja District: base of Turkana Scarp, Apr. 1960, *J. Wilson* 908! & *J. Wilson* 1023! & Napau Pass road, 16 Nov. 1953, *Dale* U289!
KENYA. Northern Frontier Province: Wajir, Catholic Mission Compound, 27 Apr. 1978, *Gilbert & Thulin* 1128!; Turkana District: Lokitaung, 23 May 1953, *Padwa* 211!; Masai District: Tsavo National Park (West), near Ngulia, Tsavo R., 25 Dec. 1957, *Verdcourt* 2070!; Teita District: Tsavo National Park (East), Voi Gate to Lugard's Falls road, 27.2 km., 22 Dec. 1966, *Greenway & Kanuri* 12840!
TANZANIA. Masai District: Nyumba ya Mungu, Msitu wa Tembo, 13 Apr. 1968, *Ludanga* in *Mweka* 2383!; Pare District: Nyumba ya Mungu Power Station, 27 Apr. 1975, *Backéus* 1299!
DISTR. U 1; K 1–4, 6, 7; T 2, 3; Sudan, Somalia and Ethiopia, also Arabia
HAB. *Acacia, Commiphora* bushland on stony hillsides, also dense mixed sometimes riverine thicket and woodland, also granite outcrops; 0–1200 m.

SYN. *Holochiloma resinosum* Hochst. in Flora 24: 371 (1841)*

NOTE. In Baringo District there are plants (23 July 1976, *Timberlake* 73; 24 July 1976, *Timberlake* 166; Aug. 1974, *Kenya Nat. Mus. 2nd 1974 Exped.* 37) with finely pubescent coarsely toothed leaves which appear to be a form of *P. resinosa* or a related species. *Mathenge* 50 (Kenya, Meru Game Reserve, Leopard Rock Camp, 6 June 1963) with very coarsely toothed glabrous leaves, up to 3.5 cm. long and wide, is similar; all are sterile.

subsp. **holstii** (*Gürke*) *Verdc.* comb. nov. Types: Tanzania, Tanga District, Moa, *Holst* 3079 (B, syn., K, isosyn.!) & Uzaramo District, Dar es Salaam, *Stuhlmann* s.n.** (B, syn.†)

Leaf-blades slightly succulent, larger, (4–)6.5–8 cm. long, 2.3–4 cm. wide, usually bluntly acuminate at the apex; petiole 1.2 cm. long. Inflorescences usually more extensive, 5–8.5 cm. long, up to 8 cm. wide.

KENYA. Kwale District: behind Diani Beach, 9 July 1968, *Gillett* 18642!; Kilifi, 8 Feb. 1946, *G. W. Jeffery* 462!; Lamu District: track from Mpekatoni to Kitwa Pembe Hill, 16 July 1974, *R.B. & A.J. Faden* 74/1116!
TANZANIA. Tanga District: Mtimbwani, 30 June 1960, *Semsei* 3046! & Moa, July 1893, *Holst* 3079!; Uzaramo District: Dar es Salaam, Oyster Bay, 24 June 1972, *Harris* 6509!
DISTR. K 4 (intermediates), 7; T 3, 6; not known elsewhere
HAB. Coastal evergreen forest and thicket on coral, grassland with scattered trees and shrub clumps; 0–20(–580) m.

SYN. *P. holstii* Gürke in P.O.A. C: 338 (1895); Bak. in F.T.A.: 291 (1900); T.T.C.L.: 641 (1949); K.T.S.: 591 (1961)

NOTE. *P. resinosa* and *P. holstii* have always been considered separate species but intermediates occur and I have considered *P. holstii* to be a derivation adapted to wetter conditions. Examples of intermediates are *Polhill & Paulo* 558 (Tana River District, Kurawa, 25 Sept. 1961), *Sangai* in *E.A.H.* 15713 (Kilifi District, Jilore, 17 Mar. 1973) and *Gillett* 20396 (Lamu District, 3 km. NW. of Ijara-Kiunga road fork). *Harris* 6192 (Uzaramo District, 17 km. N. of Dar es Salaam, Kunduchi) has lateral branches on the old grooved quadrangular scandent stems which are almost spine-like and apically ± regularly crenate leaves but is probably only a form of subsp. *holstii*.

---

* Cufodontis by a slip says this is a *nomen nudum* but it is accompanied by a full description
** Sheets of *Stuhlmann* 7923 from Dar es Salaam (EA!, BM!) might be isosyntypes and *Peter* 51816 (B!), a fragment from *Stuhlmann* s.n., may also be from the syntype or an isosyntype.

4. **P. gracillima** *Verdc.*, sp. nov. *P. resinosae* (Hochst.) Schauer subsp. *holstii* (Gürke) Verdc. valde affinis sed inflorescentiis gracilioribus vero terminalibus sed ut videtur lateralibus, foliis sustinentibus inflorescentiae eam superantibus plerumque grosse 2–3-dentatis. Type: Kenya, Kilifi District, Mangea Hill, *Luke & Robertson* 1820 (K, holo.!, EA, iso.)

Small brittle-stemmed erect or scandent shrub or "contorted bush" 1–3 m. tall, fragrant; stems slender, grey or purplish grey, glabrous or pubescent. Leaf-blades fragrant, ovate to oblong, 1.2–6 cm. long, 0.8–3.4 cm. wide, acuminate at the apex, rounded at the base, thin, entire to coarsely 1–3-dentate on either side, almost lobed, ± glabrous but obscurely minutely glandular or with hairs at base of costa beneath, with only 2 main nerves on either side; petiole very slender, 0.2–1 cm. long, mostly sparsely puberulous to densely pubescent. Inflorescences small, terminal but often overtopped by leaf pair from lower node, 1–1.5 cm. long, 1–3 cm. wide; peduncles 0.5–2 cm. long, slightly puberulous; pedicels up to 3 mm. long in fruit; bracts oblong to lanceolate, 2–3 mm. long. Calyx 1.5–2 mm. long, slightly lobed, glabrous, obscurely glandular. Corolla white; tube 2.5 mm. long; largest lobe 1.5 mm. long. Fruit obovoid, 3.5–4 mm. long, 2.5–3 mm. wide, ± ridged in dry state, sitting in a shallow cup 2.5 mm. wide.

KENYA. Kilifi District: Mangea Hill, 25 Mar 1989, *Luke & Robertson* 1820!
TANZANIA. Kilwa District: Selous Game Reserve, Malemba Thicket, 26 Feb. 1976, *Vollesen* in *M.R.C.* 3222!; Newala, 6 Mar. 1959, *Hay* 45!
DISTR. K 7; T 8; probably in Mozambique
HAB. Deciduous coastal thicket on sand; also lowland forest with *Cynometra, Brachylaena, Manilkara, Brachystegia,* etc.; 400–660 m.

SYN. [*P. holstii* sensu Vollesen in Opera Bot. 59: 82 (1980), *non* Gürke]

NOTE. I have not seen *M.R.C.* 1229 cited by Vollesen.

5. **P. serratifolia** *L.*, Mant. Pl. Alt.: 253 (1771); Munir in Journ. Adelaide Bot. Gard. 7: 13 (1984); Nicholson et al., Interpr. van Rheede's Hort. Malab.: 262 (1988) (detailed discussion). Type: ?SE. India, ?*Koenig, Herb. Linnaeus* 782.4 (LINN, lecto.)

A much branched shrub or tree 2.5–8 m. tall, rarely scandent or even prostrate; stems pubescent, puberulous or glabrescent, striate. Leaf-blades oblong to obovate-elliptic, 2.5–15(–17) cm. long, 2–8.5(–14) cm. wide, obtuse to subacute or shortly acuminate at the apex, cuneate or rounded to subcordate at the base, entire or crenate (serrate in some extra-African varieties), glabrous or some hairs at base of midrib and in lower nerve-axils beneath or on midrib and venation (some extra-African varieties have velvety leaves); petiole 0.4–1.5 cm. long. Flowers unpleasantly aromatic in extensive corymbs 4–7(–15) cm. long, 4–14(–23) cm. wide; peduncles 2–2.5 cm. long; axes densely puberulous; bracts ± linear, 1–3 mm. long. Calyx campanulate, adpressed pubescent; tube 1–2 mm. long, rim ± 2-lipped, very shortly lobed, the lobes ± 0.5 mm. long. Corolla greenish white, scented; tube cylindrical or ± funnel-shaped, 2.5–4 mm. long, pubescent outside above and at throat; limb 2-lipped, one lip entire, the other 3-lobed; lobes 1.2–1.5 mm. long, 1.2–2 mm. wide, glabrous or adpressed puberulous outside. Fruits black, subglobose, 4 mm. diameter, glabrous or pulverulent.

KENYA. Kwale District: Shimba Hills, 1.6 km. from Kwale–Mombasa, 15 Nov. 1968, *Magogo & Estes* 1242! & Shimba Hills, Kwale Forest Area, 19 May 1968, *Magogo & Glover* 1090!
TANZANIA. Rufiji District: Mafia I., Banja–Kanga, 16 Aug. 1937, *Greenway* 5122! & Mafia I., 2 Aug. 1932, *Schlieben* 2618!; Zanzibar I., Makunduchi, 4 Jan. 1960, *Faulkner* 2452!; Pemba I., Makongwe I., 16 Dec. 1930, *Greenway* 2736!
DISTR. K 7; T 6, 8; Z; P; Mozambique, Madagascar, Mauritius, Seychelles to Bangladesh, Thailand, S. China, Philippines, Ryukyu Is., Taiwan, Oceania and south to Australia
HAB. Sandy seashore above high-water mark with Sapotaceae, *Dodonaea* and *Sophora,* thicket and riverine forest; ± sea-level–375 m.

SYN. *Cornutia corymbosa* Burm.f., Fl. Ind.: 132, t. 41/1 (1768), *non Premna corymbosa* Rottler & Willd. Type: Sri Lanka, *Hermann* 2: 1 (BM, HERM, lecto.!)
Premna integrifolia L., Mant. Pl. Alt.: 252 (1771), *nom. illegit.* Type as for *Cornutia corymbosa*
P. corymbosa Rottler & Willd. in Ges. Naturf. Fr. Berlin, Neue Schrift 4: 187 (1803). Type: India, Madras, *Rottler* (K, iso.!)

P. obtusifolia R.Br., Prodr. Fl. Nov. Holl. 1: 512 (1810); Benth., Fl. Austral. 5: 58 (1870); Fosberg & Renvoize in Fl. Aldabra: 224, fig. 35/6 (1980); Moldenke in Rev. Fl. Ceylon 4: 334 (1983) (very extensive citation of synonymy and discussion of variation). Types: Australia, "New Holland 1770", *Banks & Solander* & Cape Grafton, *Banks & Solander* & North Coast, *R. Brown*, Prince of Wales I., *R. Brown* & Carpentaria, Coen R., *R. Brown* 2324 (BM, syn.!)
[*P. corymbosa* sensu Brenan, T.T.C.L.: 640 (1949), *non* Rottler & Willd.]

NOTE. Moldenke (Rev. Fl. Ceylon 4: 334 (1983)) includes the name *P. corymbosa* (Burm.f.) Rottler & Willd. in his long synonymy but uses the later name *P. obtusifolia* R.Br. A considerable number of references give the combination (Burm.f.) Rottler & Willd. but it is incorrect. Rottler & Willd. do not cite Burmann f.'s name; their name refers to a new species dating from 1803. Moldenke recognises several varieties and forms including a forma *serratifolia* (L.) Mold. (Phytologia 36: 438 (1977)), based on *P. serratifolia* L., Mant. Pl. Alt.: 253 (1771) although clearly it is not permissible to have the Linnean epithet used for a variety of a species bearing a much more recent name. Linnaeus cites only "*Cornutioides*", Fl. Zeyl.: 416 [i.e. 195] (1747), which has led to suggestions that *P. serratifolia* is illegitimate, being based on the same type as *Cornutia corymbosa* Burm.f. since under *Cornutioides* in Fl. Zeyl. Linnaeus cited "*Sambucus zeylanica odorata aromatica* Burm. Zeyl. 209 and Herm. Zeyl. 14" and Burmann f. cites both "*Cornutioides*" and *Sambucus* ..... 209 as the basis of *Cornutia corymbosa*. The choice of a Linnean specimen as the type of *Premna serratifolia* obviates the difficulty. Under *P. integrifolia* Linnaeus refers directly to *Cornutia corymbosa* Burm.f. and to Folium hirci, Rumph. Amb. 3: 28, t. 134 (1743). Nicholson et al. explain in detail that Munir is correct to use *P. serratifolia* L. for this common species. Nevertheless the citation by Linnaeus of what amounts to the same types for both his names is very puzzling but makes sense when one examines Hermann's herbarium where there are two specimens, one 2:1 with entire leaves and one 4:80 with crenulate leaves, but both numbered 416 (the species number of "*Cornutioides*" in Linn. Fl. Zeyl.: 195 (1747)) where it is described as "integerrima".

6. **P. velutina** *Gürke* in P.O.A. C: 338 (1895); T.T.C.L.: 641 (1949). Types: Tanzania, Uzaramo, *Stuhlmann* 6692 & 7176 (B, syn.†)

Spreading, climbing, straggling or even sometimes ± creeping shrub or small tree 1.2–4.5 m. tall; bark white, smooth; young stems very densely spreading yellowish pubescent, later glabrescent or glabrous with pale ± corky epidermis. Leaf-blades ovate, oblong, elliptic or almost round, 2–8 cm. long, 1.5–5 cm. wide, often ± asymmetric, rounded to acuminate at the apex, rounded to subcordate at the base, entire or serrate towards apex or for upper ½, sparsely yellowish pubescent above, rugulose with ultimate venation impressed, velvety golden-yellow-brown tomentose beneath, the venation closely reticulate and raised if visible; petiole 0.4–1 cm. long. Flowers in usually flattened ± spreading panicles 2–8 cm. long, the axes densely spreading yellow-brown pubescent; peduncles 0–2 cm. long; pedicels 0.5-2 mm. long; bracts filiform, 2–3 mm. long. Calyx cupular, pubescent, unequally 5-toothed; tube 1.5 mm. long; lobes ± 1 mm. long, 2 broader and rounder and 3 narrower and ± acute. Corolla white or yellow, glabrous outside; tube 2.5 mm. long, densely hairy at the throat; limb unequally 4-lobed, the lobes ± 1 mm. long. Style shortly bifid. Fruits globose, orange to black, 3–6 mm. in diameter when dry, sitting in shallowly lobed venose sparsely pubescent cups 6 m. wide.

KENYA. Kitui/Kilifi Districts: Galana Ranch South, 9 June 1975, *Bally* 16827!; Tana R. District: Kora base, Tana River I., 1 Aug. 1976, *Kibuwa* 2458!; Lamu District: Utwani Forest Reserve, Mambosasa, 18 Oct. 1957, *Greenway & Rawlins* 9370!
TANZANIA. Lushoto District: Korogwe, 4.8 km. W. of Magunga Halt, 28 Dec. 1960, *Semsei* 3131!; Morogoro District: Turiani, 31 Mar. 1953, *Drummond & Hemsley* 1921!; Kilwa District: 13 km. from Njinjo on road to Kandawale, 12 Oct. 1978, *Rose Innes & Magogo* 353!
DISTR. K ?4, 7; T 3, 6, 8; Mozambique
HAB. Grassland with scattered trees, swamp fringes, margins of *Terminalia hildebrandtii*, *Combretum schumannii*, *Blighia*, *Milicia* forest, riverine woodland, etc.; 18–500 m.

7. **P. hans-joachimii** *Verdc.*, sp. nov. affinis *P. hildebrandtii* Gürke et *P. velutinae* Gürke ab ambabus foliis supra pilis basi glanduloso-incrassatis obtectis differt; a posteriore petiolis 0.5–2 cm. longis fructibus majoribus diversa. Typus: Tanzania, Lindi District: Mlinguru, *Schlieben* 5730 (K, holo.!, BR, iso.!)

Shrub ± 1 m. tall or tree (*fide* Gillman); young stems densely spreading white pubescent, later glabrous and strongly ridged. Leaf-blades drying ± dark brown, ± round to rounded-oblong or somewhat obovate, 2–7.5 cm. long, 2–7 cm. wide, rounded to shortly acuminate at the apex, ± round to truncate at the base, entire or with obscure to ± distinct teeth towards the apex, slightly discolorous, densely softly hairy above with whitish long and short hairs with incrassated glandular bases, the surface appearing granular, also hairy on the main nerves beneath but areas between ± glabrous; petiole 0.5–2 cm. long.

Inflorescences ± 3.5 cm. long, ± 6 cm. wide, overtopped by upper leaves; peduncle and secondary branches 1–1.5 cm. long; pedicels 0.5 mm. long; all axes spreading white pubescent; bracts linear, 7 mm. long; secondary bracts and bracteoles filiform, 0.5–2 mm. long. Calyx adpressed white pubescent; tube 2–2.5 mm. long; lobes ovate-triangular, 0.5 mm. long. Corolla white; tube cylindrical, 3–4 mm. long, sparsely hairy above and throat pubescent; 3 short lobes ovate, 1.2 mm. long, 1.5 mm. wide and 1 longer lobe 2 mm. long, 1.5 mm. wide; stamens exserted. Style 4.5 mm. long. Fruit 6 mm. in diameter in calycine cup 7 mm. in diameter.

TANZANIA. Lindi District: Mlinguru, 17 Dec. 1934, *Schlieben* 5730! & Ngongo, 25 June 1943, *Gillman* 1403!
DISTR. T 8; not known elsewhere
HAB. Bushland; 275 m.

SYN. [*P. velutina* sensu Brenan, T.T.C.L.: 641 (1949), pro parte, *non* Gürke]

NOTE. *Bidgood et al.* 1403 (Tanzania, Lindi District, Rondo Plateau, Rondo Forest Reserve, 8 Feb. 1991) appears to be the same species having the same characteristic leaf-indumentum but the leaves attain 12 × 7 cm. the petioles 5.5 cm. and the densely gland-dotted fruits 8 mm. in diameter; it came from 800 m. and collectors record unpleasant smell.

8. **P. hildebrandtii** *Gürke* in E.J. 18: 165 (1893) & in P.O.A. C: 338 (1895); Bak. in F.T.A. 5: 291 (1900); T.T.C.L.: 640 (1949); K.T.S.: 591 (1961); Vollesen in Opera Bot. 59: 82 (1980). Type: Kenya, Mombasa I., *Hildebrandt* 2008 (B, syn.†, BM, K, isosyn.!) & Tanzania, Pangani, *Stuhlmann* 35 (B, syn.†)

Climbing shrub to 12 m. or small straggling tree or bush 3–6 m. tall; young ± slender leafy shoots brownish red, lenticellate, densely covered with spreading pale brownish pubescence; main stems leafless with purplish brown peeling epidermis, up to 10 cm. in diameter (*fide Ament & Magogo* 57), striate, pubescent but at length glabrous or with scattered pubescence and often densely woolly buds in axes of fallen leaves. Leaf-blades ovate to ± oblong-elliptic, 3–15 cm. long, 2.5–8.5(–10) cm. wide, narrowly distinctly acuminate at the apex or ? abnormally rounded, truncate to rounded or subcordate at the base, usually entire but ± repand or obscurely toothed in some southern specimens, thin, sometimes discolorous, thinly velvety above with very obvious multicellular ± adpressed hairs and more thickly similarly velvety beneath but finely so in both cases; petiole slender, 0.9–5 cm. long, similarly pubescent to the stems. Flowers very sweetly scented in extensive ± dense panicles of cymes 6–10 cm. long and wide, the axes similarly pubescent to the stems; bracts linear to lanceolate, 3–8 mm. long; pedicels to 1 mm. long. Calyx obconic, 2.2 mm. long, densely adpressed white pubescent; lobes short, very rounded or triangular, 0.5–1 mm. long, less pubescent and drying brownish. Corolla white or yellow-green; tube 3 mm. long, ± glabrous; lobes rounded, 1–1.5 mm. long and wide, densely adpressed pubescent outside. Fruit black, ellipsoid or ± globose, 6–7 mm. long, 5–6 mm. wide, ± minutely glandular, sitting in a ± flat or shallow pubescent persistent calyx.

KENYA. Kitui District: Endau Forest, 2 Feb. 1962, *Mbonge* 32!; Teita District; Voi R., 31 Jan. 1953, *Bally* 8639! & Tsavo National Park East, Ndololo, 5 Jan. 1967, *Greenway & Kanuri* 12990!
TANZANIA. Tanga District: Steinbruch Forest Reserve, June 1951, *Eggeling* 6139!; Morogoro District: Mtibwa Forest Reserve, Aug. 1952, *Semsei* 899!; Rufiji, 21 Jan. 1931, *Musk* 165!; Pemba I., Fundo I., Nov. 1929, *Vaughan* 935!
DISTR. K 4,7; T 3, 6, 8; P; not known elsewhere
HAB. Evergreen swamp forest, *Rhaphia, Milicia,* etc., also dry coastal evergreen forest, often on coral limestone, *Azima, Cordia, Dobera, Ficus, Albizia* woodland, riverine thicket and even derived grassland; 0–915 m.

NOTE. Probably partly pollinated by butterflies to which it is very attractive. *Semsei* 670 (Lindi District, Rondo Plateau, Mchinjiri, Mar. 1952) is a coarser plant with leaves to 15 × 11 cm. and petioles to 10 cm. and fruit (detached) up to 1 cm. long with very rugose pyrene. With so little material it is not clear if this is a recognisable taxon.

9. **P. richardsiae** *Mold.* in Phytologia 32: 335 (1975), as '*richardsii*'. Type: Tanzania, Iringa District, Ruaha National Park, Kimiramatonge Hill, *Richards* 21049 (NY holo., EA, K, iso.!)

Spreading shrub ± 4 m. tall, with spreading crown; stems with grey longitudinally furrowed bark, sparsely pubescent, glabrescent, but with dense pale yellow ± conspicuous glands. Leaves opposite; blades narrowly to broadly ovate, 6.5–12 cm. long, 3.5–9.5 cm. wide, acuminate at the apex, rounded to widely cuneate at the base, ± glabrous above save for minute pubescence on main nerves and midrib, and with scattered pale yellow glands, densely pubescent on midrib beneath and on main nerves and with distinct pale yellow glands scattered over the surface; only the main nerves raised beneath but no other venation obvious on either surface; petiole 1–7 cm. long. Inflorescences corymbose, many-flowered, 4–12(–15) cm. long, slender, the main branches opposite, all axes densely shortly pubescent. Calyx just over 1 mm. long, truncate or shallowly toothed. Corolla white or cream, tomentose outside; tube 2.5 mm. long; lobes 0.5 mm. long. Style 3 mm. long, plainly exserted. Fruit ovoid or ± obovoid, ± 5–6 mm. long and 5 mm. wide, sitting in a loose shallow calycine cup, 3–3.5 mm. long, which is distinctly covered with yellow glands.

TANZANIA. Singida District: Manyoni Kopje, 3 May 1932, *B.D. Burtt* 3661!; Mpwapwa District: Kongwa Ranch, 18 Feb. 1966, *Leippert* 6295! & Mpwapwa, 20 Feb. 1930, *Hornby* 186!; Iringa District: Ruaha National Park, Kimiramatonge Range, N. end, 6 Mar. 1970, *Greenway & Kanuri* 14034!
DISTR. T 5, 7; not known elsewhere (see note)
HAB. Granite rocks; 1200–1380 m.

SYN. [*P. angolensis* sensu T.T.C.L.: 640 (1949), *non* Gürke]

NOTE. This has been confused with *P. angolensis* which differs totally in its habit and venation and in the corolla being glabrous outside. *Woodburn* 48 (Mbulu District, NW. edge of Yaida Plain, ? Ronbemake, 13 Aug. 1967), a poor specimen, probably belongs here.

10. **P. schimperi** *Engl.*, Hochgebirgsfl. Trop. Afr.: 356 (1892); E.P.A.: 796 (1962). Type: Ethiopia, in the valley of the R. Reb at Gerra-Abuna-Tekla-Haimanot, *Schimper* 1131 (B, holo.†, K, iso.!)

Shrub 1.5–6 m. tall; bark pale brown; young stems densely velvety pubescent, later glabrous, chestnut and ± corky. Leaf-blades with camphor-like smell, ovate to rounded-elliptic, 2.5–12(–14.7) cm. long, 1.8–8.5(–14) cm. wide, acuminate or ± rounded at the apex, rounded at the base, closely shallowly dentate-crenate or ± entire, ± discolorous, densely pubescent above on midrib and main nerves, velvety pubescent beneath or sparsely pubescent only, occasionally almost glabrous, with dense bright yellow often almost crystalline glands on both surfaces, often drying yellow-green; petiole 0.3–4 cm. long, pubescent. Flowers in dense corymbose panicles 1.5–8 cm. long, 3–8 cm. wide; peduncle 1.5–2.5 cm. long; pedicels very short; axes pubescent; bracts up to 8 mm. long. Calyx with yellow glands; tube 1.5 mm. long, 2-lipped, one lip 2 mm. long, bifid, the lobes 0.5 × 1 mm., rounded and ciliate, the other lip oblong, 2 × 1 mm.; lobes often pale. Corolla aromatic, green or greenish white; tube 2 mm. long, puberulous outside; limb 2-lipped, the lobes 1.5–2 mm. long, ciliate. Style 3.5–4.5 mm. long. Fruit purple, greenish or ? white, globose, 4.5–5 mm. diameter, glabrous to puberulous and minutely glandular, sitting in a lacerate calycine cup up to 6.5 mm. wide; said to be edible.

UGANDA. Toro District: Mpanga R., 20 June 1906, *Bagshawe* 1058!; Kigezi District: Ruhinda, Jan 1951, *Purseglove* 3557!; Masaka District: Kyotera, Nov. 1945, *Purseglove* 1901!
TANZANIA. Ngara District: Busubi, Nyakisasa, 18 Nov. 1948, *Ford* 881! & Muganza, Rusengo, 8 Mar. 1960, *Tanner* 4761! & Busubi, Keza, 25 Mar. 1960, *Tanner* 4802!
DISTR. U 1, 2, 4; T 1; Sudan, Ethiopia, Somalia (N.)
HAB. Secondary scrub, riverine forest and grassland; 1200–1890 m.

SYN. *P. viburnoides* A. Rich., Tent. Fl. Abyss. 2: 171 (1850); Bak. in F.T.A. 5: 292 (1900), pro parte, *non* Schauer (1847), *nom. illegit.* Type: Ethiopia, Choa, *Petit* (P, holo.)
  P. viburnoides A. Rich. var. *schimperi* (Engl.) Pichi Serm. in Webbia 7: 336 (1950)

NOTE. As constituted here the species is very variable and might perhaps be divisible into subspecies according to the presence or absence of woolly indumentum on the leaves beneath. Populations with more densely woolly indumentum and fewer glands on the leaves occur in Ethiopia as well as East Africa. *Schimper* 566 was identified by Engler as true *P. viburnoides* and is densely velvety. The *Grant* specimen from Unyamwesi and *Kirk* specimen from the lower Shire valley cited by Baker under *P. viburnoides* are not that species.

11. **P. oligotricha** *Bak.* in F.T.A. 5: 292 (1900); K.T.S.: 591 (1961); Verdc. in K.B. 45: 702 (1990). Type: Kenya, Teita District, Ndi Mt., *Scott Elliot* 6202 (K, holo.!, BM, iso.!)

Small aromatic shrub or sometimes somewhat scandent, 1–3.6 m. tall; bark grey; stems slender, dull almost purplish brown, the youngest innovations densely spreading white pubescent and often densely glandular but soon glabrous and ± longitudinally ridged; older stems roughened with remains of short shoot-bases, the lateral shoots mostly very short. Leaves green, drying ± blackish, sometimes ± viscid, opposite, small, elliptic or

FIG. 10. *PREMNA OLIGOTRICHA* — **1**, habit, × ⅔; **2, 3**, leaves, × ⅔; **4**, part of lower leaf surface, × 6; **5**, flower, × 8; **6**, corolla, opened out, × 8; **7**, ovary and style, × 10; **8**, longitudinal section of ovary, × 14; **9**, fruit, × 4; **10**, pyrene, × 6. 1, 4–8, from *Verdcourt* 1783; 2, from *Gillett* 19128; 3, 9, 10, from *Bally* 16790. Drawn by Mrs. M.E. Church.

rounded-elliptic, 0.7–5(–7) cm. long, 0.5–3(–6) cm. wide, rounded to subacute at the apex, rounded at the base, entire to crenate or coarsely toothed with rounded lobes, ± densely covered above with short flattened several-celled hairs and sessile glands, similarly pubescent beneath but the hairs and glands denser, the surface often greyish velvety and with some white spots due to aggregations of hairs but occasionally glabrescent; petiole 0.3–2(–3.2) cm. Inflorescences terminal, 2–5.5 cm. long and wide, somewhat lax; peduncles 1–3 cm. long; pedicels 0.5–2.5 mm. long; bracts linear, 1.5 mm. long; all axes spreading puberulous. Calyx cupular, puberulous to pubescent and glandular; tube 1.5 mm. long; lobes ovate to triangular or ± linear, 0.5–1 mm. long. Corolla greenish cream with purple throat or white, cream or yellow with brown lines or midlobe white and rest green; tube ± 2 mm. long, glabrous outside; limb 2-lipped, one lip 3-lobed, the lobes round to elliptic, 1.2–1.5 mm. long, 1–1.3 mm. wide, puberulous outside and densely ciliate. Style exserted, not divided at the apex. Fruit globose or obovoid, 4–6 mm. long, 3.5–5 mm. wide, glandular at the apex; pyrene rugose; calycine cup black, or paler and venose, glandular, 5 mm. wide, the teeth triangular up to 2.5 mm. long. Fig. 10.

KENYA.   Northern Frontier Province: Dandu, 8 Apr. 1952, *Gillett* 12728!; Machakos District: 46.5 km. Thika–Garissa, near Kithimani, 24 May 1957, *Verdcourt* 1783!; Teita District: Tsavo National Park (East), Voi Gate to Sobo road, km. 34.5, 20 Dec. 1966, *Greenway & Kanuri* 12807!
TANZANIA.   Pare District: NW. spur of North Pare Mts., above Kifaru Estate, 19 May 1968, *Bigger* 1843! & Mkomazi Game Reserve, Maore Junction, 25 July 1972, *Mbano* in *C.A.W.M.* 5750!; Lushoto District: Lake Manka, July 1967, *Procter* 3709!
DISTR.   K 1, ?2, 4, 6, 7; T 3; Somalia, Ethiopia
HAB.   *Commiphora, Acacia* bushland with scattered *Delonix*, etc., *Commiphora* thicket, *Combretum, Acacia* bushland with thin grassland, *Adenia, Sterculia, Lannea* woodland, often on lava ridges; (0–)360–1650(–1750) m.

SYN.   *P. somaliensis* Bak. in F.T.A. 5: 289 (1900); E.P.A.: 796 (1962). Type: Somalia (N.), Ahl Mts., Mait, *Hildebrandt* 1526 (K, holo.!, BM, iso.!)
       Clerodendrum kibwesense Mold. in Phytologia 4: 48 (1952) & 61: 404 (1986). Type: Kenya: Kibwesi, *Scheffler* 62 (BR, holo. not found, K!, S, iso.)

NOTE.   The stems were once used by the Southern Kamba for arrow shafts. Moldenke wrongly refers Kibwesi to Tanzania and also cites *Hornby* 2499 from Mozambique but this specimen is *P. senensis*. *Kuchar* 11837 (Kilifi District, Mnarani Club, 11 July 1979) appears to be *P. oligotricha* and must be from near sea-level. Some states of *P. oligotricha* with very small leaves and very condensed inflorescences appear quite different.

12. **P. senensis** *Klotzsch* in Peters, Reise Mossamb., Bot. 1: 263 (1861); Gürke in P.O.A. C: 338 (1895); Bak. in F.T.A. 5: 292 (1900); T.T.C.L.: 641 (1949); E.P.A. 796 (1962); F.F.N.R.: 370 (1962); Coates Palgrave, Trees S. Afr.: 805 (1977); Vollesen in Opera Bot. 59: 83 (1980). Type: Mozambique, Rios de Sena, *Peters* (B, holo.†)

Several-stemmed much-branched shrub 2–5 m. tall or a small tree, rarely scandent; bark grey with scattered pale lenticels, sometimes peeling in small flakes; stems usually dark grey-brown or purplish but sometimes pale, at first with spreading fawn pubescence, soon glabrous and ridged. Leaves very variable, ± thin, smelly, often in whorls of 3; blades oblong or ovate, 2–8.5(–12) cm. long, 1.3–5.5(–10) cm. wide, usually distinctly acuminate (sometimes very long-acuminate) at the apex but occasionally rounded, truncate then shortly cuneate at the base or subcordate, entire or obscurely to distinctly coarsely crenate, distinctly toothed or even lacerate on some parts of the margin, sparsely to densely velvety pubescent on both sides and with very obscure gland dots, sometimes almost glabrous; petiole 0.5–3.5 cm. long. Inflorescences mostly small on short lateral shoots, 1–6 cm. wide; peduncles slender, 0–1(–3) cm. long; pedicels 0.5–1.5 mm. long; axes densely shortly spreading pubescent or with dense longer pubescence and scattered white glands; bracts oblanceolate or ± linear, 2–4 mm. long. Calyx cupular, ciliate but otherwise glabrous to densely white pubescent all over, glandular; tube 1.5 mm. long; lobes triangular or rounded, (0.5–)1–1.5(–3 in fruit) mm. long. Corolla white, glabrous or sparsely hairy on outside of lobes; tube 3–3.5 mm. long; limb 2-lipped, 3-and 2-lobed, the lobes elliptic-oblong, 3 mm. long, 1.5–2 mm. wide; throat and lower part of lobes inside densely white pubescent. Ovary glandular above; style 5 mm. long. Fruit black or violet, globose, 3.5–6 mm. in diameter, sitting in a ribbed distinctly lobed or almost entire cup usually covered with small glands; pyrene rugose.

KENYA.   Kwale District: Shimba Hills, Giriama Point, 27 Mar. 1968, *Magogo & Glover* 530!
TANZANIA.   Mbulu District: Lake Manyara National Park, Chem Chem R., 21 Nov. 1963, *Greenway & Kirrika* 11061!; Ufipa District: Rukwa valley, Milepa, 17 Dec. 1946, *Pielou* 37!; Kondoa District: near Serya [Salia], 28 Dec. 1927, *B.D. Burtt* 1127!

DISTR. **K** 7; **T** 1–8; Burundi, Zaire, ?Somalia (see note), Mozambique, Malawi, Zambia, Zimbabwe, Botswana and Caprivi Strip

HAB. *Brachystegia spiciformis*, *Combretum zeyheri* woodland, *Acacia* woodland, thicket and riverine woodland; (100–)300–1000(–1500) m.

SYN. *P. sp.* sensu Oliv. in Trans. Linn. Soc., Bot., 29: 132 (1875) as 'possibly allied to *senensis*', quoad *Grant* s.n.

    Ehretia tetrandra Gürke in E.J. 28: 311 & 461 (1900); Bak. in F.T.A. 4(2): 22 (1905). Type: Tanzania, by Ruaha R., *Goetze* 471 (B, holo.†, BR, K, iso.!)

    [*Premna viburnoides* sensu Bak. in F.T.A. 5: 292 (1900), quoad *Grant* s.n., *non* A. Rich.]

    [*P. schimperi* sensu T.T.C.L.: 641 (1949), quoad *Grant* s.n., *non* Engl.]

    [*P. holstii* sensu T.T.C.L.: 641 (1949), quoad *Hornby* 145, *non* Gürke]

NOTE. Very similar to *P. richardsiae* which also occurs in **T** 5 but that has very distinctive yellow glands on the leaves and calyx and the fruiting calycine cup is shallow and unlobed. Material of *P. senensis* from **T** 8 has glabrous inflorescences and is more generally glabrescent. The Kenya plant cited seems to be a form of this rather protean species but no other material has been seen. It is one of the species used for fire sticks. I have not seen *Gorini* 229 from Somalia, Transjuba, which Chiovenda refers to this species; Cufodontis suggests it needs confirmation.

    The delimitation of *P. senensis* is unsatisfactory. Some material from **T** 5 has very short inflorescences, leaves drying grey-green rather than discolorous and calyx-lobes 1.5 mm. long, lengthening to 2–3 mm. in fruit, e.g., *Newman* 94 (Kondoa District, Farkwa, 14 Jan. 1966) and *Peter* 45663 (Mpwapwa District, Ugogo, near Gulwe, 8 Dec. 1925). Peter suggested it was a new species and routine namers have suggested it was identical with *Ehretia tetranda* but that has the short calyx-lobes of typical *P. senensis* and the variation is too erratic to separate taxa. *Greenway & Kirrika* 11061 (Mbulu District, Lake Manyara National Park, Chem Chem R., 21 Nov. 1963), for instance, has most lobes long, narrowly triangular and acuminate, but on the other hand **T** 4 material, e.g., *Richards* 18338 (Mpanda District, Rukwa Sonta Woodland, 3 Nov. 1963) has the calyx almost truncate. Other material has large, almost obtuse leaves and appears very diverse, e.g., *Richards* 21059 (Iringa District, Ruaha Nat. Park, track to airfield near Mwangusi R., 27 Jan. 1966) but I have been unable to satisfactorily distinguish taxa; further work may indicate that it is necessary.

    Apart from this variation some poor or sterile material has been referred here temporarily as *P. senensis* in a broad sense, e.g., Pangani District, Useguha, Hale, 21 May 1926, *Peter* 40421. The species would certainly repay field study throughout its range.

13. **P. discolor** *Verdc.* sp. nov. nulla affinitate arcte obvia, typice ob folia valde discoloria supra viridia glabra praeter costam puberulam subtus alba, indumentum foliorum duplicem ex pilis brevibus patentibus et tomento microscopico constans, axes inflorescentiae aurantiaco-brunneos insignis. Typus: Kenya, Kilifi District, Cha Simba, *Luke & Robertson* 1886 (K, holo.!, EA, iso.)

Climber or shrub 3–10 m. tall with arching branches, with glabrous pale grey-brown striate lenticellate corky stems; young shoots glabrous to pubescent. Leaves borne on rather short lateral shoots, fragrant; blades elliptic-oblong, very slightly obovate, 4–15 cm. long, 2.5–7.5 cm. wide, acute, very shortly acuminate or ± rounded at the apex, cuneate at the base, slightly to very discolorous beneath, pubescent and tomentose or ± glabrous, entire or very obscurely undulate, both surfaces with raised finely reticulate venation when dry; petiole 0.8–1.8 cm. long. Dichasia pale yellowish green in life, drying orange-brown, 2–6 cm. long, 3–8 cm. wide, glabrous or pubescent; peduncle 1–2 cm. long; pedicels 1 mm. long. Calyx not lobed or slightly so. Corolla white; tube 2 mm. long, densely hairy inside; lobes oblong-ovate, 1.5–1.8 mm. long, 1.2 mm. wide. Fruit white turning purple-black, 4–6 mm. in diameter; calycine cup 2 mm. tall, 4–5 mm. wide, obscurely veined or very veined except near the entire margin.

var. **discolor**

Leaves very white beneath, very finely arachnoid-tomentose, the individual hairs visible only under a microscope, and also with pubescence on the venation. Venation of calycine cup obscure or not evident.

KENYA. Kilifi, Cha Simba, 14 Aug. 1989, *Luke & Robertson* 1886! & Vitengeni Gorge, *Luke* 1324

DISTR. **K** 7; not known elsewhere

HAB. Jurassic limestone outcrop with *Pandanus*, *Euphorbia*, *Gyrocarpus*, *Ficus* and *Cola*; 260 m.

var. **dianiensis** *Verdc.* a var. *discolori* foliis subtus leviter albescentibus fere glabris, calyce fructifero valde venoso differt. Typus: Kenya, Kwale District, Diani Forest, *Gillett & Kibuwa* 19854 (K, holo.!, BR!, EA, iso.)

KENYA. Kwale District: Diani Forest, 22 June 1970, *Friis* 121! & 13 July 1972, *Gillett & Kibuwa* 19854! & 24 Aug. 1989, *Robertson & Luke* 5893!

DISTR. **K** 7; not known elsewhere
HAB. Very mixed dry evergreen forest on coral with little soil; 0–12 m.

NOTE. This had been identified as a *Clerodendrum* but the small 3–4-locular globose fruit indicates *Premna*. The leaf-shape and texture and entirely glabrous inflorescences distinguish it from *P. serratifolia* although in Australia that name is allowed to cover a very wide range of plants.

14. P. sp. A

Shrub to 4.5 m. tall; young stems greenish, striate, glabrous save for a very few sparse hairs, later darker with pale lenticels, slender, nodular with persistent petiole-bases. Leaves paired or judging by scars possibly in 4's; blades oblong to oblong-obovate, 6–13 cm. long, 3–7 cm. wide, acuminate or cuspidate at the apex, truncate to rounded at the base, entire to very coarsely crenate, ± discolorous, minutely sparsely pubescent on midrib and main nerves above, glabrous beneath; petiole 1–3 cm. long. Inflorescences lax, terminal on short lateral branches and main branch but overtopped by the leaves, 4–5 cm. long and wide; axes glabrous but gland-dotted. Flowers not seen. Fruits globose, 5 mm. in diameter when dry, in an entire or lobulate shallow calycine cup 4–5 mm. wide; pyrene rugulose.

TANZANIA. Lindi District: Rondo Plateau, Mchinjiri, Feb. 1952, *Semsei* 648!
DISTR. **T** 8; not known elsewhere
HAB. Probably evergreen forest; 810 m.

NOTE. Probably close to *P. senensis*.

15. P. tanganyikensis *Mold.* in Phytologia 7: 83 (1959). Type: Tanzania, Newala District, Kitangari, *Gillman* 1328 (K, holo.!, EA, iso.!)

Scandent shrub or small tree to 4 m.; young stems densely orange-ferruginous velvety with dendritically branched hairs, later glabrous, with obscurely ridged pale grey-brown thinly corky epidermis; lateral shoots abbreviated, 0.5–3 cm. long. Leaves elliptic or obovate to ± obovate-oblanceolate, 3–10 cm. long, 1–4 cm. wide, acuminate at the apex, cuneate at the base, densely stellate ferruginous pubescent on both surfaces when young, later glabrescent with indumentum much more scattered, subentire to distinctly fairly distantly toothed especially towards the apex; petioles 0.5–1 cm. long, densely ferruginous stellate pubescent with branched hairs. Inflorescences short dense terminal and axillary ovoid clusters mostly on the short lateral shoots, 1–2 cm. long, ± 1.5 cm. wide; bracts linear, 2–3 mm. long; pedicels obsolete. Calyx with dense ferruginous stellate hairs; tube tubular, ± 3 mm. long, the teeth triangular, 1 mm. long. Corolla yellowish brown or yellowish cream, 3–3.5 mm. long; lobes rounded-oblong, 2 mm. long, 1.2 mm. wide, stellate pubescent outside; stamens exserted. Fruit fawn coloured, obovoid to globose, 3.5 mm. in diameter, ridged below.

TANZANIA. Lindi District: Rondo Plateau, Mchinjiri, Mar. 1952, *Semsei* 664! & Rondo Forest Reserve, 15 Feb. 1991, *Bidgood et al.* 1582!; Newala District: Kitangari, 23 Mar. 1943, *Gillman* 1328!
DISTR. **T** 8; Mozambique
HAB. "Makonde thicket" on orange sands; also semi-evergreen forest with *Milicia, Albizia, Dialium* and *Pteleopsis*; 450–700 m.

NOTE. Moldenke gives the height as 2–5 m., but on what authority I do not know; there is no such data on the label of *Gillman* 1328 the only sheet he cites. He saw and annotated *Semsei* 664 but did not cite it.

16. P. chrysoclada (*Boj.*) *Gürke* in E.J. 33: 293 (1903); Chiov., Racc. Bot. Misc. Consol. Kenya: 98 (1935); T.T.C.L.: 640 (1949); K.T.S.: 589 (1961). Type: Kenya, Mombasa I., *Bojer* (?P, holo.!, K, iso.!)

Much-branched shrub or small tree 1–6(–8) m. tall, occasionally somewhat scandent; branchlets coarsely velvety with very dense rusty or orange-brown multibranched hairs, sometimes quite long; older stems glabrescent, eventually glabrous, pale brown, ridged and ± corky. Leaves variable, opposite, aromatic; blades elliptic, oblong, oblong-obovate or obovate-lanceolate, 1.5–9.5 cm. long, 0.8–6 cm. wide, acute to acuminate at the apex, rarely ± rounded, narrowly rounded, broadly cuneate or rounded at the base, entire or obscurely to distinctly crenate or toothed, thinly pubescent to velvety above with small stellate hairs and longer simple hairs which may, however, be missing, and similarly pubescent or velvety beneath particularly on the raised nerves; gland-dots dense but obscure; petiole 0.5–1.2 cm. long. Flowers small in dense mostly elongate much-branched complicated terminal panicles 1–7 cm. long; bracts filiform, 4 mm. long; pedicels 0.5 mm. long; axes pubescent like the stems. Calyx cupular, 1.8 mm. long, truncate or with 5 short teeth to 0.5 mm. long, glandular and densely stellate-pubescent; corolla greenish white or cream to yellow, glandular and slightly puberulous outside; tube 2.5–4 mm. long; limb 2-lipped, the lobes rounded to oblong, 1.2–2 mm. long, 1.3–1.5 mm. wide. Style green, forked at tip. Drupes black or purplish black, ellipsoid to obovoid, 4.5–6 mm. long, 4–4.5 mm. wide, pubescent, sitting in a cup 3.5 mm. wide.

KENYA. Kwale District: Shimba Hills Development Scheme, Kidongo N., 17 Dec. 1968, *Mwangangi* 1277!; Mombasa, *Boivin*!; Kilifi District: Malindi, Oct. 1951, *Tweedie* 993!
TANZANIA. Tanga District: Mtimbwani, 30 June 1960, *Semsei* 3049!; Uzaramo District: Pugu Forest Reserve, Aug. 1953, *Semsei* 1308!; Lindi District: 9.5 km. S. of R. Mbemkuru on Kilwa–Lindi road, 6 Dec. 1955, *Milne-Redhead & Taylor* 7566!; Zanzibar I., without locality, *Mrs Taylor* 525!
DISTR. K 7; T 3, 6, 8; Z; not known elsewhere
HAB. Coastal *Brachystegia* woodland, scrub and thicket, bushland and sometimes in riverine woodland and grassland, also abandoned cultivations; 0–450 m.

SYN. *Vitex chrysoclada* Boj. in Ann. Sci. Nat., sér 2, 4: 268 (1835); Schauer in DC., Prodr. 11: 694 (1847); Bak. in F.T.A.: 317 (1900)
Premna zanzibarensis Vatke in Linnaea 43: 531 (1882); Gürke in P.O.A. C: 338 (1895). Type: Zanzibar I., Kokotoni, *Hildebrandt* 1167 (B, holo.†, BM, iso.!)

NOTE. The leaves are used in hot water as an inhalent for fever, and the roots boiled in water with frankincense for dysentery.
Frontier Tanzania 49 (Tanzania, Uzaramo District, Pande Forest, 23 Oct. 1989) is sterile; the young stems and undeveloped leaves have ferruginous stellate pubescence but the more adult leaves are almost glabrous save for a very few stellate hairs towards the base of the leaf and denser ones on the petioles. I suspect it may be only a state of *P. chrysoclada*. Specimens with glabrescent leaves do occur.

17. **P. schliebenii** *Werderm.* in N.B.G.B. 12: 89 (1934); T.T.C.L: 641 (1949). Type: Tanzania, Ulanga District, Mahenge, Lipindi, *Schlieben* 1668 (B, holo.†, BM, BR, iso.!)

Tree 15–16 m. tall, laxly branched with 1–3 stems and flattish spreading crown; bark smooth, buffish yellow-brown, finely longitudinally fissured; young shoots densely yellowish-rusty stellate-tomentose and with longer branched hairs. Leaf-blades oblong, 5–29 cm. long, 2–14 cm. wide, rounded to shortly acuminate at the apex, ± cuneate or slightly auriculate at the base, entire, repand or occasionally denticulate, with small yellowish stellate hairs above and on the prominent venation beneath to completely velvety tomentose beneath; petiole 1–1.5 cm. long, densely tomentose. Inflorescence paniculate, many-flowered, up to 7 cm. long, 12 cm. wide, the axes yellowish ferruginous tomentose with dense mixed stellate and branched hairs; bracts lanceolate, 0.3–1.2 cm. long; pedicels 1 mm. long; peduncles to 1.5 cm. long. Calyx cyathiform, 2 mm. long, with stellate and glandular hairs intermixed; teeth 5, very short. Corolla white or green; tube cylindric, 3–4 mm. long, glabrous; limb 4-lobed, 3 mm. wide; 3 lobes small and rounded but 1 more elongate, recurved or deflexed, all with minute glandular hairs. Stamens scarcely exserted. Drupes green, subglobose, up to 6 mm. long, 4.5 mm. wide, pubescent and gland-dotted.

TANZANIA. Lushoto District: E. Usambaras, Misima, 27 Jan. 1947, *Greenway* 7922! & Sigi Singali, *Zimmermann*! & Lunguza, 30 Jan. 1917, *Zimmermann* in *Herb. Amani* 8185!; Ulanga District: Lipindi, 26 Jan. 1931, *Schlieben* 1668!
DISTR. T 3, 6; Mozambique (intermediates)
HAB. Rocky gully in scattered tree grassland and dry evergreen forest; (300–)360–800 m.

NOTE. Schlieben reported it yielded good timber for house-building. This is very close to *P. chrysoclada* and before seeing the Schlieben material I had annotated that from the Usambaras as a variety. More evidence is needed but it is clearly more or less a case of a savanna/forest pair having different habits but ± identical floral structure. Many such pairs are known and always cause controversy over treatment which varies between non-recognition and specific rank. More details of the exact habitat are also needed. The label of the BM isotype does not give some of the information given by Werdermann. *Schlieben* 5750 (Tanzania, Lindi District, Mlinguru, 19 Dec. 1934) and other specimens from Lindi are rather intermediate but considerably smaller in habit.

18. **P. sp. B**

Shrub 3 m. tall; branches slender, dull purplish grey with pale lenticels, longitudinally striate and bark somewhat peeling, at first ± bifariously pubescent, later glabrescent;

lateral shoots very short, borne in the axils of nodular petiole-bases of fallen leaves, pubescent. Leaves narrowly ovate, 1–2.5 cm. long, 0.5–1 cm. wide, acute to acuminate at the apex, rounded at the base, sparsely pubescent above and on costa beneath; venation closely reticulate but not raised beneath; petiole 3–4 mm. long. Inflorescences very short, ± 1 cm. long, terminal on the short shoots but usually appearing immediately above the nodules with very reduced leaves; pedicels ± 1.5 mm. long. Calyx cupular, 1.8 mm. long, minutely rugulose, divided up to half way into obtuse lobes. Corolla green with lower lobe white; tube 2.5 mm. long; lobes unequal. Fruit not seen.

TANZANIA.   Newala, 8 Dec. 1968, *Jahl* 187!
DISTR.   **T** 8; not known from elsewhere but presumably in Mozambique
HAB.   Thicket, once cleared but now covered with shrubs; 750 m.
NOTE.   Probably allied to *P. resinosa* and *P. oligotricha*.

11. **KAROMIA**

Dop in Bull. Mus. Hist. Nat. Paris, sér. 2, 4: 1052 (1932); R. Fernandes in Garcia de Orta, sér. Bot. 7: 36 (1988)

Shrubs or trees, glabrous or pubescent, unarmed or with small axillary spines; stems sometimes ± subscandent. Leaves deciduous, petiolate, entire or toothed. Inflorescences cymose, axillary or crowded at apices of shoots and appearing terminal; bracts mostly small, deciduous. Calyx often brightly coloured; tube very short, urceolate; limb broadly expanded, very accrescent to form an entire or 5-lobed usually venose rotate plate-like structure in fruit. Corolla-tube gibbous, very asymmetric, contracted beneath the limb, split dorsally almost to the base; limb 5-lobed and ± 1-lipped, very irregularly formed of 2 separated posterior lobes, 2 lateral lobes and a larger median anterior labelliform cucullate lobe. Stamens inserted near throat, the filaments enrolled into a ring in bud, distinctly exserted at anthesis and curved; filaments not prolonged beneath the point of insertion. Style enrolled in a ring with the filaments in bud. Fruit dry and hard, turbinate or turbinate-obpyramidal, rounded above or ± flat and with 4 small horns in a cross, at length dividing into ?2–4 pyrenes. Fig. 11.

Nine species, all but one formerly placed in *Holmskioldia*, occurring in Vietnam, E., SE. and S. Africa and particularly Madagascar. Apart from one species occasionally cultivated in East Africa a recently described but imperfectly known endemic species may belong to this genus. R. Fernandes has given good reasons for recognising *Karomia*; she places the Madagascan and Vietnam species in sect. *Karomia* and the African species in sect. *Cyclonemoides* R. Fernandes. *K. gigas* had not then been described although she was aware of its existence and thought it unlikely to be a *Holmskioldia*.
K. speciosa (Hutch. & Corbishley) R. Fernandes, a native of South Africa (Natal, Transvaal), Swaziland and Mozambique, is cultivated in Tanzania at Amani and probably elsewhere. It is a shrub or small tree 3–6 m. tall, with ovate entire coarsely toothed or almost lobed leaves ± 4.5 cm. long, 2.5 cm. wide; fruiting calyx pink or pinkish purple, 2–2.5 cm. wide and corolla blue or purplish pink (Tanzania, Amani, 20 May 1950, *Greenway* 8452! & 15 Mar. 1973, *Ruffo* 640!).

1. Cultivated plants with fruiting calyx 1.5–2.5 cm. wide   2
   Wild plant with fruiting calyx 4.5–7 cm. wide   . . . .   *K. gigas*
2. Fruiting calyx ± entire, flatter and more plate-like   see *Holmskioldia sanguinea* (p. 4)

   Fruiting calyx distinctly obtusely lobed and more funnel-
   shaped   . . . .   *K. speciosa*

K. gigas (*Faden*) *Verdc.*, comb. nov. Type: Kenya, Kilifi District, Chonyi–Ribe road, *Faden et al.* 77/439 (US, holo., EA, F, K!, P, iso.)

Tree about 12 m. tall, with a narrow crown, resembling teak; bark tan-coloured, peeling, prominently lenticellate. Leaf-blades rounded-ovate, 15.5–23 cm. long, 11.5–19 cm. wide, acute or shortly acuminate at the apex, rounded then shortly cuneate at base, glabrous above, extremely finely puberulous beneath and with minute stipitate glands and larger sessile ones; venation closely reticulate and prominent beneath; petiole stout, 1.3–3 cm. long, margin of abscission surface with long inwardly directed hairs hiding the surface. Inflorescences not known. Fruits obconic, ± 1 cm. wide at top of calyx-tube, puberulous; fruiting calyx-limb probably shallowly bowl-shaped in life, 4–7 cm. in

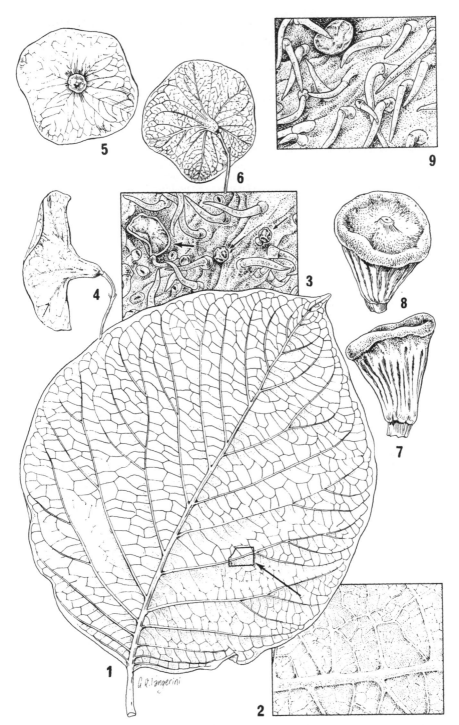

Fig. 11. *KAROMIA GIGAS* — **1**, leaf, lower surface, × ½; **2**, leaf venation detail, × 6; **3**, leaf, lower surface showing hairs, stalked glands (small arrows) and sessile gland (large arrow), × 155; **4–6**, fruiting calyx × ½; **4**, side view, **5**, front view, **6**, back view; **7**, sterile fruit, side view, × 2; **8**, sterile fruit, showing top, × 2; **9**, top of fruit showing hairs and sessile gland (arrow), × 155. All from *Faden et al.* 77/439; 3, 9, drawn from SEM. Drawn by Alice R. Tangerini. Reprinted with permission from the Smithsonian Institution.

diameter, entire or very faintly (4–)5-lobed, closely reticulate-venose, sparsely puberulous or ± glabrous; upper superior discoid part of fruit cushion-shaped, ± 1 cm. in diameter; persistent pedicel slender, 1–1.3 cm. long; fruit with 4 trigonous nutlets. Fig. 11.

KENYA. Kilifi District: just N. of Mwarakaya on Chonyi–Ribe road, 4.8 km. after turn-off from Kilifi–Kaloleni road, 16 Feb. 1977, *Faden et al.* 77/439! & same tree, 15 Feb. 1980, *Gillett* 22775
DISTR. **K** 7; not known elsewhere
HAB. Relict forest on limestone rocks with *Sterculia appendiculata, Gyrocarpus americanus, Milicia excelsa,* etc.; 140 m.

SYN. *Holmskioldia?sp.*; Moldenke in Phytologia 48: 322, 385 (1981); R. Fernandes in Garcia de Orta, sér. Bot. 7: 45 (1988).
H. gigas Faden in K.B. 43: 660, fig. 1 (1988)

NOTE. Only one tree and a sapling were noted and on a later visit in 1985 it was found by H. Beentje that these had been cut down. It is possible it may occur in other forest remnants in 'Kayas' on rocky hills but a search in the vicinity has so far produced nothing. I have preferred to place it in *Karomia* but until flowers are available its true position will remain uncertain.

12. CLERODENDRUM*

L., Sp. Pl.: 109 (1753) & Gen. Pl., ed. 5: 285 (1754); Thomas in E.J. 68: 1–106 (1936); Moldenke in Rev. Fl. Ceylon 4: 407 (1983) (very comprehensive generic synonymy etc.)**

Trees, shrubs or lianes or occasionally perennial herbs or pyrophytes with large woody rootstocks, sometimes with persistent spinescent petiole-bases often acting as scrambling aids. Leaves simple, opposite or in whorls of 3–4, entire or toothed. Flowers in lax cymes, extensive panicles or dense corymbs or heads, axillary or terminal, often appearing axillary but actually terminating abbreviated shoots, rarely solitary. Calyx often coloured and showy, campanulate to tubular, truncate to 5-toothed, often accrescent in fruit. Corolla white, blue, violet (etc.) or red, almost regular or quite irregular; tube narrowly cylindrical, straight or curved, often widened towards throat, sometimes very long or only as long as calyx; limb spreading or reflexed, the anterior lobes larger than the posterior pairs. Stamens 4–5, didynamous, inserted in the corolla-tube, alternate with the lobes, usually long-exserted. Ovary imperfectly 4-locular, each locule with one lateral ovule. Style terminal, elongate; stigma shortly 2-fid, exserted. Drupes obovoid or globose, often 4-grooved and ± (2–)4-lobed; mesocarp ± fleshy; endocarps bony or crustaceous, smooth or rugose, separating into 4 pyrenes or 2 pairs of pyrenes; seeds with no albumen.

A large genus of about 400 species (584 taxa fide Moldenke) mostly in the tropics and subtropics of the Old World but some in America and a few in temperate regions. Linnaeus recognised several genera which have now been combined: *Ovieda* L., *Volkameria* L. and *Siphonanthus* L.; Moldenke suggests that *Cyclonema* Hochst. is perhaps distinct enough to separate. Similarly *Kalaharia* Baillon has been kept distinct by several workers, including Gürke, Briquet, Moldenke and R. Fernandes (adnot.), but if this is split then one might be justified in suggesting that other subgenera should be raised to generic rank. This should be done only after a detailed world study. I have mainly followed Thomas's adequate classification*** of the genus, to which reference should be made for more detailed information on the subgenera and sections. Unfortunately none of his new names for supraspecific or infraspecific taxa is validly published since Latin was by then obligatory; his new species are validly described in an appendix to his paper (op. cit.: 99–106). Moldenke has accepted the classifications of Briquet and Thomas and produced a hybrid of the two with a key (Phytologia 57: 338–342 (1985)) but he has not validated the Thomas names. Several of Thomas's species I have been unable to identify, despite his excellent descriptions, even when from well-collected areas. These I have retained in his sequence although it is just possible they have been badly misplaced. In all cases there appear to be no isotypes known. *C. micranthum* Gilli, the type of which I have seen, is *Hoslundia opposita* Vahl (Labiatae) (see K.B. 45: 382 (1990)). *C. racemosum* (I.T.U., ed. 2: 442 (1952)) is an error for *Combretum racemosum.*
This account was produced with some difficulties with much of the K and BM material away on loan for over ten years. Certain types reported to be at K have not actually been seen.

---

* Often spelt *Clerodendron,* presumably to preserve the entirely Greek derivation 'kleros' — fortunate and 'dendron' — tree, but Linnaeus used the -um ending in 1753 (and 1737).
** Moldenke has published exhaustive bibliographic notes on this genus in Phytologia from part 57(2) (1985) to 64(3) (1988) — 38 parts in all. Vast amounts of ill-digested information is repeated verbatim and uncritically. I have given some references under individual species.
*** It should be pointed out that *C. eupatorioides* Bak., on which Thomas based his sect. *Stenocalyx,* is *Eremomastax speciosa* (Hochst.) Cuf. (Acanthaceae) (see Verdc. in K.B. 44: 697 (1989)).

A synopsis of the classification used is as follows, with Thomas's sections validated; his revision of the African species has served well over fifty years and not been replaced. The account is undoubtedly that of a 'splitter' but he was reluctant to sink taxa described by his peers, although often he suggests they should be. His arrangement is, I think, very reasonable.

subgen. **Clerodendrum** (subgen. *Euclerodendrum* (Schauer) Thomas, *nom. illegit.*)
sect. **Clerodendrum** (sect. *Paniculatae* Schauer, *nom. illegit.*)*

Cymes in capitate subumbelliform, often very amply branched, spreading terminal panicles; leaves mostly cordate, pubescent to velvety but not squamulose. Corolla hypocrateriform with unequal subsecund lobes. Calyx enlarging in fruit. Type: *C. infortunatum* L. Species 1.

sect. **Densiflorae** *Schauer* in DC., Prodr. 11: 664 (1847)

Cymes in a terminal compact panicle or 1 or more heads, axillary or pseudoterminal. Leaves usually cordate, pubescent. Corolla narrowly funnel-shaped to hypocrateriform with unequal often subsecund lobes. Calyx enlarging in fruit. Type: *C. fragrans* (Vent.) R.Br. = *C. chinense* (Osbeck) Mabb. Species 2.

sect. **Oxycalyx** *Verdc.*, sect. nov. Calyx 0.5–1 cm. longus per $\frac{1}{2}$–$\frac{2}{3}$ longitudinem divisus, lobis lanceolatis vel ± subulatis plerumque acuminatis vel raro subobtusis. Corollae tubus 1.2–3 cm. longus. Typus: *C. pleiosciadium* Gürke

subsect. **Oxycalyx** (subsect. *Acuminata* Thomas, *nom. illegit.*). Species 3, 4.

subsect. **Apiculata** *Verdc.*, subsect. nov. Calycis lobi apiculati. Typus: *C. splendens* G. Don. Species 5; *C. splendens* (cult.) — see p.87.

subsect. **Fallax** *Verdc.*, subsect. nov. (subsect. *Obtusata* Thomas, *nom. invalid.*). Calycis lobi acuti vel subacuti. Typus: *C. speciosissimum* C. Morren. *C. speciosissimum* (cult.) — see p.87.

sect. **Macrocalyx** *Verdc.*, sect. nov. Calyx 1–3 cm. longus per $\frac{2}{5}$–$\frac{4}{5}$ longitudinem divisus; inflorescentiae laxae; corolla 1.5–17 cm. longa. Typus: *C. thomsoniae* Balf. Species 6, 7; *C. thomsoniae* (cult.) — see p.87.

sect. **Capitata** *Verdc.*, sect. nov. (subsect. *Spicata* Thomas et subsect. *Capitata* Thomas, *nom. invalid.*). Inflorescentiae capitatae vel subspicatae, valde congestae; calycis lobi ± petaloidei vel foliacei, saepe colorati, ovati, elliptici vel lanceolati. Typus: *C. capitatum* (Willd.) Schumach. & Thonn. Species 8–13.

sect. **Cylindrocalyx** *Verdc.*, sect. nov. Rami haud spinosi. Calycis tubus cylindricus, 5-angulatus, 1 cm. longus, per $\frac{1}{3}$ longitudinem divisus. Corollae tubus 3–4 cm. longus. Typus: *C. hildebrandtii* Vatke. Species 14.

sect. **Siphonocalyx** *Verdc.*, sect. nov. Frutices vel scandentes; folia plerumque opposita; petioli demum spiniformes; flores in cymas foliaceas vel nudas vel inflorescentias capitatas dispositi. Calyx anguste tubulosus, 0.4–1 cm. longus; lobi triangulares, haud patentes; corolla alba. Typus: *C. silvanum* Henriques. Species 15–18.

sect. **Eurycalyx** *Verdc.*, sect. nov. Fructices vel scandentes; folia plerumque opposita, petiolata; petioli demum spiniformes; rami haud cavi. Inflorescentiae foliaceae. Calyx expansus, 3–5 mm. longus, per $\frac{1}{3}$ longitudinem fissus, plerumque coloratus, in fructu accrescens et expansus. Fructus membranaceus. Typus: *C. volubile* P. Beauv. Species 19, 20.

---

* This section is habitually called *Paniculatae* by Moldenke and others but includes the type of the genus.

sect. **Microcalyx** *Verdc.*, sect. nov. Frutices vel scandentes, ramulis teretibus vel 4–6-angulatis partim cavis. Folia plerumque 3–4-verticillata; petioli demum spiniformes. Inflorescentiae cymosae, capitatae vel capitibus in pseudoumbellas dispositis. Calyx campanulatus, 2–3 mm. longus, apice expansus, per ⅕ longitudinem fissus, lobis ± patentibus, in fructu cupuliformis. Tubus corollae usque 1.5 cm. longus, lobis 4–5. Typus: *C. formicarum* Gürke. Species 21–23.

sect. **Odontocalyx** *Verdc.*, sect. nov. Frutices vel arbores; rami haud cavi; folia plerumque verticillata, margine ± revoluta, glanduloso-punctata; petioli haud spiniformes. Inflorescentiae cymosae, ± parvae, plerumque subumbelliformes. Calyx breviter campanulatus, lobis angustis acuminatis vel subulatis. Tubus corollae parvus, usque ± 1 cm. longus, lobis parvis. Placentae integrae haud furcatae. Typus: *C. glabrum* E. Mey. Species 24–26.

sect. **Cornacchinia** *(Savi) Briq.* in E. & P. Pf. 4, 3A: 176 (1895)* (*Cornacchinia* Savi in Mem. Soc. Ital. Mod. 21: 184 (1837); *Clerodendrum* L. subgen. *Cornacchinia* (Savi) Thomas in E.J. 68: 23 (1936))

Shrubs with verticillate petiolate leaves, the blades entire, pubescent to velvety beneath; petioles not becoming spinescent; stems not hollow. Inflorescences axillary, umbellate-cymose, long-pedunculate; bracts rather large and leafy. Calyx 4–5-lobed, up to 8 mm. long, the lobes acute. Corolla white; tube narrowly cylindrical, ± 2 cm. long, not dilated. Ovary imperfectly 4-locular; placentas placed in angles, not forked. Pericarp spongy, very rugose, becoming hardened. Fruiting calyx accrescent. Fruit eventually dividing into 4 parts. Type: *C. acerbianum* (Vis.) Benth. in Benth. & Hook.f. Species 27.

sect. **Konocalyx** *Verdc.*, sect. nov. Herbae lignosae, suffrutices vel frutices. Folia opposita vel verticillata, sessilia vel breviter petiolata, plerumque grosse serrata vel lobata; petioli haud spiniformes. Inflorescentiae cymosae et foliaceae usque capitatae. Calyx anguste infundibuliformis, 3–9 mm. longus, per ¼–⅓ divisus. Limbus alabastri plerumque irregularis. Tubus corollae 3–15 cm. longus. Antherae 3–4 mm. longae; filamenta cum stylo per revolutiones tres involuta. Ovarium glabrum vel hirsutum, placentis haud furcatis. Typus: *C. incisum* Klotzsch. Species 28–31.

subgen. **Cyclonema** *(Hochst.) Thomas* in E.J. 68: 22 (1936) (*Spironema* Hochst. in Plantae Schimperianae No. 330 (1840), *non* Raf. (1838) *nec* Lindl. (1840); *Cyclonema* Hochst. in Flora 25: 225 (1842); *Clerodendrum* L. sect. *Cyclonema* (Hochst.) Gürke in P.O.A. C: 341 (1895)).

Shrubs, subshrubs, or pyrophytic herbs, more rarely scandent or decumbent, or even small trees; stems not hollow. Leaves usually verticillate, sessile or shortly petiolate, often distinctly crenate or coarsely serrate, but sometimes quite entire. Inflorescences cymose, usually leafy, mostly lax, terminal and/or axillary. Calyx divided for up to half its length but often less, campanulate to subspherical, frequently red or purple; lobes 4–6, rounded to narrowly triangular, acute to usually perfectly rounded at the apex. Corolla pale to deep blue or a mixture of both or occasionally partly white or yellow, zygomorphic; tube 0.6–1.2 cm. long; basally saccate in front, deeply divided behind at apex; lobes very unequal, forming a ± 2-lipped limb, the anterior lip larger and curved. Ovary glabrous, glandular or hairy; placentas not furcate. Type: *C. myricoides* (Hochst.) Vatke.

sect. **Cyclonema** (sect. *Chaunocymosa* Thomas, *nom. invalid.*)

Inflorescences leafy, mostly lax. Species 33–50.

sect. **Stacheocymosa** *Verdc.*, sect. nov. Inflorescentiae haud foliaceae, ± spicatae, ramulis lateralibus valde abbreviatis, tempore fructificantis elongatae auctae. Typus: *C. kissakense* Gürke. Species 51, 52.

* This has invariably been looked on as a distinct isolated section or subgenus but is so very similar in general appearance to *C. glabrum* and its allies that I feel too much weight has been given to the curious wrinkled pericarp. Gürke (P.O.A. C: 341 (1895)) included *C. eriophyllum* and *C. tricholobum* together with *C. acerbianum* in sect. *Cornacchinia* but the fruits of the first two were not known to him.

subgen. **Kalaharia** (*Baillon*) *Thomas* in E.J. 68: 23 (1936) (*Kalaharia* Baillon, Hist. Pl. 11: 111 (1892); Briq. in E. & P. Pf. 4, 3A: 172 (1895))

Decumbent shrubs; branches armed with axillary curved spines and indurated spine-like pedicel-remnants. Stem never hollow. Leaves opposite. Flowers solitary in the axils. Calyx regularly 5-lobed, hairy inside and out. Corolla reddish yellow; tube short, ± 1 cm. long, widened above but not split. Ovary cylindrical, glandular, 4-locular above. Placenta split to margin. Fruiting calyx scarcely accrescent. Type: *C. spinescens* (Oliv.) Gürke (= *Kalaharia spinipes* Baillon). Species 53.

A number of species have been cultivated in East Africa and the following list is probably not complete. All the species have been included in the general key to species since some native species are occasionally cultivated and some cultivated ones can appear to be wild. *C. bungei* Steud. (*C. foetidum* Bunge, *non* D. Don (Jex-Blake, Gard. E. Afr., ed. 4: 108 (1957)) from China is a pubescent shrub up to 2 m., with broadly ovate acuminate stalked coarsely toothed leaves up to 30 cm. long and dense corymbose heads, 10–20 cm. wide, of rosy red flowers about 2 cm. wide; corolla-tube 3–4 times as long as the calyx; the stems are sometimes said to be spiny but in material I have seen the old wood is roughened with raised lenticels. *C. calamitosum* L. from Java (e.g. Nairobi City Park, May 1965, *Greensmith* in *E.A.H.* 13159!) is a shrub 1–2 m. tall with oblong to obovate stalked coarsely undulate-toothed leaves 4–14 × 2–8 cm. and axillary and terminal 4–10-flowered very fragrant cymes; calyx ± 1 cm. long, finely shortly pubescent, divided into narrow lobes; corolla white with tube greenish outside, shortly pubescent, ± 3.5 cm. long with lobes ± 1–1.25 cm. long. Jex-Blake (Gard. E. Afr., ed. 4: 109 (1957)) reports a *C. mastocanthum* to be tall, quick growing and mauve-flowered, but I have been unable to trace the name, nor work out of what it could be a misspelling. *C. chinense* (Osbeck) Mabb. (*C. philippinum* Schauer, *C. fragrans* (Vent.) R. Br. in Ait.f. (Jex-Blake, Gard. E. Afr., ed. 4: 109 (1957)) from ?China or S. tropical Asia, is treated fully on p. 94 since it has now escaped from cultivation and occurs as a ruderal. *C. speciosissimum* C. Morren (*C. buchananii* (Roxb.) Walp. var. *fallax* (Lindl.) Mold., *C. fallax* Lindl., — Jex-Blake, Gard. E. Afr., ed. 4: 108 (1957); Bailey, St. Cyclop. Hort., ed. 2, 1, fig. 997 (1939)) from Java, is an erect shrub to ± 4 m., with broadly ovate entire or repand-denticulate cordate pubescent leaves 10–35 × 8.5–26 cm.; petioles 1.4–21 cm. long; inflorescences terminal lax many-flowered panicles 15–45 × 14–25 cm.; calyx red, 5–9 mm. long, with ovate lobes 3 mm. long; corolla red, scarlet or vermilion with slender tube 2–2.5 cm. long and limb 4–5 cm. wide with ± reflexed lobes 1.5–1.8 cm. long. *C.* × *speciosum* Bull (*thomsoniae* × *splendens*) (e.g. Uganda, Mengo District, Kampala, Makerere University Hill, 5 May 1971, *Lye* 6051!; Kenya, Nairobi, Oct. 1961, *Grahame Bell Ltd.* in *E.A.H.* H 260/61/2!) is a scrambler with bright red corolla and purple calyces very similar indeed to *C. umbellatum* but the specimens match with authentic material of the hybrid.* *C. splendens* G. Don (Jex-Blake, Gard. E. Afr., ed. 4: 134 (1957); Huber in F.W.T.A., ed. 2, 2: 444 (1963); Moldenke in Rev. Fl. Ceylon 4: 435 (1983) (very full account); Everett, Encycl. Hort. 3: 793 (1981)) from W. Africa (e.g. Nairobi Arboretum, 16 June 1952, *G.R. Williams Sangai* 448! & Block 33, 15 Apr. 1952, *Dyson* 349! & Nairobi, Kilimani, Caledonian Shopping Centre, 7 Feb. 1977, *Kahurananga* 824!; Tanga, Aug. 1950, *Bally* in *C.M.* 16735!; Lushoto, Magharib, 21 Apr. 197?, *Mtali* 50; Dar es Salaam, State House, 30 Aug. 1972, *Ruffo* 469!) Shrub or climber 1.2–5 m. long, with elliptic to round, oblong or even lanceolate, stalked glabrous to pubescent cordate leaves 5–18 × 3–8 cm., inflorescences showy, many-flowered, axillary and terminal corymbose cymes, 7–11 × 6–8 cm.; peduncle 1.5–3.5 cm. long; calyx 0.7–1 cm. long, divided into broadly triangular connivent lobes 3–4 mm. long; corolla deep red to crimson, the tube 2 cm. long and lobes 10–15 × 6–7 mm. Moldenke records this as wild from Tanzania but I have seen no material. *C. thomsoniae* Balf. — usually spelt 'thomsonae', (Jex-Blake, Gard. E. Afr., ed. 4: 134 (1957); Huber in F.W.T.A., ed. 2, 2: 442, fig. 307 (1963); Moldenke, Rev. Fl. Ceylon 4: 433 (1984), very extensive treatment) from W. Africa, very extensively cultivated in East Africa but only one herbarium specimen seen

---

* Bull definitely states his *speciosum* is a hybrid between *C. balfourii* Hort., i.e. *C. thomsoniae* Balf., and *C. splendens* G. Don; his name is accompanied by an adequate description (Cat. 44:4 (1869)) and is apparently the earliest publication of this epithet. Moldenke says the name *C. speciosum* Teijsm. & Binn. (1869 but name only) has mostly been used for a scarlet-flowered cultivated form of *C. umbellatum* Poir. but is actually a form of *C. thomsoniae* (forma *speciosum* (Teijsm & Binnend.) Voss) and repudiates it as a hybrid. The epithet has also been used by Carrière (1873) and D'Ombrain in Fl. Mag. 8, t. 432 (1869). I have not had time to pursue the matter.

(Tanzania, Amani, 24 Apr. 1973, *Magogo* 444!). Climber to 7 m. or shrub 1–2 m. tall, with elliptic stalked entire ± glabrous leaves 6–14 × 3–7 cm.; cymes axillary towards the ends of the twigs, very lax, 5–9 × 4–8.5; calyx white or greenish, becoming yellow or pink, rarely reddish (in forms and hybrids), 1.8–2 cm. long with ovate lobes 8–10 mm. long; corolla rose to deep red, scarlet or crimson; tube 2.5 cm. long; lobes 6–10 × 3.5–4 mm. *C. villosum* Blume (*C. velutinum* Thomas) is treated fully on p. 94 since it appears it may have been naturalized around Amani early this century. It was certainly cultivated then but no recent specimens have been seen. *C. wallichii* Merr. (*C. nutans* Don, *non* Jack) (Tanzania, E. Usambaras, Magila, 7 June 1932, *F.M. Rogers* H28/32) is a shrub to ± 3 m., with narrowly oblong, oblong-lanceolate or -oblanceolate leaves 7.5–28 × 1.4–4 cm. and lax terminal panicles 10–50 × 6–14 cm.; calyx pink to dark red, up to 1.4 cm. long divided into ovate lobes ± 5–10 mm. long; corolla greenish white or cream, the narrow tube ± 1.5 cm. long and lobes up to ± 2 cm.

Some native species have also been cultivated, e.g. *C. capitatum* (Willd.) Schumach. & Thonn. (Kenya, Nairobi Arboretum, 16 June 1952, *G.R. Williams Sangai* 446! & Nairobi, Oct. 1961, *Grahame Bell* 20!)

Some species known from incomplete material are omitted from the key (32, 41, 42, 47, 49 and 50). The *C. capitatum* and *C. myricoides* groups of species are very difficult and some may prefer to take a very wide view, in which case species 8–13 will be included under the former and species 33 to 50 under the latter but the range of variation then admitted is enormous.

1. Axillary curved thorns and spine-like petiole-bases present; corolla scarlet or orange-red; rambling shrub or subshrub forming tangled masses (subgen. *Kalaharia*)   53. *C. uncinatum*
   Plant unarmed or with spine-like petiole-bases but then corolla not red (but calyx may be) 2
2. Corolla ± regular or slightly irregular, tube not as below and often longer 3
   Corolla distinctly irregular, the tube inflated in front, slit at back, 0.2–1.2 cm. long; limb essentially 2-lipped with anterior lobe the largest, usually blue or dark blue or mauve, etc. (subgen. *Cyclonema**) 36
3. Flowers in very tight capitate inflorescences, the bracts and calyx-lobes usually coloured (red, mauve, etc.), broadly elliptic or ovate, 0.8–3.5 × 0.4–2.5 cm., acuminate, often venose, much longer than the calyx-tube; corolla long and tubular, (3.5–)4–17.5 cm. long (sect. *Capitata*) 50
   Flowers in various inflorescences, lax to densely capitate, but if capitate then calyx-lobes quite different 4
4. Subshrubs with essentially herbaceous shoots under 1 m. tall from a woody rootstock; calyx-teeth triangular; corolla-tube 1.3–11.5 cm. long 5
   Trees, shrubs or climbers, rarely as low as 1 m. and then with linear or lanceolate calyx-lobes 2.2–10 mm. long, or small, intricately branched shrubs 8
5. Calyx 8–10 mm. long; corolla 10–11.5 cm. long 6
   Calyx 2.5–3 mm. long; corolla 1.3–9 cm. long 7
6. A true pyrophyte 13–20 cm. tall, with capitate inflorescence   29. *C. pusillum*
   Larger subshrub 30–80 cm. tall, with spicate inflorescence   30. *C. lutambense*
7. Corolla 5–9 cm. long; leaves elliptic to lanceolate, 1.7–13.5 × 1.2–6 cm., acuminate, usually very coarsely toothed or lobed, rarely entire . . . .   28. *C. incisum*
   Corolla 1.3–4.2 cm. long; leaves linear-lanceolate to oblanceolate, 0.5–7 × 0.15–1.5 cm., acute or subacute, less coarsely toothed . . . .   31. *C. ternatum*

---

* Specific delimitation within *Cyclonema* is very difficult; see notes on p. 136.

8. Corolla (4–)6.3–11.3 cm. long; shrub with rather lax
   inflorescences; calyx-tube 3–5 mm. long, with lanceolate
   lobes 1–1.7 cm. long . . . .   7. *C. rotundifolium*
   Corolla under 4 cm. long or if ± 4 then other characters not
   the same 9
9. Foliage and young stems, etc., distinctly drying black;
   corolla white or yellow, inky-blue when bruised and
   also drying black, the tube 7 mm. long and lobes 3 × 2
   mm.; western in distribution (**U** 2, 4; **K** 5; **T** 1)   19. *C. melanocrater*
   Foliage not so distinctly drying black 10
10. Leaves small, 0.3–1.8 × 0.15–1 cm.; corolla-tube 1.1–1.3 cm.
    long, the lobes unequal, the largest 4 × 6 mm., the
    smaller 3 × 1.8 mm. (**K** 1) . . . .   18. *C. robecchii*
    Leaves larger or only young foliage of similar size, never all
    so small 11
11. Ovary spongy and wrinkled, divided into 4 deeply grooved
    lobes, the fruit similar to a blackberry, but with low
    spongy-corky processes; flowers in dense terminal
    clusters; corolla-tube 1.8–3 cm. long (sect.
    Cornacchinia) . . . .   27. *C. acerbianum*
    Ovary and fruit not as above, mostly smooth 12
12. Corolla red, crimson, etc., or marked with it (many of the
    species cultivated) 13
    Corolla white 19
13. Flowers in compact heads 14
    Flowers in lax or fairly dense panicles 15
14. Corolla-tube 3–4 times as long as calyx; calyx-lobes
    triangular (cult.) . . . .   *C. bungei*
    Corolla-tube not much longer than the calyx; calyx-lobes
    linear-lanceolate; corolla often double (cult. and
    naturalized) . . . .   2. *C. chinense*
15. Calyx mostly red, 0.5–1.2 cm. long 16
    Calyx white to pink, longer, 1.6–3 cm. long 18
16. Calyx-lobes ovate or triangular, 1–3 mm. long; leaves
    deeply cordate; inflorescence many-flowered; corolla-
    tube 2–2.5 cm. long (cultivated) . . . . . .   *C. speciosissimum*
    Calyx-lobes ovate to lanceolate, 3–10 mm. long; leaves
    rounded to shallowly cordate; corolla-tube 1–1.7 cm.
    long 17
17. Calyx-lobes triangular to lanceolate, 3–7 mm. long (usually
    longer, 5–7 mm.); corolla-tube 1–1.7 cm. long (wild)   5. *C. umbellatum**
    Calyx-lobes similar but usually smaller, 3–4 mm. long;
    corolla 1.5–2 cm. long (cultivated) . . . .   *C. splendens*
18. Leaves cordate at the base; calyx cream; corolla yellowish
    or white with red inside 3 upper lobes; tube (2–)2.5–3
    cm. long (wild, **U** 2) . . . .   6. *C. fuscum*
    Leaves rounded to cuneate at the base; calyx white, yellow
    or pink; corolla deep red, scarlet or crimson, the tube
    2–2.5 cm. long (cultivated) . . . .   *C. thomsoniae*
19. Corolla-tube over 2.5 cm. long 20
    Corolla-tube mostly under 2 cm. long, rarely up to 2.4 cm.
    (see species 3) 23
20. Calyx-tube 3–6 mm. long, widened at throat with triangular
    recurved lobes 1 mm. long; liane or scrambling shrub
    with inflorescences mostly borne on the old leafless
    stems but sometimes terminal; corolla-tube 2.5–3.6
    cm. long . . . .   17. *C. schweinfurthii*
    Calyx-tube longer, with teeth or lobes 0.25–2.2 cm. long 21
21. Cultivated shrub with coarsely undulate-toothed leaves   *C. calamitosum*
    Wild plants with entire or slightly undulate leaves 22

---

* The hybrid *C.* × *speciosum* will key near here with calyx-lobes up to 10 mm. long and rounded to
feebly cordate leaf-bases rather than shallowly but distinctly cordate.

22. Scandent subshrub; calyx-tube campanulate, slightly
    inflated below, 1 cm. wide; corolla usually with red
    marks but possibly sometimes pure white; plant ±
    pubescent (**U** 2) . . . .   6. *C. fuscum*
    Shrub or tree 1.2–9 m. tall, sometimes subscandent; calyx-
    tube ± narrowed at base, 4 mm. wide; corolla white, ±
    mauve on fading; plant usually glabrous but a rare
    pubescent variant in Tanzania (**K** 1, 7; **T** 3, 6, 8)   14. *C. hildebrandtii*
23. Inflorescences elongate, mostly borne on leafless old
    wood; liane with calyx-lobes shorter than tube; corolla-
    tube 0.7–1.8 cm. . . . .   15. *C. silvanum*
    Inflorescences terminal or from upper axils 24
24. Calyx-lobes distinctly longer than the tube 25
    Calyx-lobes about equalling or shorter than the tube 30
25. Leaves velvety or woolly-pubescent beneath 26
    Leaves glabrous, glabrescent or more sparsely pubescent
    beneath 28
26. Leaves 15–18 × 7–9 cm.; calyx-lobes 5 × 2 mm.; corolla-tube
    2 cm. long (**T** 6, known only from description)   4. *C. polyanthum*
    Leaves 3–13 × 2–11 cm. 27
27. Leaves in whorls of 3, elliptic to ± round, 3–8 × 2–5 cm.;
    flowers in axillary capitate cymes ± 2 cm. wide; calyx-
    lobes linear from triangular base, 2.2–3.5 mm. long,
    without large glands; corolla-tube 1–1.8 cm. long   26. *C. tricholobum*
    Leaves opposite, broadly obovate, 10–17.5 × 4–13 cm.;
    flowers in elongate lax leafy panicles, the lowest cymes
    9–12 cm. long, the upper 0.5–3 cm. long; calyx-limb
    spreading, ± 1 cm. long, divided for ± ⅔ its length into
    ovate-triangular teeth 6 mm. wide at base with often
    sparse but conspicuous large glands (**T** 3, cultivated
    and perhaps naturalized formerly) . . . . .   1. *C. villosum*
28. Inflorescences narrow elongate thyrsoid panicles, 10–50 ×
    6–14 cm.; leaves narrowly oblong-oblanceolate or -
    lanceolate, 7.5–28 × 1.4–4 cm., glabrous; calyx-lobes
    ovate, 5–10 mm. long; corolla-tube ± 1.5 cm. long
    (cultivated) . . . .   *C. wallichii*
    Without above characters combined 29
29. Calyx usually red or purple; flowers in terminal compact
    many-flowered cymes 3–6 cm. long, 3.5–9 cm. wide,
    with leafy lanceolate bracts 1.5–3 × 0.3–1.2 cm.
    scattered through the cymes; calyx-lobes 4–10 mm.
    long; corolla-tube 1.5–2 cm. long (but flowers often
    double) (cultivated and naturalized) . . . . .   2. *C. chinense*
    Calyx green; flowers in terminal often condensed many-
    flowered composite cymes 4–10(–12) cm. wide, often
    some in lower 5–8 axils from apex; calyx-lobes (4–)5.5–
    8 mm. long; corolla-tube 1.2–2.4 cm. long . . . .   3. *C. pleiosciadium**
30. Calyx-lobes distinctly shorter than calyx-tube or calyx-tube
    plus undivided part of limb 31
    Calyx-lobes about equalling the tube, sometimes slightly
    shorter or slightly longer 34
31. Calyx-tube subovoid, 2 mm. long, with whitish, yellow or
    green ± funnel-shaped or rather flat limb 5–7 mm. wide
    divided into broadly triangular teeth 1–2 mm. long, 2
    mm. wide; flowers in compound corymbs 6–9(–12) cm.
    long, 7–10.5 cm. wide; corolla yellow-green; tube 5–7
    mm. long (**T** 4) . . . .   20. *C. volubile*
    Calyx structure not as above 32

---

* If not satisfactorily run down above proceed to couplet 34; the ratio of calyx-tube to lobes is
variable in some species.

32. Leaves distinctly 3-nerved from the base, 1.2–9.5 × 0.8–6
cm.; branchlets 5–8-ridged; calyx-tube 2.5 mm. long,
lobes 1.5–2 mm. long; corolla-tube (3–)6–8 mm. long,
lobes 2–4 × 1.5 mm.; branches often hollow . . . 22. *C. formicarum*
Leaves not so distinctly 3-nerved from the base and
branchlets less ridged 33
33. Corolla-tube (4–)6–8 mm. long, with lobes 3–4 × 2–3 mm.;
calyx-tube 2.2 mm. long, with lobes 1–1.5 mm. long;
leaves acute to obtuse, glabrous and very densely
punctate (**K** 7; **T** 3, 6, 8; **Z**; **P**) . . . . 24. *C. glabrum*
Corolla-tube 1.4–2 cm. long, with lobes 4–5 × 3–4 mm.;
calyx-tube 3–5.5 mm. long, with triangular lobes 2 mm.
long; leaves acuminate, sparsely pubescent beneath or
glabrous, not punctate (**U** 2; **T** 4) . . . . 16. *C. tanganyikense* *
34. Flowers very numerous, in large pyramidal or oval panicles
24 × 12 cm.; leaves ± glabrous; calyx-tube 2.5 mm. long,
with ovate-triangular lobes 2–2.5 mm. long (**T** 6; known
only from the type) . . . . 21. *C. myrianthum*
Flowers not so numerous, or if in almost as large
inflorescences then leaves densely velvety 35
35. Inflorescences axillary and terminal, forming large
complicated dense corymbose panicles of cymes 7–20
cm. wide; leaves 6.5–19.5 × 3.5–14.5 cm., densely ±
adpressed velvety with fawn to ferruginous or ± red
indumentum; calyx-tube 2 mm. long, with lobes 1.5–2
mm. long; corolla-tube (3.5–)9–10 mm. long, with lobes
3–4 mm. long and wide 23. *C. johnstonii*
Inflorescences usually small lax cymes 4–6 cm. long or very
condensed heads 2–4.5 cm. diameter; leaves 1–12 ×
0.5–7 cm., mostly densely pubescent and very densely
punctate; calyx-tube 2–2.5 mm. long, with linear-
lanceolate lobes 2–3 mm. long; corolla-tube 0.8–1.7 cm.
long, with lobes 3.5 mm. long, 1.5 mm. wide . . . 25. *C. eriophyllum*
36. Inflorescences almost spiciform, leafless with congested
lateral branches; corolla-tube 3–4 mm. long (sect.
Stacheocymosa) 37
Inflorescences lax or much less congested, usually ± leafy
cymes; corolla-tube over 5 mm. long except in species
38 (sect. *Cyclonema***)) 38
37. Scrambling or suberect slightly fleshy shrub 0.6–3 m. tall,
usually flowering when leafless or when leaves are very
undeveloped; adult leaves 1.5–6 × 1.2–4 cm. (**K** 4, 6, 7; **T**
3, 6) . . . . 51. *C. makanjanum*
Erect herb or subshrub 20–60 cm. tall from a woody stock;
leaves 4–28 × 1–7 cm., present at flowering time (**T** 6, 8) 52. *C. kissakense*
38. Liane to 5 m., with leafless older stems bearing leafy shoots
(in Flora area); inflorescences terminal on main and
lateral shoots (**U** 2; forest) . . . . 33. *C. violaceum*
Herbs, subshrubs, shrubs or small trees or small climbers 39
39. Pyrophytic subshrubby herb 30–40(–100) cm. tall from a
thick woody rootstock; leaves in whorls of 3, drying
pale green or brownish, oblanceolate to elliptic-
oblong, 4–13(–16) × 1.2–5 cm.; inflorescences terminal,
narrow, 7–12(–23) cm. long (**T** 7) . . . . . . 39. *C. prittwitzii*
Habit very various but more shrubby 40

---

* If still not satisfactorily run down carry on to next couplet.
** Specific delimitation within *Cyclonema* is very difficult; see notes on p. 136.

40. Calyx-lobes distinctly elongate, triangular to triangular-lanceolate; inflorescences few-flowered simple axillary or terminal dichasia or borne on lateral suppressed branches; leaves small, ± subsessile 41

    Calyx-lobes broadly rounded, ± semi-circular, or if with a tendency to be broadly triangular then inflorescences many-flowered complicated cymes running together at apices of branches to form often dense terminal inflorescences with supporting leaves reduced to bracts; leaves small to very large, sessile to long-petiolate 43

41. Leaves and inflorescences congested on very reduced side shoots; calyx-lobes more elongate and densely glandular within; adapted to more arid areas (**K** 1)   45. *C. rupicola*

    Leaves and inflorescences mostly on normal shoots (note sp. 46 has very short side shoots but the calyx-lobes are short and ± semicircular or ovate and not or sparsely glandular within) 42

42. Calyx-lobes equilaterally triangular but not elongate; inflorescences axillary, 1–3-flowered cymes, 1–3 per node, often 3 per axil; peduncles 1.5–3 cm. long; corolla small, the tube 5 mm. long and longest lobe 5 × 3 mm.; lateral lobes yellow; leaves small, elliptic, 1–5 × 0.4–2 cm. (**T** 1, 4) . . . .   43. *C. tanneri*

    Calyx-lobes distinctly elongate-triangular, mostly lanceolate-triangular; inflorescences much as in last but poorly known; corolla probably mostly larger; leaves 1.3–4.5 × 0.7–2.5 cm. (or 7.5 × 4.5 in a variety) (**T** 4) . . . .   44. *C. taborense*

43. Leaves sessile, 7–25 × 3–21 cm., narrowed to a square ± cordate base but can be so reduced as to appear merely attenuate, practically glabrous, very thin (**T** 1, 7; forest)   34. *C. bukobense*

    Leaves cuneate to rounded at base, sessile or petiolate, or if narrowed to a broad base (as in *C. myricoides* subsp. *austromonticola*) then leaves thicker and more densely hairy 44

44. Leaves 3–5, clustered on very short abbreviated side-shoots (the internodes suppressed), oblanceolate, obcuneate or narrowly elliptic, 1.2–6 × 0.6–2.2 cm.; inflorescences appearing lateral but from apices of side-shoots, 1–2-flowered; peduncles 2.5–4.5 cm. long, very slender (T7)   46. *C. commiphoroides*

    Leaves and inflorescences not on short abbreviated side-shoots and without other characters combined 45

45. Leaves thickly white or ferruginous pubescent on both surfaces, 10–13 × 5–7 cm., acuminate at the apex; corolla bud-limbs glabrous except for marginal ciliations of lobes (**T** 6; Uluguru Mts.) . . . . .   36. *C. suffruticosum**

    Leaves glabrous to densely pubescent or velvety but if so then grassland or woodland species with smaller leaves 46

46. Inflorescences terminal many-flowered elongate thyrsoid panicles; leaves much longer than broad 47

    Inflorescences not so elongate or if so then leaves not much longer than wide 49

47. Leaves oblanceolate or elliptic to oblanceolate-oblong, typically 5.5–25(–30) × 1.3–7 cm. and glabrescent; inflorescences 13–30 cm. long of rather small flowers; corolla-tube 5 mm. long, lobes 8.5 × 3.5–4.5 mm. (a single atypical specimen seen from **T** 4 has broadly elliptic leaves) . . . .   37. *C. alatum*

    Leaves up to 12 × 3 cm.; inflorescences 20 cm. long; corolla apparently 4 mm. long including lobes . . . . .   38. *C. sp. B*

---

* Very imperfectly known and probably not distinct from *C. sansibarense*.

48. Branches of dichasia short, 2–2.5 cm. long with subsecund
    flowers; pedicels ± 1 mm. long; leaves small, at least
    upper ones under 3 cm. long (T 5, Itigi) . . . .   48. *C. sp. F*
    Inflorescence mostly longer, with pedicels much longer 49
49. Evergreen forest plants with essentially larger leaves up to
    25 × 11 cm. and longer peduncles up to 8 cm.; flowers
    often larger; habit usually more shrubby . . . . 35. *C. sansibarense*
    Grassland or woodland plants with essentially smaller
    leaves 2–15 × 0.4–6 cm. and shorter peduncles 0–7 cm.
    long (usually short); flowers often smaller; habit
    usually more subshrubby except in several subspecies 40. *C. myricoides*
50. Scandent shrubs very often climbing by petiole-bases or
    abbreviated hook-like branchlets, or if ± erect then
    calyx and corolla glabrous or nearly so 51
    Erect shrubs or pyrophytic subshrubs or herbs without
    modified petiole-bases or abbreviated hook-like
    branchlets; calyx and corolla rarely glabrous, usually
    distinctly hairy 52
51. Calyx-lobes glabrous or only shortly obscurely ciliate, acute
    or subacute, usually not so distinctly acuminate, the
    venation often rather obscure, scarcely to fairly
    distinctly raised and reticulate, often distinctly densely
    punctate; corolla-tube glabrous to sparsely pubescent
    or with sessile glands . . . . 13. *C. cephalanthum*
    Calyx-lobes hairy and long-ciliate, distinctly acuminate, the
    venation usually very distinctly raised and reticulate,
    not distinctly punctate (except in *C. cephalanthum*
    subsp. *montanum*); corolla-tube usually distinctly
    pubescent (glabrous in a few variants not occurring in
    Flora area) 55
52. Leaves cordate and ± long-petiolate, the petioles attaining
    12(–30) cm. 53
    Leaves not cordate or scarcely so and usually only shortly
    petiolate; petioles up to 5(–8.5) cm. but mostly quite
    short, 0.3–3 cm. 54
53. Leaves large, more round, 11–36(–60) cm. long, 4–32 cm.
    wide; petioles 1–12(–30) cm. long; corolla 12–14.5(–
    17.5) cm. long (U 2; T 1, 6) . . . . 8. *C. poggei*
    Leaves usually smaller and more oblong or elliptic, 3.5–22
    cm. long, 2.5–13.5 cm. wide; petioles 0.7–7.5 cm. long;
    corolla 5–7 cm. long . . . . 11. *C. frutectorum*
54. Subshrubby pyrophyte with several strictly erect unbranched
    stems 0.15–1 m. tall from an erect woody rootstock;
    corolla-tube 5–7 cm. long (T 1; 4)   10. *C. buchneri*
    Larger, mostly branched, shrubs (0.3–)2–6 m. tall; corolla-
    tube 3.5–15 cm. long 55
55.* Corolla-tube 7–15 cm. long, usually over 10 cm.; always
    shrubby; calyx-lobes 2–2.8 cm. long (K 7; T 3, 6, 8)   9. *C. robustum*
    Corolla-tube 3.5–8.5 cm. long; mostly climbing but
    occasionally erect; calyx-lobes 0.5–1.8 cm. long (U 1–4;
    K 3, 5; T 1, 7) 56
56. Calyx-lobe cilia short, scarcely 0.5 mm. long; calyx-tube
    sometimes punctate; corolla-tube ± 3.5 cm. long (T 7)   13. *C. cephalanthum*
                                        subsp. *montanum*

    Calyx-lobe cilia longer, 1–1.5 mm. long; calyx-tube not
    punctate; corolla-tube (4–)6–8.5 cm. long (U 1–4; K 5;
    T 1) . . . . 12. *C. capitatum*

---

* Note there are two leads to this couplet. See also note at beginning of key.

1. **C. villosum** *Blume,* Bijdr.: 811 (1826); Lam, Verben. Malay Arch.: 289 (1919); Moldenke in Fifth Summ. Verbenac. 1: 361 (1971) & in Rev. Fl. Ceylon: 475, 476 (adnot. in text) (1983). Type: W. Java, Bogor [Buitenzorg], Bantam, *Blume* (L, holo.)

Shrub 2.5–3 m. tall; branches obtusely angled or terete; densely tomentose at the apex, velvety or pilose beneath; internodes rather long; accessory buds present. Leaves opposite, ovate to broadly obovate, 10–17.5 cm. long, 4–13 cm. wide, acute at the apex, rounded to slightly cordate at the base, coarsely and irregularly crenate or toothed, velvety or adpressed pubescent above, hispid beneath; venation plane above but very prominent beneath; petiole 4–7 cm. long. Panicles lax, elongate, to ± 17 cm., leafy at base with elongate internodes, the leaves decreasing upwards with uppermost bract-like; axillary cymes opposite, the lower 9–12 cm. long, the upper 0.5–3 cm. long; peduncles 2–5 cm. long. Calyx 1.3 cm. long, densely villous; tube cupulate-campanulate, 3 mm. long and in diameter; limb widely spreading, ± 1 cm. long, divided for ± ⅔ the length into 5 ovate-triangular acute-ciliate teeth 6 mm. wide at the base, nerved, sparsely pubescent and with very characteristic discoid-cupular glands (sessile glands *fide* Thomas). Corolla villous; tube 1–1.5 cm. long, 3 mm. wide, not dilated at the throat; limb almost 2-lipped; lobes unequal, the smallest 7 mm. long, the largest 1 cm. long. Stamens 4–4.5 cm. long, long-exserted. Style 5 cm. long. Ovary black, 4-lobed, 2 mm. tall. Mature fruit globose.

TANZANIA.   Lushoto District: E. Usambaras, Amani, Oct. 1912, *Grote in Herb. Amani* 3935 & cultivated near Amani, 5 Sept. 1916, *Peter* 17691! & without any data, *Peter* 51800!
DISTR.   T 3; India, Burma, Thailand, Indochina, Malaya, Philippines, Sumatra to Lesser Sunda Is. (Banka, Billiton & Lepar)
HAB.   Rain-forest, cultivated and possibly formerly naturalized; 900 m.

SYN.   *C. velutinum* Thomas in E.J. 68: 60, 99 (1936); T.T.C.L.: 634 (1949). Type: Tanzania, E. Usambaras, Amani, *Grote in Herb. Amani* 3935 (B, holo.†)

NOTE.   There seems no doubt from Thomas's description that his *C. velutinum* is nothing more than *C. villosum* Blume. It is possible that *Grote in Herb. Amani* 3935 was obtained from a naturalized plant — the data is simply given as 'Urwald'. No specimens have been seen from the area since the Peter collections were made. Moldenke states it is very close to *C. viscosum* Vent. but the corolla is only as long as or slightly longer than the calyx.

2. **C. chinense** (*Osbeck*) *Mabb.*, The Plant Book, corr. reprint: 707 (1990). Type: China, near Huang-Pu (Whampoa), *Osbeck* (S, lecto., FHO, photo.)

Subshrub, or ± herbaceous, 0.9–1.8 m. tall; stems ± square, shortly pubescent. Leaves broadly ovate, 6–29 cm. long, 5–28 cm. wide, acute or acuminate at the apex, truncate to widely cordate at the base, usually coarsely serrate, sometimes 1–3-lobed, rather sparsely pubescent; petiole 2–23.5 cm. long. Flowers strongly scented, in terminal compact many-flowered cymes 3–6 cm. long, 3.5–9 cm. wide; bracts lanceolate, leafy, scattered throughout the cymes, 1.5–3 cm. long, 0.3–1.2 cm. wide, acuminate. Calyx usually red or purple, campanulate, 1–1.5 cm. long; lobes lanceolate, 4–10 mm. long, acuminate. Corolla white or pink, funnel-shaped in common 'flore pleno' form but hypocrateriform in the single wild form, 1.5–2 cm. long.

KENYA.   Kiambu District: Kabete, May 1960, *Verdcourt* 2772!; Nairobi District: Eastleigh section 1, 10 Oct. 1971, *Mwangangi* 1848!
TANZANIA.   Lushoto District: Amani, Malaria Research Institute, 18 Apr. 1973, *Magogo* 424!
DISTR.   K 4; T 3; probably a native of China or S. tropical Asia but origin now not ascertainable; widespread cultivated and naturalized throughout tropical Asia, Oceania (where it is now a bad weed) and much of Africa and America; probably cultivated in East Africa
HAB.   Waste ground, rocky places between experimental plots, etc.; 1650–1740 m.

SYN.   *Cryptanthus chinensis* Osbeck, Dagbok Ostind. Resa: 215 (1757)
    Volkmannia japonica Jacq., Hort. Schoenbr. 3: 48, t. 338 (1798), *non Volkameria japonica* Thunb. (1784)
    Volkameria fragrans Vent., Jard. Malmaison 2, t. 70 (1804). Type: plant grown in France, Jardin de Malmaison
    Clerodendrum fragrans Willd., Enum. Pl. Hort. Berol.: 659 (1809), *nom. illegit.*
    C. fragrans (Vent.) R.Br. in Ait.f., Hort. Kew, ed. 2, 4: 63 (1812); Jex-Blake, Gard. E. Afr., ed. 4: 109 (1957), *nom. illegit.*
    C. fragrans (Vent.) R.Br. var. *multiplex* Sweet, Hort. Brit.: 322 (1827). Type: Bot. Mag. 43, t. 1834 (1816)
    C. fragrans (Vent.) R.Br. var. *pleniflora* Schauer in DC., Prodr. 11: 666 (1847), *nom. superfl.*. Type: Bot. Mag., t. 1834 (1816) (lecto.)

C. philippinum Schauer in DC., Prodr. 11: 667 (1847), as *"Clerodendron"*; Howard & Powell in Taxon 17: 54 (1968); Moldenke in Rev. Fl. Ceylon 4: 469 (1983) (extensive synonymy). Type: Philippines, Luzon, Prov. Albay*, *Cuming* 1096 (B, holo.†, GH, lecto., K, iso.!, KIEL, iso.†)
[*C. japonicum* sensu Huber in F.W.T.A., ed. 2, 2: 443 (1963), *non* (Thunb.) Sweet]

NOTE. Moldenke has discussed the naming of the single and double forms but his argument seems confused. Osbeck's type is the double-flowered form (Mabberley *in litt.*) and according to the Kew syntype the name *C. philippinum* refers to the double variety as stated by Howard & Powell. If names were needed for these forms then it is the single form which needs a name.

3. **C. pleiosciadium** *Gürke* in E.J. 18: 177 (1893) & in P.O.A. C: 341 (1895); Bak. in F.T.A. 5: 203 (1900); Thomas in E.J. 68: 55 (1936); T.T.C.L.: 364 (1949); Haerdi in Acta Trop., suppl. 8: 151 (1964). Type: Tanzania, Handeni District, Pangani, Kiwanda, *Fischer* 467 (B, holo.†)**

Erect herb or shrub with angular stems 0.3–1.5 m. tall. Leaves in whorls of 3(–4), mostly from very reduced branchlets which are almost spinous after petioles fall, ovate to oblong, 2–13 cm. long, 1.5–8.6 cm. wide, acuminate at the apex, rounded, obtuse or truncate at base or attenuate into the petiole, entire, very coarsely irregularly crenate-dentate or lobulate, ± discolorous, sparsely ± adpressed pilose on the nerves, also densely punctate beneath with ± red gland-dots; petiole 0.7–8 cm. long. Flowers in terminal many-flowered composite cymes, often condensed, 4–10(–12) cm. wide including accessory lateral cymes, mostly aggregated at apex but often some in lower 5–8 axils; peduncles 0–2 cm. long; pedicels (3–)5–7 mm. long; axes pubescent; bracts filiform or linear, 4–6 mm. long, pubescent. Calyx ± glabrous; tube 2.5–3 mm. long; lobes lanceolate, (4–)5.5–8 mm. long, sparsely pubescent or almost glabrous. Corolla white, sweet-scented (but plant sometimes said to be offensively smelling), glabrous or puberulent and with red glands outside; tube (1.2–)1.5–1.7(–2.4) cm. long; lobes very rounded, 3.5–4 mm. long, 2 mm. wide, with red glands and minute pubescence outside. Stamens green turning ± black, together with style exserted, 9 mm. Ovary 4-lobed. Fruit deep red or purplish black, subglobose, 0.8–1 cm. long and wide, 4-lobed or 1-lobed by abortion or 1–2 lobes ± aborted and small, the lobes separating. Seeds very pale, very compressed, pip-shaped, 7 mm. long, 4 mm. wide, rugulose. Fig. 12/1–3, p. 96.

TANZANIA. Lushoto District: Korogwe, 23 June 1970, *Archbold* 1236! & Korogwe, Segera, 7 Oct. 1966, *Faulkner* 3872!; Morogoro District: Uluguru Mts., 20 Apr. 1935, *E.M. Bruce* 1126!; Ulanga District: Mahenge, 30 Apr. 1932, *Schlieben* 2149!
DISTR. T 3, 6, 8; Malawi, Mozambique and Zimbabwe
HAB. Forest, woodland, cultivated land and regeneration; 200–1200 m.

SYN. ?*C. longipetiolatum* Gürke in E.J. 18: 178 (1893); Bak. in F.T.A. 5: 304 (1900); Thomas in E.J. 68: 55 (1936); T.T.C.L.: 634 (1949). Type: Tanzania, Uluguru Mts., Morogoro, *Stuhlmann* 71 (B, holo.†)
    C. syringiifolium Bak. in K.B. 1898: 160 (1898) & in F.T.A. 5: 300 (1900), as *"syringaefolium"*. Type: ? Malawi, between Mpata and Nyasa/Tanganyika Plateau, *Whyte* (K, holo.)
    ?*C. ulugurense* Gürke in E.J. 28: 294, 465 (1900); Bak. in F.T.A. 5: 518 (1900); Thomas in E.J. 68: 55 (1936); T.T.C.L.: 634 (1949). Type: Tanzania, Uluguru Mts., eastern foothills near Tununguo, *Stuhlmann* 8691 (B, holo.†)
    C. congestum Gürke in E.J. 28: 296, 466 (1900); Bak. in F.T.A. 5: 517 (1900). Type: Tanzania, Uluguru Mts., *Goetze* 213 (B, lecto.†)***
    C. pleiosciadium Gürke var. *dentata* Thomas in E.J. 68: 56 (1936); T.T.C.L.: 634 (1949). Type: Tanzania, Morogoro, *von Prittwitz & Gaffron* 340 (B, holo.†)
    C. pleiosciadium Gürke var. *bussei* Thomas in E.J. 68: 56 (1936); T.T.C.L.: 634 (1949). Type: Tanzania, Kilwa District, Donde, Kwa Mpanda, *Busse* 624 (B, holo.†, EA, iso.!)

NOTE. Thomas's varieties with strongly toothed leaves (var. *dentata*) or very short petioles (var. *bussei*) do not seem worthy of retention.
    I have not seen type material of either supposed synonym marked with a '?' but the descriptions seem to fit this species. *C. longipetiolatum* and *C. ulugurense* are separated from *C. pleiosciadium* by inflorescence characters and from each other by various measurements all of which occur in the much greater material now available — the 6–7 mm. calyx of *C. ulugurense* is extreme but has been seen. Brenan has, however, cited *"Schlieben* 3655!" as *C. longipetiolatum* indicating that he considered

* Information from Kew isotype.
** *Wallace* 213 (Tanzania, Morogoro District, 26 Oct. 1932) bears the note in ? Miss Bruce's handwriting 'compared with type *Fischer* 467'.
*** By transferring the three syntypes *Stuhlmann* 8214, 8219 and 8237 to var. *bussei* Thomas has effectively lectotypified this name.

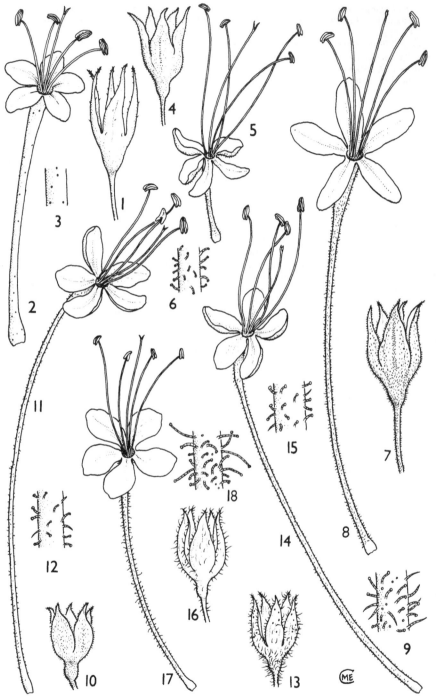

Fig. 12. *CLERODENDRUM PLEIOSCIADIUM* — **1**, calyx, × 3; **2**, corolla, × 3; **3**, part of corolla-tube, × 6. *C. FUSCUM* — **4**, calyx, × 1; **5**, corolla, × 1; **6**, part of corolla-tube, × 6. *C. ROTUNDIFOLIUM* — **7**, calyx, × 1; **8**, corolla, × 1; **9**, part of corolla-tube, × 6. *C. POGGEI* — **10**, calyx, × 1; **11**, corolla, × 1; **12**, part of corolla-tube, × 6. *C. ROBUSTUM* var. *FISCHERI* — **13**, calyx, × 1; **14**, corolla, × 1; **15**, part of corolla-tube, × 6. *C. BUCHNERI* — **16**, calyx, × 1; **17**, corolla, × 1; **18**, part of corolla-tube, × 6. 1–3, from *Faulkner* 1220; 4–6, from *Rogers & Gardner* 319; 7–9, from *Greenway & Kanuri* 12098; 10–12, from *Pirozynski* 477; 13–15, from *Faulkner* 3832; 16–18, from *Lloyd* 13. Drawn by Mrs. M.E. Church.

it distinct; it had been sent out from Berlin under that name but the duplicates I have seen (BM, BR) do not appear separable from *C. pleiosciadium*. *Peter* 32021 (Morogoro, 12 Nov. 1925) has spreading hairs on the stems and inflorescences. *Leach* 9891 from Malawi has a quite different calyx.

4. **C. polyanthum** *Gürke* in E.J. 28: 295 (1900); Bak. in F.T.A. 5: 517 (1900); Thomas in E.J. 68: 56 (1936); T.T.C.L.: 634 (1949). Type: Tanzania, Uzaramo [Usaramo] District, *Stuhlmann* 8274 (B, holo.†)

Shrub or tree; young stems finely pubescent. Leaves described as alternate, elliptic, 15–18 cm. long, 7–9 cm. wide, acute at the apex, obtuse at the base, finely pubescent above, velvety pubescent beneath, entire; petiole 4–6 cm. long, pubescent. Inflorescences terminal many-flowered cymes. Calyx broadly campanulate, ± 7 mm. long, glabrous; lobes narrowly lanceolate, ± 5 mm. long, 2 mm. wide at the base, long acuminate at the apex. Corolla probably white; tube very slender, 2 cm. long, glabrous outside; lobes ovate, obtuse.

TANZANIA.   Uzaramo [Usaramo] District, July 1894, *Stuhlmann* 8274
DISTR.   T 6; known only from the description
HAB.   Damp wooded valleys

NOTE.   Gürke states that this stands very near to *G. pleiosciadium:* 'the many flowered thickly crowded inflorescences, calyx and corolla' showing few differences but *C. polyanthum* differs widely in its almost velvety acute (not acuminate) entire leaves. I have seen nothing that matches this description. The large velvety leaves should render it immediately recognisable.

5. **C. umbellatum** *Poir.*, Encycl. Méth. Bot. 5: 166 (Jan. 1804); Huber in F.W.T.A., ed. 2, 2: 442 (1963). Type: Sierra Leone, *Smeathman*, a specimen given to Lamarck by de Beauvois* (P, holo. microfiche!)

Erect single-stemmed or shrubby herb or climber from a woody rootstock, 0.3–5 m. long or tall; stems brown with dense pubescence, eventually ± glabrescent. Leaves opposite, oblong-ovate, 3.3–18 cm. long, 1.5–11 cm. wide, shortly acutely acuminate at the apex, truncate to slightly or distinctly cordate at the base, ± discolorous, slightly hairy on venation above and shortly pubescent beneath, mainly on the raised venation and minutely punctate. Inflorescences terminal and from subtending axils, together up to 10 cm. wide; axes shortly spreading pubescent; pedicels ± 7 mm. long. Calyx becoming plum-coloured or maroon, shortly pubescent; tube funnel-shaped, 5 mm. long; lobes pinkish, lanceolate to triangular, 3–7 mm. long, acute. Corolla sweetly scented, white or cream with red throat, one lobe often with crimson or red markings on a white ground or ? all mauve; tube slender, 1–1.7 cm. long; limb ± asymmetrically globose in bud; lobes elliptic, 6–10 mm. long, 4 mm. wide, rounded. Stamens exserted, 3–4 cm. long. Fruit shiny, blackish grey, 1–1.2 cm. long, 7 mm. wide; seed white, ellipsoid, compressed, 8 mm. long, 5 mm. wide; calyx persistent. Fig. 13, p. 98.

UGANDA.   W. Nile District: Koboko, Mar. 1939, *Hazel* 720!; Bunyoro District: Hoima, 27 Dec. 1905, *Dawe* 800!; Busoga District: 16 km. S. of Bugiri, Igwe mutalla, 17 Oct. 1950, *G.H.S. Wood* 22!
KENYA.   W. Suk District: Suk Escarpment, Sept. 1963, *Tweedie* 2732! & Kacheliba Escarpment, May 1932, *Napier* 2007!
TANZANIA.   Buha District: Kasulu, Heru Juu, 16 Aug. 1954, *F.G. Smith* 1232! & Kibondo, 7 Aug. 1950, *Bullock* 3100! & Kasulu district, July 1956, *Procter* 496!
DISTR.   U 1–4; K 2; T 4; Senegal to Nigeria, Cameroon, Bioko [Fernando Po], Zaire, Burundi, Central African Republic, Sudan and Ethiopia; has been cultivated, e.g. at Kew
HAB.   Grassland with scattered trees, scrub of *Lophira, Butyrospermum, Annona* etc., riverine thickets; 900–1800 m.

SYN.   *C. scandens* P. Beauv., Fl. Owar. 2: 6, t. 62 (1808); F.W.T.A. 2: 274 (1931); Bak. in F.T.A. 5: 304 (1900); Thomas in E.J. 68: 58 (1936). Type: Nigeria, near Oware, *Palisot de Beauvois* (?G, holo., ?P-JU, iso.)
    Volkameria cordifolia Hochst. in Flora 25: 227 (1842), adnot.; Schauer in DC., Prodr. 11: 657 (1847). Type: Ethiopia, no locality or collector's number given in original reference but clearly near R. Tacazze, *Schimper* 1132 (TUB, holo., BM, iso.!)
    Clerodendrum cordifolium (Hochst.) A. Rich., Tent. Fl. Abyss. 2: 170 (1850); Oliv. in Trans. Linn. Soc., Bot. 29: 132 (1875); Engl., Hochgebirgsfl Trop. Afr.: 357 (1892); Gürke in P.O.A. C: 341 (1895); Bak. in F.T.A. 5: 304 (1900); Thomas in E.J. 68: 56 (1936); T.T.C.L.: 633 (1949)

* The microfiche makes it clear the specimen was collected by Smeathman and given to Lamarck by de Beauvois — Poiret speaks of Citoyen Beauvais in error.

Fig. 13. *CLERODENDRUM UMBELLATUM* — **1**, habit, × ⅔; **2**, flower, × 1; **3**, calyx, × 2; **4**, corolla, opened out, × 2; **5**, ovary, × 8; **6**, stigma, × 8; **7**, longitudinal section of ovary, × 10; **8**, part of fruiting branchlet, × 1; **9**, **10**, 2 views of pyrene, × 2. 1, from *Lye* 2040; 2–7, from *Tweedie* 2732; 8–10, from *Tweedie* 2332. Drawn by Mrs. M.E. Church.

NOTE. Although *C. cordifolium* has usually been kept separate and is more often ± erect than climbing the habit is very variable and it is not possible to maintain it distinct from *C. umbellatum* which displays a similar variation of habit in W. Africa but is predominantly climbing. Brenan (T.T.C.L.: 633 (1949)) comments on the uncertainty of Thomas's record "ostseite des Tanganyika-Gebirges, 2500 m., *von Trotha* 23".

6. **C. fuscum** *Gürke* in E.J. 18: 175 (1893) & P.O.A. C: 341 (1895); Bak. in F.T.A. 5: 304 (1900); Thomas in E.J. 68: 61 (1936); T.T.C.L.: 634 (1949); Moldenke in Phytologia 59: 475 (1986) & 63: 119 (1987). Type: Uganda, Kigezi District, Kayonza Mt. [Kajonsa], *Stuhlmann* 3061 (B, lecto.†)

Scandent subshrub, with young stems pubescent with spreading hairs. Leaves opposite, ovate, 6–10(–15.5) cm. long, 4–7(–15.5) cm. wide, abruptly acuminate at the apex, cordate at the base, entire or slightly undulate, sparsely pubescent above, more densely brown-pubescent beneath; petiole 0.5–1(–5.5) cm. long, pubescent. Flowers in lax terminal pubescent panicles; peduncles 5–9 cm. long; pedicels slender, 1.5–2 cm. long; bracts linear, (0.5–)1 cm. long. Calyx cream, 1.6–2(–2.8) cm. long, pubescent, 5-lobed to the middle; tube campanulate, slightly inflated below, 1 cm. wide; lobes equal, lanceolate, 1(–2.2) cm. long, 6–7 mm. wide at base, acute. Corolla yellowish white, purple-red outside (*fide* Stuhlmann), and with red or purplish red splashes inside 3 upper lobes; tube (2–)2.5–3 cm. long, 1 mm. wide, shortly pubescent; lobes narrowly obovate or oblanceolate, (0.5–)1 cm. long, obtuse or subacute, spreading or 3 upper reflexed. Stamens 4, recurved, much exserted. Style filiform, exserted 3 cm. Fruit black, shining, 1.2 cm. tall, 1.2–1.3 cm. wide. Fig. 12/4–6, p. 96.

UGANDA. Kigezi District: S. end of Lake Mutanda, 10 Jan. 1933, *Rogers & Gardner* 319! & Echuya Forest Reserve, 8 Aug. 1960, *Paulo* 671! & Kayonza [Kanyonsa] Mt., 28 Jan. 1892, *Stuhlmann* 3061 & 30 Jan. 1892, *Stuhlmann* 3096
DISTR. U 2; Zaire, Rwanda, Burundi, Gabon
HAB. Bamboo forest; secondary growth near lakeside; 1860–2346 m.

SYN. *C. macrocalyx* De Wild. in B.J.B.B. 7: 172 (1920). Types: Zaire, Bomili, *Bequaert* 1637, Avakubi, *Bequaert* 2031 & Masisi–Walikale, *Bequaert* 6492 (all BR, syn.)
C. grandicalyx Bruce in K.B. 1934: 306 (1934). Type: Uganda, Kigezi District, Lake Mutanda, *Rogers & Gardner* 319 (K, holo.!, EA, iso.!)

NOTE. Some of the parenthetical measurements are from De Wildeman's description.
    Brenan included this species in T.T.C.L. believing Stuhlmann's material to have come from NW. Tanzania.

7. **C. rotundifolium** *Oliv.* in Trans. Linn. Soc., Bot. 29: 132, t. 89 (1875); Gürke in P.O.A. C: 341 (1895); Bak. in F.T.A. 5: 308 (1900); Thomas in E.J. 68: 61 (1936); F.P.N.A. 2: 143 (1947); T.T.C.L.: 634 (1949); Brenan in Mem. N.Y. Bot. Gard. 9: 36 (1954); Jex-Blake, Gard. E. Afr., ed. 4: 225 (1957); K.T.S.: 585 (1961); U.K.W.F.: 614 (1974); Blundell, Wild Fl. E. Afr.: 398, t. 131 (1987). Type: Tanzania, Bukoba District, Karagwe, *Grant* 461 (K, holo.!)

Erect shrub 0.75–3 m. tall or sometimes climbing and even reported to be a small tree; stems with dense short spreading pubescence, later glabrescent and lenticellate. Leaves opposite or in whorls of 3 or rarely alternate, ovate to ± round, (2–)5–28 cm. long, (0.9–)4.5–26 cm. wide, shortly acuminate at the apex, rounded to cordate or less often broadly cuneate at the base, entire, crenate or crenate-dentate, pubescent to densely pilose above, usually densely so or velvety beneath or sometimes indumentum restricted to the nervation; petiole 2–9(–11.5) cm. long. Flowers fragrant, in terminal, not very dense, subumbellate dichasial cymes and with some in axils beneath, usually all ± aggregated in one inflorescence; pedicels 1.4–4 cm. long, pubescent with short soft hairs, glandular or not; bracts linear, ± 4 mm. long. Calyx densely pubescent; tube subglobose, 3–5 mm. long; lobes lanceolate, 1–1.7 cm. long, 3.5–7 mm. wide, acuminate, becoming more triangular in fruit, 1.5–2.5 cm. long, 0.6–1.2 cm. wide. Corolla white; tube slender, (4–)6.3–11.3 cm. long, glandular-pubescent outside; lobes broadly elliptic or oblong, 1–2.2 cm. long, 7–8.5 mm. wide, obtuse. Stamens and style exserted 3.5 cm.; stigma purplish. Fruit red, drying black and shiny, depressed subglobose, 1–1.5 cm. long, 1.2–1.8 cm. wide, 4-lobed. Fig. 12/7–9, p. 96.

UGANDA. Karomoja District: Moroto, Lia R., Mar. 1959, *J. Wilson* 662!; Ankole District: Kichwamba, 26 June 1945, *A.S. Thomas* 4174!; Mengo District: Kampala, King's Lake, Oct. 1935, *Chandler & Hancock* 103!

KENYA. Trans-Nzoia District: Moi's [Hoey's] Bridge, 15 Aug. 1963, *Heriz-Smith & Paulo* 832!; Meru, 5 Aug. 1914, *Battiscombe* 847!; Kericho, Sotik, 15 June 1953, *Verdcourt* 957!
TANZANIA. Bukoba District: Karagwe, Nyaishozi [Nyashozi], Oct. 1931, *Haarer* 2314!; Mbulu District: Lake Manyara National Park, Ndabaka R. Drift, 31 May 1965, *Greenway & Kanuri* 12098!; Lushoto District: W. Usambaras, Lushoto–Mombo road, 6.4 km. SE. of Lushoto, 10 June 1953, *Drummond & Hemsley* 2884!
DISTR. U 1–4; K 3–6 (7, Teita Hills, fide Jex-Blake); T 1–3, 5, 6; Zaire, Burundi, Sudan, Malawi and Mozambique, also occasionally cultivated
HAB. *Acacia* savanna, *Combretum* bushland, rocky places, stream-banks, scrub, anthills in dried parts of swamps, short grassland on hillsides, also cultivations in cleared rain-forest; 360–2130 m.
SYN. *C. stuhlmannii* Gürke in E.J. 18: 173 (1893) & in P.O.A. C: 341 (1895); Bak. in F.T.A. 5: 309 (1900). Types: Tanzania, Bukoba District, Ihangiro, *Stuhlmann* 916 & Lake Ikimba, *Stuhlmann* 1631 & Bweranyange [Weranyanye], *Stuhlmann* 1777 (all B, syn.†)
  C. guerkei Bak. in F.T.A. 5: 308 (1900). Type: Tanzania, W. Usambaras, Kwa Mshusa [Mshuza], *Holst* 8908A (K, holo.!)
  C. zambesianum Bak. in F.T.A. 5: 309 (1900); Thomas in E.J. 68: 62 (1936). Type: Malawi, Msope, *Buchanan* 359 (K, lecto.!)
  Siphonanthus rotundifolius (Oliv.) S. Moore in J.L.S. 37: 198 (1905)
  Clerodendrum rotundifolium Oliv. var. *keniense* T.C.E. Fries in N.B.G.B. 8: 701 (1924), as *"keniensis"*. Type: Kenya, Embu District, E. Mt. Kenya, Nithi R., *R.E. & T.C.E. Fries* 1643a (UPS, holo., BR, iso.!)
  C. rotundifolium Oliv. var. *stuhlmannii* (Gürke) Thomas in E.J. 68: 62 (1936)
  C. hildebrandtii Vatke var. *pubescens* Mold. in Phytologia 53: 197 (1983). Type: Tanzania, Musoma District, Serengeti National Park, Banagi, Ikoma, along Seronera [Serenela] R., *Tanner* 1664 (MICH, holo., K, iso.!)
NOTE. *C. guerkei* Bak. is a kind of weak subspecies with longer indumentum and occurs in T 3 (lowland, not W. Usambaras) and T 6. I have not formally recognised it but it is a geographical entity.

8. **C. poggei** *Gürke* in E.J. 18: 171 (1893); Bak. in F.T.A. 5: 309 (1900); F.P.S. 3: 195 (1956). Type: Zaire, Lualaba, *Pogge* 1116 (B, holo.†)

Shrub or small tree 0.9–2.5 m. tall, mostly with a single stem and with leaves restricted to the apex but also reported to be a climber in Angola. Leaves ovate or rounded-ovate, 11–36(–60) cm. long, 3.8–32 cm. wide, acute to subacuminate at the apex, obtuse to subcordate at the base, the margin entire to very coarsely toothed, slightly bristly pilose above but scarcely scabrid, with short ± soft spreading pubescence beneath or glabrescent; venation very reticulate beneath; petiole 1–12(–30) cm. long. Cymes aggregated into very dense ovoid heads 15–17.5(–20) cm. long, 6–9 cm. wide; pedicels 5 mm. long; bracts lanceolate or oblanceolate, 1.7–3 cm. long, 1.5–2.5 cm. wide, acuminate, pubescent; bracteoles 1–1.3 cm. long. Calyx reddish purple, densely pubescent; tube campanulate, 0.6–1.6 cm. long, deeply 5-lobed; lobes lanceolate, ± 1.5–2 cm. long, 3–7 mm. wide, acuminate. Corolla cream; tube slender, curved, 12–14.5(–17.5) cm. long, pubescent; lobes unequal, 1.7 cm. long, 7 mm. wide. Stamens and style exserted 2–3 cm. Fruit dark greenish black, 1.3 cm. long, 1.2 cm. wide, shiny. Fig. 12/10–12, p. 96.

UGANDA. Toro District: Kikonda [Kikonde], Dec. 1925, *Maitland* 1367!; Ankole District: Kalinzu Forest, June 1938, *Eggeling* 3703!; Kigezi District: Ishasha Gorge, Mar. 1947, *Purseglove* 2406!
TANZANIA. Buha District: 16 km. N. of Kigoma, Kakombe, 7 July 1959, *Newbould & Harley* 4276! & Kakombe valley, 1 Mar. 1964, *Pirozynski* 477! & Lugomela stream valley, 20 Mar. 1964, *Pirozynski* 574!
DISTR. U 2; T 4; Cameroon, Gabon, Central African Republic, Zaire, Burundi, Sudan, Ethiopia and Angola
HAB. Forest (including gallery), sometimes on rock faces; 760–1500 m.
SYN. *C. speciosum* Gürke in E.J. 18: 171 (1893), *non* Bull.* (1869), *nom illegit.* Types: Angola, Pungo Andongo, *Mechow* 121 (B, syn.†) & *Soyaux* 230 (B, syn.†, K, isosyn.!)
  C. angolense Gürke in E.J. 28: 291 (1900); Bak. in F.T.A. 5: 518 (1900); Thomas in E.J. 68: 62 (1936); Moldenke in Phytologia 57: 478 (1985) & 58: 180 (1985). Type as for *C. speciosum*
  C. hysteranthum Bak. in F.T.A. 5: 306 (1900). Type: Angola, Pungo Andongo, *Welwitsch*** (K, holo.)
  C. megasepalum Bak. in F.T.A. 5: 306 (1900). Type: Angola, Pungo Andongo, Cazella, *Welwitsch* 5705 (K, holo.)

* *C. speciosum* Bull is said to be a hybrid of *C. splendens* G. Don and *C. balfourii* Hort., i.e. *C. thomsoniae* Balf. See p. 87 for possible occurrence of this hybrid in East Africa.
** No number cited but presumably same gathering on which Hiern partly based his *Siphonanthus sanguinea*.

C. orbiculare Bak. in F.T.A. 5: 307, 518 (1900). Type: Angola, Catete & Quilango, *Welwitsch* 5688 (K, holo.!)

Siphonanthus sanguinea Hiern, Cat. Afr. Pl. Welw. 1: 839 (1900), *nom. superfl.* Types: Angola, Pungo Andongo, Cazella, *Welwitsch* 5705 (LISC, syn., BM, K, isosyn.!) & Catete & Quilango, *Welwitsch* 5688 (LISC, syn., BM, K, isosyn.!)

Clerodendrum capitatum (Willd.) Schumach. & Thonn. var. *butayei* De Wild. in Ann. Mus. Congo, Bot., sér. IV, 1: 117 (1903) & sér. V, 3: 131 (1909). Type: Zaire, 'Bas Congo', *Butaye* in *Gillet* 2245 (BR, holo.)

C. euryphyllum Mildbr. in N.B.G.B. 11: 679 (1932); Thomas in E.J. 68: 63 (1936); T.T.C.L.: 635 (1949). Type: Tanzania, Ulanga District, Mahenge, *Schlieben* 2048 (B, holo.†, BM, BR, HBG, P, iso.!)

C. euryphyllum Mildbr. var. *glabrum* Thomas in E.J. 68: 63 (1936), *nom. invalid.* Type: Cameroon, *Tessman* 2076 (B†) & *Nolde* 293 (B†)

NOTE. The material from Ethiopia is atypical but clearly very close to this species and I consider it conspecific. The species shows some resemblance to *C. fischeri* but the leaves are distinctly subcordate.

9. **C. robustum** *Klotzsch* in Peters, Reise Mossamb., Bot. 1: 259 (1861). Type: Mozambique, Querimba I. and 'Mainland of Mozambique', *Peters* (B, syn.†)

Shrub (0.3–)2–3 m. tall, with unbranched or branched stems presumably from a woody stock, but none seen and presumably not so compact as in the next species; young stems shaggy with yellowish brown hairs. Leaves broadly elliptic, oblong or slightly obovate, 7–17 cm. long, 4–11 cm. wide, shortly acuminate at the apex, broadly cuneate to rounded at the base, venation mostly conspicuously impressed above and strongly raised beneath, ± bullate, entire or less often coarsely undulate, densely pilose above and especially beneath, particularly on the venation; petiole 0.5–1(–2.5) cm. long. Inflorescences subglobose, 3–5 cm. long, 4–6 cm. wide, rarely a second smaller inflorescence present at lower (leafless) node just beneath; bracts green or purplish, lanceolate to broadly ovate, 1.5–2.5 cm. long, 0.5–1.8 cm. wide, acuminate and apiculate, long-ciliate and pilose or glabrous. Calyx-lobes similar, elliptic-lanceolate, 2–2.8 cm. long, (0.2–)6–8 mm. wide. Corolla white, 7–15 cm. long, usually over 10 cm., glabrous to more usually covered with glandular hairs and sessile glands, lobes oblong up to 0.8–1.7 cm. long, 3.5–8 mm. wide. Stamens and style exserted 3.5 cm. Fruit not seen.

1. Calyx-lobes and bracts very narrow, 5 mm. wide; corolla-
   tube ± 15 cm. long, with short glandular hairs or mostly
   sessile glands . . . . var. c. **mafiense**
   Calyx-lobes and bracts broader 2
2. Corolla practically glabrous . . . . var. a. **robustum**
   Corolla with mostly dense glandular hairs and sessile
   glands or shorter hairs 3
3. Bracts and calyx-lobes more lanceolate; corolla mostly
   densely hairy . . . . var. b. **fischeri**
   Bracts and calyx-lobes more broadly ovate; corolla less
   hairy . . . . var. d. **latilobum**

a. var. **robustum**

Similar in bracts and sepals to the much better known var. *fischeri* but corolla-tube glabrous or with very few sparse hairs or sessile glands.

TANZANIA. Tunduru District: 96 km. from Masasi, 19 Mar. 1963, *Richards* 17956!; Lindi District: Rondo [Muera] Plateau, May 1903, *Busse* 2611! & Nachingwea District, 8 Mar. 1956, *B.D. Nicholson* 41!
DISTR. **T** 8; Mozambique
HAB. Wooded grassland; 900–1000 m.

SYN. *C. fischeri* Gürke var. *robustum* (Klotzsch) Thomas in E.J. 68: 64 (1936); T.T.C.L.: 635 (1949); Moldenke in Phytologia 59: 414 (1986), incorrect name

NOTE. Curiously Thomas sank Klotzsch's name as a variety of the much younger name *fischeri* and Brenan and Moldenke both repeat this without comment. Careful reading of Klotzsch's description supports Thomas's conclusion that *robustum* and *fischeri* are conspecific. Thomas also cites *Braun* in *Herb. Amani* 1237 (Amani) as var. *robustum* but I have not seen this specimen; it must, I think, be a form of *C. cephalanthum.*

b. var. **fischeri** (*Gürke*) *Verdc.*, comb. nov.

Bracts and sepals relatively much more lanceolate, particularly the upper ones (some lower ones can be distinct broadly elliptic), usually over 5 mm. wide, usually greenish. Corolla-tube ± 7–12(–14) cm. long, with glandular hairs and sessile glands. Leaves entire. Fig. 12/13–15, p. 96.

KENYA. Kilifi District: Kibarani, 20 Mar. 1945, *Jeffrey* K140! & Sokoke Forest, Aug. 1938, *Moggridge* 138!; District uncertain: 'coastal grasslands' 24 June 1909, *Battiscombe* 50! & *Gardner* in *F.D.* 1454!
TANZANIA. Lushoto District: Segoma Forest, 30 July 1966, *Faulkner* 3832!; Ulanga District: Ifakara, Oct. 1959, *Haerdi* 347/0!; Lindi District: Rondo Plateau, Mchinjiri, Mar. 1952, *Semsei* 693!
DISTR. K 7; T 3, 6, 8; Mozambique, Malawi and Zimbabwe
HAB. Wet grasslands, woodland (including *Brachystegia*) and forest clearings; 0–450 m.

SYN. *C. fischeri* Gürke in E.J. 18: 172 (1893); Bak. in F.T.A. 5: 306 (1900); Thomas in E.J. 68: 64 (1936); T.T.C.L.: 635 (1949); K.T.S.: 584 (1961); Moldenke in Phytologia 59: 412 (1986). Type: Tanzania, Handeni District, Kiwonda, *Fischer* 483 (B, holo.†)

NOTE. *R.M. Graham* in *F.D.* 2003 (Kwale, Aug. 1929), in grassland, has very ciliate calyx-lobes, narrow bracts and distinctly pubescent corolla-tubes and is approaching var. *fischeri*. It is not, I think, anything to do with *C. cephalanthum*, which also occurs nearby in the Shimba Hills. In Zimbabwe some specimens merge with *C. buchneri* but the two are widely distinct throughout most of their ranges.

c. var. **mafiense** *Verdc.* var. nov. a var. *fischeri* bracteisque lobis calycis angustioribus corollae tubo ± 15 cm. longo differt. Typus: Tanzania, Rufiji District, Mafia I, *Wallace* 706 (K, holo.!, EA, iso.)

Bracts and calyx-lobes narrow, 2.5 cm. long, 5 mm. wide. Corolla-tube ± 15 cm. long, with short glandular hairs or mostly sessile glands.

TANZANIA. Rufiji District: Mafia I., 24 Mar. 1933, *Wallace* 706!
DISTR. T 6; not known elsewhere
HAB. Not known for type, probably coastal bushland at near sea-level; *Greenway* 5163 is from *Parinari*, *Uapaca*, *Vitex*, *Maprounea* association at 9 m.

NOTE. The only other members of this group collected on the island are *Greenway* 5234, certainly very different, which I have referred to *C. cephalanthum* Oliv. subsp. *swynnertonii* (S. Moore) Verdc., var. *schliebenii* (Mildbr.) Verdc. and *Greenway* 5163 (Kirongwe [Kerongwe], 25 Aug. 1937) which is a form of var. *mafiense* with more hairy corolla.

d. var. **latilobum** *Verdc.*, var. nov. a var. *fischeri* bracteis sepalisque inferioribus et medianis ovatioribus; glandulis corollae tubi sessilibus vel subsessilibus differt. Typus: Tanzania, Lindi District, Nachingwea, Namanga Hill, *Anderson* 1237 (K, holo.!, EA, iso.!)

TANZANIA. Bagamoyo District: Wami, 30 May 1921, *Swynnerton* 1263!; Ulanga District: Msolwa Camp, 23 May 1976, *Vollesen* in M.R.C. 3676! & Ifakara, Oct. 1959, *Haerdi* 347/0!
DISTR. T 6, 8; not known elsewhere
HAB. Woodland, sometimes on termite mounds; ?–250–?400 m.

SYN. [*C. fischeri* sensu Thomas in E.J. 68: 64 (1936) & T.T.C.L.: 635 (1949) quoad *Schlieben* 6119, *non* Gürke sensu stricto]
      [*C. capitatum* sensu Haerdi in Acta Tropica, Suppl. 8: 151 (1964), *non* (Willd.) Schumach. & Thonn.]
      [*C. buchneri* sensu Vollesen in Opera Bot. 59: 82 (1980), *non* Gürke]

NOTE. One sheet of *Migeod* 673 had huge leaves, 21 × 16 cm., with very impressed venation above, very prominent beneath.

10. **C. buchneri** *Gürke* in E.J. 18: 172 (1893); Bak. in F.T.A. 5: 305 (1900); Thomas in E.J. 68: 64 (1936); T.T.C.L.: 635 (1949); Moldenke in Phytologia 58: 300, 350 (1985). Type: Angola, Malange, Feira, *Buchner* 572 (B, lecto.†)

A strict subshrubby herb (0.15–)0.3–1 m. tall, with mostly several unbranched stems from a thick woody rootstock; stems densely shaggy with spreading purplish or ± ferruginous hairs. Leaves oblong-elliptic or ± ovate to ± obovate, 4.5–11.5(–18) cm. long, 3–6.5(–7.8) cm. wide, acuminate at the apex, broadly cuneate to rounded at the base, venation slightly to usually very impressed above and very prominent beneath, ± bullate, entire to obscurely undulate, rarely lobed, pilose above, ± densely pubescent beneath, chiefly on the venation; petiole 1.2–3 cm. long. Inflorescences capitate, subglobose or somewhat elongate, 3.5–9 cm. long and wide but sometimes with extra clusters in lower whorls, then up to 10 cm. long; bracts maroon or purplish, 1.2–3.5 cm. long, 0.5–1.5 cm. wide, long-ciliate and pubescent. Calyx-lobes coloured as bracts, elliptic-lanceolate, 1–1.5 cm. long, 4–6 mm. wide, long-apiculate, long-ciliate and glabrous to pilose on the faces. Corolla white; tube (4–)5–7 cm. long, with spreading glandular hairs or ± sessile glands or both; lobes oblong-elliptic, 11–12 mm. long, 7 mm. wide, rounded at apex. Fruits 1.3 cm. long, 0.8–1.3 cm. wide. Fig. 12/16–18, p. 96.

TANZANIA.   Buha District: 6.4 km. from Kibondo, 15 Nov. 1962, *Verdcourt* 3315!; Tabora District: Kakoma, 12 Feb. 1936, *Lloyd* 13!; Mpanda District: Kibwesa Point, 13 Sept. 1958, *Newbould & Jefford* 2210!
DISTR.   T 1, 4, 8; Zaire, Zambia, Botswana and Angola
HAB.   Woodland of bamboo, *Syzygium*, *Parinari*, *Protea*, *Brachystegia*, *Combretum* etc., also *Julbernardia* and *Brachystegia* woodland, rounded hills with long grassland mostly on sandy soil, also old cultivations; (600–)800–1275(–1350) m.

SYN.   *C. strictum* Bak.* in F.T.A. 5: 305 (1900). Type: Angola, Pungo Andongo, by the R. Caghuy between Caghuy and R. Cuanza, *Welwitsch* 5685 (K, holo.!, BM!, LISC, iso.)
  C. cuneifolium Bak.* in F.T.A. 5: 305 (1900). Type: Angola, Pungo Andongo, without precise locality, *Welwitsch* 5684 (K, holo.!, BM!, LISC, iso.)
  Siphonanthus stricta Hiern*, Cat. Afr. Pl. Welw. 1: 840 (1900). Type: Angola, Pungo Andongo, *Welwitsch* 5685 (LISC, holo., BM, K, iso.!)
  Siphonanthus cuneifolia Hiern* in Cat. Afr. Pl. Welw. 1: 840 (1900). Type: Angola, Pungo Andongo, *Welwitsch* 5684 (LISC, holo., BM, K, iso.!)
  Clerodendrum hockii De Wild. in B.J.B.B. 3: 266 (1911). Type: Zaire, Shaba, valley of the little Luemba, *Hock* (BR, holo.)
  C. humile Chiov. in Nuov. Giorn. Bot. Ital., n.s. 29: 117 (1923). Type: Zaire, Shaba, Biano [Bianos] Plateau, *Bovone* s.n. (FT, holo.)
  C. impensum Thomas var. *buchneroides* Thomas in E.J. 68: 66 (1936); T.T.C.L.: 636 (1949), *nom. invalid*. Type: Tanzania, Tabora, *Braun* 5397 (B, holo.†, EA, iso.!)

NOTE.   I had at first considered this a subspecies of *C. fischeri* but the distinctive habit, shorter corolla-tube and smaller lobes as well as the different geography indicate it is best kept distinct. Only very few intermediates have been seen out of a large number of specimens. *Haerdi* 347/0 (Ulanga District, Ilingera, 30 Oct. 1953) has a short corolla but a branched shrubby habit. The T 8 record is based on a single specimen with no basal parts (*Jahl* 107, Masasi, Chidya, 12 Feb. 1968, 600 m.). One apparently prostrate specimen has been seen from outside the Flora area.

11. **C. frutectorum**** *S. Moore* in J.B. 57: 249 (1919); Thomas in E.J. 68: 63 (1936); Moldenke in Phytologia 59: 472 (1986). Type: Zaire, Shaba, Shiwele, *Kassner* 2473 (BM, holo., K, iso.!)

Shrub, possibly somewhat climbing; young stems with long obviously multicellular hairs and also short pubescence, later glabrescent; older stems with nodular petiole-bases. Leaves oblong-ovate, 3.5–22 cm. long, 2.5–13.5 cm. wide, acuminate at the apex, rounded to distinctly cordate at the base, entire to coarsely crenate, sparsely slightly bristly pubescent, more densely pubescent beneath, particularly on the venation; petiole 0.7–9 cm. long, hairy as the stems. Inflorescences subglobose to semi-globose, 2.5–3.5 cm. long, 3.5–6 cm. wide; bracts lanceolate to elliptic, 1.2–2.5 cm. long, 0.45–1.5 cm. wide (outer ones wider), acuminate at the apex, strongly narrowed to the base, densely rather shortly pubescent, the cilia and hairs ± 0.5 mm. long. Calyx-lobes similar to the narrow bracts. Corolla white; tube 5–7 cm. long, with short glandular hairs outside; limb not seen expanded, 9 mm. long, 6 mm. wide, similarly hairy. Fruit not known.

TANZANIA.   Kigoma District: Uvinza, W. of Lugufu, km. 1176.5 to 1175.8 on central line, 8 Feb. 1926, *Peter* 36522! & Uvinza, SW. of Malagarasi, 30 Jan. 1926, *Peter* 35876!
DISTR.   T 4; Zaire and Zambia
HAB.   Presumably *Brachystegia* woodland; 1060–1100 m.

NOTE.   I am not sure that this is not just a form or state of *C. poggei* or in some cases intermediates with *C. capitatum*.

12. **C. capitatum** (*Willd.*) *Schumach. & Thonn.*, Beskr. Guin. Pl.: 61 (1827); Schauer in DC., Prodr. 11: 673 (1847); Bak. in F.T.A. 5: 305 (1900); F.W.T.A. 2: 275 (1931); Junghans in Bot. Tidsskr. 58: 94 (1962); Huber in F.W.T.A., ed. 2, 2: 443 (1963); Hepper, W. Afr. Herb. Isert & Thonning: 128 (1976); Moldenke in Phytologia 58: 417, 432 (1985). Type: Guinea, Aquapim, *Isert* (B-WILLD 11682, holo., microfiche 831!)

---

* These names were published independently, Hiern taking up Baker's unpublished names which were, however, published first.
** This curious epithet is presumably genitive plural of frutectum — a thicket.

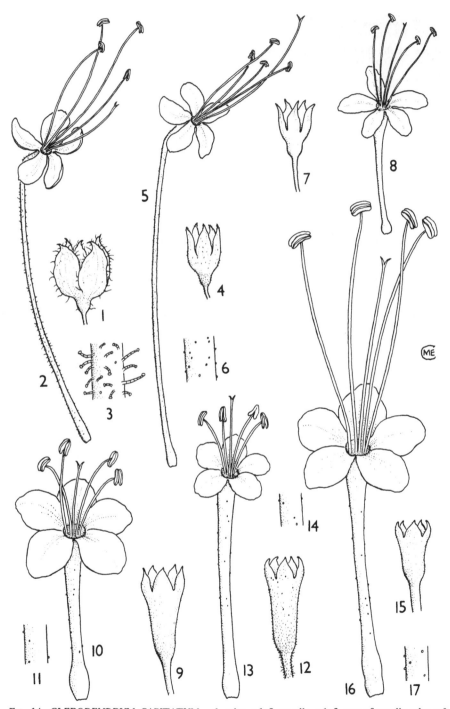

FIG. 14. *CLERODENDRUM CAPITATUM* — **1**, calyx, × 1; **2**, corolla, × 1; **3**, part of corolla-tube, × 6.
C. CEPHALANTHUM subsp. *CEPHALANTHUM* — **4**, calyx, × 1; **5**, corolla, × 1; **6**, part of corolla-
tube, × 6. *C. HILDEBRANDTII* var. *HILDEBRANDTII* — **7**, calyx, × 1; **8**, corolla, × 1.
C. SILVANUM var. *NUXIOIDES* — **9**, calyx, × 3; **10**, corolla, × 3; **11**, part of corolla-tube, × 6.
C. TANGANYIKENSE — **12**, calyx, × 3; **13**, corolla, × 3, **14**, part of corolla-tube, × 6.
C. SCHWEINFURTHII — **15**, calyx, × 3; **16**, corolla, × 3; **17**, part of corolla-tube, × 6. 1–3, from
Tweedie 3715; 4–6, from *Faulkner* 2323; 7, 8, from *Musyoki & Hansen* 978; 9–11, from *G.H.S. Wood*
757; 12–14, from *Purseglove* 572; 15–17, from *Carcasson* in B11409. Drawn by Mrs. M.E. Church.

Mostly a climbing or scrambling shrub to 6* m., the older stems usually with straight or hook-like persistent petiole-bases up to 2.5 cm. long, or sometimes erect or ± spreading shrub 1.8–6 m. tall, or even rarely (in W. Africa) ± herbaceous, 30 cm. tall. Stems ± densely covered with long ± ferruginous spreading hairs, later glabrescent and pale. Leaves oblong-elliptic or ± ovate, 1.5–17(–20.5) cm. long, 1–9(–11) cm. wide, acuminate at the apex, rounded to subcordate or cuneate at the base, entire to distinctly coarsely undulate, sparsely pubescent on venation above, sparsely to ± densely pubescent beneath; petiole 0.3–5(–8.5) cm. long, hairy like the stems. Inflorescences subglobose or hemispherical, 3–5.5 cm. wide, with occasional cymes from nodes immediately beneath. Calyx-tube 2.5 mm. long; lobes elliptic, green or often mauve or purple, 1.2–1.8 cm. long, 4–8(–10) mm. wide, venose, very acuminate at the apex, long ciliate and ± pilose on face or only sparsely hairy, rarely glabrous (in W. Africa). Corolla sweetly scented, white; tube very narrow, (4–)6–8.5 cm. long, densely long hairy or with rather short gland-tipped hairs, less often glabrous (in W. Africa); lobes oblong, 1.1–1.4 cm. long, 6–6.5 mm. wide. Stamens and style exserted 2.5–3.5 cm. Fruit shiny black, subglobose, 1.2 cm. long and wide, 4-lobed, the lobes ± oblong, grooved and granular on inner face. Fig. 14/1–3.

UGANDA. W. Nile District: 1 km. SE. of Metu rest camp, 15 Sept. 1953, *Chancellor* 269!; Teso District; Serere, Aug. 1934, *Chandler* 875!; Mengo District: Kirerema, Oct. 1914, *Dummer* 265!
KENYA. Trans-Nzoia District: Kapretwa, Oct. 1969, *Tweedie* 3715!; N. Kavirondo District: near Bungoma, Sept. 1968, *Tweedie* 3581! & W. of Bungoma, Nov. 1971, *Tweedie* 4145!
TANZANIA. Biharamulo District: Rusumo Falls, 23 Mar. 1960, *Tanner* 4790!; Bukoba District: Kiziba, Bugandika, 19 Apr. 1948, *Ford* 366! & Nyakato, 6 Jan. 1935, *Gillman* 211!
DISTR. U 1–4; K 3, 5; T 1; Gambia and Senegal to Cameroon, Zaire, Central African Republic, Rwanda, Sudan, Zambia and Angola; also cultivated in Nairobi
HAB. Forest, dense riverine thicket, seasonally wet wooded grassland; 780–1950 m.

SYN. *Volkameria capitata* Willd., Sp. Pl. 3(1): 384 (1800)
    Clerodendrum conglobatum Bak. in F.T.A. 5: 296 (1900). Type: Angola, Pungo Andongo, *Welwitsch* 5629 (K, holo.!, BM!, LISC, iso.)
    Siphonanthus conglobata Hiern, Cat. Afr. Pl. Welw. 1: 840 (1900). Type: Angola, Pungo Andongo, Cabondo, *Welwitsch* 5629 (LISC, holo., BM, iso.!)
    S. capitata (Willd.) S. Moore in J.L.S. 37: 198 (1905)
    Clerodendrum talbotii Wernham in Cat. Talbot's Nigerian Pl.: 90 (1913). Type: Nigeria, Oban, *Talbot* 341 (BM, holo.!, K, iso.!)
    C. capitatum (Willd.) Schumach. & Thonn. var. *talbotii* (Wernham) Thomas in E.J. 68: 65 (1936) (see note)
    C. capitatum (Willd.) Schumach. & Thonn. var. *conglobatum* (Bak.) Thomas in E.J. 68: 65 (1936)

NOTE. I have taken a narrower view of this species than most recent routine namers at Kew. Some West African specimens have very short (± 4 cm.) slender corolla-tubes and come near to *C. capitatum* var. *talbotii* in this respect but the Kew isotype of that has much broader calyx-lobes than any other material I have seen and appears rather distinctive; Thomas may be correct in maintaining it as a variety. Plants with glabrous calyx-lobes referred by Huber to var. *cephalanthum* (Oliv.) H. Huber are not I think that plant (maintained here as a separate species) but a distinct variety. A specimen from E. Usambaras, Amani, Oct. 1945, *Fernie* in *Herb. Amani* 9728 does appear to be *C. capitatum* and is obviously a climber. I suspect it was cultivated — the label gives no details of the exact habitat. No other wild members of this group, which grow in the E. Usambaras, have long ciliate calyx-lobes, merely very short hairs or quite glabrous. *A.S. Thomas* 905A (Uganda, Masaka District, Sese Is., Fumve, 27 Feb. 1933) was originally determined as an abnormal form of *C. capitatum* and I can only agree with this; the leaves and calyces are normal but the corollas (detached) described as greenish are quite unlike the description given above, the tube being coarse, 3–4.5 mm. wide, rather than 0.7–1.5 mm. (dry) and the limb more zygomorphic with lobes up to 2 cm. long, 7.5 mm. wide; the reason for this is not evident and no other material has been seen from the Sese Is.

13. **C. cephalanthum** *Oliv.* in Hook., Ic. Pl. 16, t. 1559 (1887) & in Gard. Chron., ser. 3, 3: 652 (1888); Hemsl. in Bot. Mag. 129, t. 7922 (1903); Thomas in E.J. 68: 66 (1936); T.T.C.L.: 635 (1949). Type: Zanzibar I., *Kirk* (K, holo.!)

Usually scandent or sometimes ± erect shrub 0.6–6 m. long or tall, climbing by spine-like persistent petiole-bases 1–1.5 cm. long. Young stems finely to densely pubescent but later glabrous, or entirely glabrous with pale brown corky bark or sometimes persistently spreading pubescent. Leaves oblong-elliptic, 2.8–25.4 cm. long, 1.3–9(–16) cm. wide, acute to acuminate or subacute at the apex, often rather less

---

* A claim by Tanner that it climbs to 30 m. seems exaggerated.

acuminate than in species 9, cuneate to rounded at the base, entire or very widely crenate-dentate near apex, usually coriaceous, ± bullate, venation usually distinctly impressed above and raised beneath, ± glabrous or sparsely hairy on the midrib beneath or in one subspecies pubescent to ± velvety beneath; petiole 0.8–4 cm. long, pubescent. Inflorescences few–many-flowered, capitate or subglobose to ovoid, 3–6 cm. long and wide, often run together for up to 12 cm. long or 25 cm. diameter; pedicels up to 7 mm. long. Bracts linear to elliptic then similar to but usually smaller than the calyx-lobes. Calyx green, pink, purple or maroon; tube 3 mm. long; lobes ovate or narrowly ovate, 0.8–1.7 cm. long, 6–7.5 mm. wide, ± accrescent, acute, subacute or even ± obtuse, usually less distinctly acuminate than in related species but sometimes mucronulate to tapering, glabrous or only shortly obscurely ciliate and puberulous at tip, the venation often rather obscure, scarcely to distinctly raised and reticulate, the lamina very often distinctly densely punctate. Corolla scented, white; tube (3.5–)7.5–10 cm. long, ± glabrous outside or with gland-tipped hairs; lobes 1.3 cm. long, 6 mm. wide; style and stamens exserted 3.5 cm., the stigma and upper part of style often purple; anthers crimson. Fruit blackish, subglobose, 0.9–1.5 cm. long and wide, 4-lobed.

KEY TO INFRASPECIFIC VARIANTS

1. Leaves pubescent above and pubescent to almost velvety beneath; stems hairy; bracts and sepals pubescent and ciliate; corolla-tube short, ± 3.5 cm. long, with hairs and gland-tipped hairs (T 7) . . . .   e. subsp. **montanum**
   Leaves mostly ± glabrous or venation pubescent beneath; stems glabrous or hairy; bracts and sepals puberulous or mostly glabrous; corolla-tube 3.5–10 cm. long, glabrous or hairy and glandular (widespread) 2
2. Corolla-tube glabrous or with obscure sessile glands and rarely a few hairs; calyx-lobes often ± obtuse 3
   Corolla-tube hairy and/or with gland-tipped hairs; calyx-lobes mostly acute to acuminate 4
3. Corolla-tube 5.5–10 cm. long; leaves often coriaceous (**K** 7; **T** 6, 8; **Z**) . . . .   a. subsp. **cephalanthum**
   Corolla-tube ± 3.5–5 cm. long (T 3, 6, 7, 8) . . . . .   d. subsp. **swynnertonii** var. **schliebenii**
4. Stems ± glabrous; leaves glabrous 5
   Stems hairy; leaf-venation ± hairy beneath; corolla-tube 3–4(7.5) cm. long with gland-tipped hairs and ± sessile glands; usually shrubby (**T** 3, W. Usambaras) . . .   c. subsp. **impensum**
5. Usually a branched shrub; calyx-lobes acute to acuminate; corolla-tube 4.5–5(–6.5) cm. long (**T** 3, E. Usambaras)   b. subsp. **mashariki**
   Usually a distinct climber; calyx-lobes ± obtuse to acute; corolla-tube 3–5.5 cm. long . . . .   d. subsp. **swynnertonii** var. **swynnertonii**

a. subsp. **cephalanthum**

Leaves often distinctly coriaceous (particularly in Zanzibar material), glabrous, with venation impressed above. Calyx-tube and pedicels finely puberulous; lobes glabrous or finely puberulous. Corolla-tube 5.5–10 cm. long, glabrous or with obscure sessile glands or rarely a few hairs. Usually scandent. Figs. 14/4–6, p. 104; 15.

KENYA. Kwale District: Lungalunga–Msambweni road, Marenge Forest, 18 Aug. 1953, *Drummond & Hemsley* 3879! & Shimba Hills, Lango ya Mwagandi [Longmwagondi] Forest, 13 Nov. 1968, *Magogo & Estes* 1223!; Kilifi, *R.M. Graham* in *F.D.* 1945!

TANZANIA. Uzaramo District: Oyster Bay, 19 June 1968, *Batty* 143!; Zanzibar I.: Chukwani, 7 Aug. 1959, *Faulkner* 2323! & Fumba, 12 Sept. 1951, *R.O. Williams* 90!

DISTR. **K** 7; **T** 6, 8; **Z**; formerly cultivated at Kew for many years

HAB. Coastal forest and bushland on coral, etc; 0–330(–450) m.

SYN. *C. capitatum* (Willd.) Schumach. & Thonn. var. *cephalanthum* (Oliv.) Bak. in F.T.A. 5: 306 (1900)
     C. mossambicense Klotzsch var. *glabrum* Thomas in E.J. 68: 63 (1936); T.T.C.L.: 636 (1949). Type: Tanzania, Lindi, *Busse* 2335 (B, holo.†, BR, EA, iso.!), *nom. invalid.*
     [*C. swynnertonii* sensu Thomas in E.J. 68: 66 (1936), quoad *Stuhlmann* 804! *non* S. Moore]

NOTE. *Webster* in *E.A.H.* 8959, said to be from Cherangani, 2000 m. [6500 ft.], is undoubtedly, as determined by Greenway, *C. cephalanthum* but I am convinced it is wrongly localized and came from the coast. Huber (F.W.T.A., ed. 2, 2: 443 (1963)) records *C. capitatum* var. *cephalanthum* from

FIG. 15. *CLERODENDRUM CEPHALANTHUM* subsp. *CEPHALANTHUM* — **1**, habit, × ⅔; **2**, calyx, × 2; **3**, upper part of corolla, × 2; **4**, ovary and stigma, × 6; **5**, longitudinal section of ovary, × 8; **6**, fruit with persistent calyx, × 1; **7**, **8**, 2 views of pyrene, × 2. 1, from *Faulkner* 2347; 2–5, from *Faulkner* 2323; 6–8, from *R.O. Williams* 95. Drawn by Mrs. M.E. Church.

West Africa but the material involved is I think a distinct variety of *C. capitatum; Busse* 2335, the type of *C. mossambicense* var. *glabrum*, has rather large bracts and calyx-lobes but is I think *C. cephalanthum*. It had originally (?by Gürke) been labelled as *C. robustum*. Despite a careful study of Mozambique material and comparison with the descriptions of *C. stenanthum* Klotzsch (types: Mozambique, R. Zambesi (Rios de Sena), Boror and Querimba I., *Peters* (B, syn.†)) and *C. mossambicense* Klotzsch (type: Mozambique "Festland von Mossambique", *Peters* (B, holo.†)) I am not certain that either is undoubtedly *C. cephalanthum*, which is of course a much younger name than either. Thomas sinks *C. stenanthum* into *C. mossambicense* but the descriptions are by no means the same. It is possible both are forms of *C. robustum* Klotzsch which, as pointed out (p. 101), I feel sure should include *C. fischeri* as a variant.

b. subsp. **mashariki** *Verdc.*, a subsp. *cephalantho* foliis plerumque tenuioribus, corollae tubo glandulo-pubescenti differt. Typus: Tanzania, E. Usambaras, Sangerawe, *Greenway* 4674 (K, holo.!, EA, iso.!)

Leaves usually thinner, glabrous, with venation scarcely to very impressed above, the margin entire or often more distinctly undulate. Calyx as in subsp. *cephalanthum*. Corolla-tube 4.5–5(–6.5) cm. long, usually with short gland-tipped hairs, but these may be sparse. Usually a much branched shrub.

TANZANIA. Lushoto District: E. Usambaras, Sangerawe, 14 Oct. 1936, *Greenway* 4674! & above Magunga, 23 Sept. 1953, *Faulkner* 1258! & Amani, Ngomeni road, 22 Sept. 1960, *Paulo* 825!; Ulanga District: Uzungwa Mts., Sanje, 6 Sept. 1984, *Bridson* 631!
DISTR. **T** 3, 6; not known elsewhere
HAB. Evergreen forest; 300–1050(–1650) m.

SYN. [*C. cephalanthum* Oliv. var. *coriaceum* sensu Thomas in E.J. 68: 66 (1936) quoad *Braun* 869!, *non* Thomas sensu stricto]

NOTE. *Bridson* 631 cited above has leaves to 27 × 17 cm. with 2–3 big teeth on either side; it is in fruit only and may represent a distinct form. Some material from the Uluguru Mts. probably also belongs here, e.g. *Mabberley* 1327 (Uluguru North [Northern Forest Reserve], valley of R. Mwere, 21 July 1952).

c. subsp. **impensum** (*Thomas*) *Verdc.*, comb. et stat. nov. Type: Tanzania, W. Usambaras, Kwai, *Eick* 396 (B, holo.†)

Similar to subsp. *mashariki* but young stems distinctly hairy and leaf-venation ± hairy beneath or almost glabrous. Corolla-tube 3–4(–7.5) cm. long, with gland-tipped hairs and ± sessile glands.

KENYA. Teita District: Mbololo Forest, 13 May 1985, *Faden et al.* 359!
TANZANIA. Lushoto District: W. Usambaras, Shume Forest Reserve, Gologolo, Dec. 1951, *Parry* 96! & Shume Forest Reserve, Aug. 1955, *Semsei* 2226! & Shagayu Forest, May 1935, *Procter* 219!
DISTR. **K** 7; **T** 3; not known elsewhere
HAB. Evergreen forest; 1500–1950 m.

SYN. *C. cephalanthum* Oliv. var. *coriaceum* Thomas in E.J. 68: 66 (1936); T.T.C.L.: 635 (1949), pro parte. Type: Tanzania, W. Usambaras, without locality, 1902, *Engler* 1175 (B, holo.†)
      C. impensum Thomas in E.J. 68: 100, 66 (1936); T.T.C.L.: 635 (1949)

NOTE. Thomas cites two E. Usambara specimens as well as the W. Usambaras one under his var. *coriaceum* but in general neither subsp. *mashariki* nor subsp. *impensum* has leaves as coriaceous as the subsp. *cephalanthum*. B.D. Burtt 4889 (Handeni, Mgera, Lukigura [Likagura] valley, 28 Sept. 1933) has flowers with the corolla-tube up to 7.5 cm. but the hairy stems of subsp. *impensum*. *Stuhlmann* 574 is puzzling and would appear to belong near here but was collected in "Usegua" at "Haliboma" (19 Sept. 1888) which has not been traced; on 22 Sept. he was near Korogwe. A specimen from Pare District — N. Pare, between Kilomeni and Kisangara [Kissangara], 22 June 1915, *Peter* 11563!, 1550 m. — is a variant of *C. cephalanthum* with micropuberulous punctate calyx and subsessile glands on the corolla-tube but is only in young bud. It comes close to the two Usambara subspecies. Similar specimens from the same area, Kilomeni to Lembeni, 29 June 1915, *Peter* 11706! and Kilomeni to Kisangara, 28 June 1915, *Peter* 11656, are close to subsp. *impensum*. The Kenya Mbololo population is not typical but included mainly on geographical grounds.

d. subsp. **swynnertonii** (*S. Moore*) *Verdc.*, comb. et stat. nov. Type: Zimbabwe, Chirinda Forest, *Swynnerton* 85 (BM, lecto.!, K, isolecto.!)

Leaves thin or ± thinly coriaceous, with or without impressed nerves, mostly quite glabrous; stems glabrous, puberulous or ± pubescent in intermediates with subsp. *montanum*. Calyx ± 1–1.2 cm. long, glabrous or slightly puberulous. Corolla-tube 3.5–5.5 cm. long, very slender, glabrous, pubescent or with stalked or sessile glands. Shrub or more usually distinctly climbing by prominent curved petiole-bases.

var. **swynnertonii**

Corolla-tube with distinct hairs or gland-tipped hairs.

TANZANIA. Njombe District: Likanga, 18 July 1931, *Schlieben* 1090A!
DISTR. **T** 7; Malawi, Mozambique, Zambia and Zimbabwe

HAB. Wooded grassland; 1000 m.

SYN. *C. swynnertonii* S. Moore in J.L.S. 40: 166 (1911); Thomas in E.J. 68: 66 (1936); Brenan in Mem. N.Y. Bot. Gard. 9: 37 (1954)

NOTE. Several fruiting specimens cannot be placed. In E. Zimbabwe and Mozambique this taxon is uniform and distinctive in its long climbing habit and short hairy corolla. It is likely that many would prefer to keep it distinct. If so, the short corolla taxa from the Tanzanian forests should probably also be treated as *swynnertonii* but distinctions from *C. cephalanthum* become difficult in the Flora area. There will undoubtedly be some who consider reducing all to *C. capitatum* to be the solution, but this is an oversimplification hiding much of phytogeographical interest.

var. **schliebenii** (*Mildbr.*) *Verdc.*, comb. et stat. nov. Type: Tanzania, Ulanga District, Mahenge, Sali, *Schlieben* 2260 (B, holo.†, BR, iso.!)

Corolla-tube glabrous or with minute sessile glands.

TANZANIA. Tanga District: 8 km. SE. of Ngomeni, 30 July 1953, *Drummond & Hemsley* 3548!; Rufiji District: Mafia I., Kilindoni, 10 Sept. 1937, *Greenway* 5234! & same place, 14 July 1932, *Schlieben* 2116!; Rungwe District: Kirambo, Mbaka, 3 Nov. 1912, *Stolz* 1537!
DISTR. T 3, 6, 7, 8; Mozambique,
HAB. Seashore with bushland and scattered trees, thickets in scattered tree grassland; 9–1000 m.

SYN. *C. schliebenii* Mildbr. in N.B.G.B. 11: 678 (1932); Thomas in E.J. 68: 66 (1936); T.T.C.L.: 638 (1949)
[*C. swynnertonii* sensu Thomas in E.J. 68: 66 (1936), pro parte, *non* S. Moore sensu stricto]

NOTE. The possibility that this taxon might be the same as *C. stenanthum* Klotzsch needs discussion since that name would take precedence over *C. cephalanthum*. The corolla is described as glabrous and 2½ inches (± 6.5 cm.) long but the bracts and calyx-lobes as much larger and ciliate; moreover Thomas who had the type available would doubtless have united them had they been identical. Instead he united it with *C. mossambicense* Klotzsch described in the absence of a corolla. In such a difficult group decisions are difficult without the actual types.

e. subsp. **montanum** (*Thomas*) *Verdc.*, comb. et stat. nov. Type: Tanzania, Njombe District, Ubena, *Goetze* 770 (B, holo.†, BR, iso.!)

Leaves mostly thin and smaller than in other subspecies, usually pubescent above and beneath sometimes quite velvety. Young stems hairy. Bracts and sepals pubescent and ciliate. Corolla-tube short, ± 3.5 cm. long, with hairs and stalked glands.

TANZANIA. Iringa District: Mufindi, Ngwazi [Nkwazi], 10 May 1973, *Shabani* 1000!; Njombe District: Makumbako, 25 June 1960, *Leach & Brunton* 10116! & about 14 km. S. of Njombe, 10 July 1956, *Milne-Redhead & Taylor* 11043!
DISTR. T 7; Malawi
HAB. Grassland and evergreen bushland; 1680–2300 m.

SYN. [*C. capitatum* sensu Bak. in F.T.A. 5: 306 (1900), pro parte, quoad *Thomson, non* (Willd.) Schumach. & Thonn. sensu stricto]
C. montanum Thomas in E.J. 68: 100, 67 (1936); T.T.C.L.: 636 (1949)

NOTE. It is with this taxon that difficulties really arise concerning the distinction between *C. capitatum* and *C. cephalanthum*, but I do not believe that these warrant the sinking of the two. This would involve the loss of much information concerning the phytogeography of the eastern rain-forest area. I have chosen to consider *montanum* a subsp. of *cephalanthum* rather than of *capitatum*, particularly to avoid the absurdity of having two 'species' occurring in one area differing only in a few hairs on the calyx.

14. **C. hildebrandtii** *Vatke* in Linnaea 43: 536 (1882); Gürke in P.O.A. C: 341 (1895); Bak. in F.T.A. 5: 302 (1900); Chiov., Fl. Somala 2: 362 (1932); T.T.C.L.: 637 (1949); K.T.S.: 584 (1961); E.P.A.: 299 (1962); Vollesen in Opera Bot. 59: 82 (1980). Type: Kenya, Teita District, Ndara, *Hildebrandt* 2389 (B, holo.†)

Shrub or tree 1.2–9 m. tall, sometimes with a distinct bole or subscandent, the branches often ± pendulous; bark straw-coloured to pale brown, fissured; branchlets nearly always glabrous, rarely pubescent, roughened with persistent petiole-bases, lenticellate. Leaves broadly ovate, 1.5–13 cm. long, 0.8–11.7 cm. wide, acuminate or apiculate at the apex, broadly cuneate, rounded or subcordate at the base, slightly fleshy, entire, almost invariably glabrous; petiole 0.7–7 cm. long. Inflorescences terminal panicles of very lax dichasial cymes, each ± 9-flowered and also some additional axillary ones; peduncle and secondary peduncles 1–3 cm. long; apparent pedicels 0.8–1.7 cm. long. Calyx dull purplish at base; tube 0.8–2 cm. long, 4 mm. wide, angled, glabrous; teeth triangular, 2.5–9 mm. long, becoming 1.2 cm. in fruit. Corolla white or cream, mauve on fading, pleasantly scented; tube 2.5–3(–?4) cm. long; lobes obovate, 7 mm. long, 4–5 mm. wide. Stamens ± 2 cm. long, filaments and style reddish to mauve above; anthers red-brown. Fruit 1–1.7 cm. long and wide.

var. **hildebrandtii**

Young shoots, leaves, flowers, etc., glabrous. Fig. 14/7, 8, p. 104.

KENYA. Kwale District: Waa, Ras Kikadini, 18 Jan. 1983, *S.A. Robertson* 3604!; Mombasa I, below Fort Jesus, Aug. 1961, *Tweedie* 2202; Kilifi District: 6.4 km. N. of Malindi, 12 Nov. 1961, *Polhill & Paulo* 743!

TANZANIA. Lushoto District: Lwengera valley, 6.4 km. ENE. of Korogwe, 27 June 1953, *Drummond & Hemsley* 3037!; Tanga District: Sawa, 10 May 1968, *Faulkner* 4109!; Bagamoyo District: Chambezi Agricultural Station, 14 June 1967, *S.A. Robertson* 729!

DISTR. K 1, 4 (rare), 7; T 2 (see note), 3, 6, 8, Somalia; also cultivated in Hawaii

HAB. Thickets in scattered tree grassland, *Acacia*-desert grass formations, lowland forest and forest edges, bushland, *Adansonia, Lannea, Markhamia, Sterculia* woodland and derived cultivations; 0–900 m. (see note)

NOTE. The T 2 record is based on *Ibrahim* 647 (Laitokitok Mt. School, 20 July 1962 at 1800 m.) No other material from such an altitude inland has been seen and it needs confirmation, particularly as one collection number is crossed out and another added.

var. **puberula** *Verdc.*, var. nov. a var. *hildebrandtii* ramulis juvenilibus, paginis inferioribus foliorum, floribus breviter pubescentibus differt. Typus: Tanzania, Handeni District, Handeni–Korogwe, *Eggeling* 6823 (K, holo.!, EA, iso.)

TANZANIA. Handeni District: Handeni–Korogwe, June 1954, *Eggeling* 6823!; Uzaramo District: Kissaki Steppe, 5 Nov. 1898, *Goetze* 77!; Kilwa District: Kingupira, 15 Apr. 1975, *Vollesen* in *M.R.C.* 2238!

DISTR. T 3, 6, 8; not known elsewhere

HAB. Little known but recorded from riverine thicket; 50–600 m.

NOTE. Lack of indumentum is one of the characteristics of this species so this unusual pubescent variant is worthy of note. *C. hildebrandtii* var. *pubescens* Mold. (see p. 100) is actually *C. rotundifolium* Oliv.

15. **C. silvanum** *Henriques* in Bol. Soc. Brot. 10: 148 (1892) as '*silvaeanum*'; Exell, Cat. Vasc. Pl. S. Tomé: 264 (1944) & in Bull. Brit. Mus. Bot. 4: 386 (1973). Type: S. Tomé, R. Contador valley, *Quintas* 1347 (COI, syn.) & R. Io Grande valley, Angolares, *Quintas* 995 (COI, syn., BM, isosyn.!) & *Quintas* 1011 (COI, syn., K, isosyn.!)

Liane often attaining the canopy and 12–18 m. long, climbing by means of persistent petiole-bases which may be slender and spine-like, retrorse and finely ridged; 1.5–2 cm. long on the long slender young shoots; main stems up to 10–15 cm. wide with fissured grey corky bark; younger shoots pale grey-brown, lenticillate; stems sometimes coiling on ground and rooting; can also be a shrub 3–3.6 m. tall. Leaves often confined to the canopy, elliptic, 2–11(–21) cm. long, 1–7(–12) cm. wide, distinctly acuminate at the apex, ± rounded at the base, ± thin, glabrous, finely reticulate; petiole 0.5–5 cm. long. Inflorescences apparently axillary or at least terminating lateral leafless branches, usually borne on main stems near the ground far below the lowest leaves or sometimes terminating leafy branchlets, 11–40 cm. long, composed of numerous 3–several-flowered dichasial cymes on 1–2 cm. long stalks; pedicels 1.5–4 mm. long; bracts narrow, 2 mm. long. Calyx finely papillate-puberulous and finely striate in dry state; tube cylindrical, 3.5–10 mm. long, with acutely triangular teeth 1–2 mm. long. Corolla fragrant, white; tube narrowly cylindrical, 0.6–1.8(–2.5) cm. long; lobes rounded-elliptic, 2.5–8 mm. long, 2–5 mm. wide. Fruit deep green, oblong, 1.3 cm. long, 1 cm. wide, not deeply lobed.

SYN. *C. preussii* Gürke var. *silvanum* (Henriques) Thomas in E.J. 68: 70 (1936) as '*silvaenum*'

var. **buchholzii** (*Gürke*) *Verdc.*, comb. et stat. nov. Type: Cameroon, Barombi Station, *Preuss* 404 (B, lecto.†)

Calyx-tube 8–10 mm. long. Corolla-tube (1.5–)1.7–2.5 cm. long; lobes up to 8 × 5 mm.

UGANDA. Toro District: Fort Portal, 20 Oct. 1906, *Bagshawe* 1269!

DISTR. U 2; Guineé to Gabon, Zaire and Angola

HAB. Forest; 1350 m.

SYN. *C. preussii* Gürke in E.J. 18: 175 (1893); Bak. in F.T.A. 5: 302 (1900). Type: Cameroon Mt., *Preuss* 1008 (B, holo.†, K, iso.!)
    C. buchholzii Gürke in E.J. 18: 176 (1893); Bak. in F.T.A. 5: 301 (1900); Thomas in E.J. 68: 69 (1936); Huber in F.W.T.A., ed. 2, 2: 443 (1963); Troupin, Fl. Rwanda 3: 272, fig. 88/1 (1985); Moldenke in Phytologia 58: 294 (1985)
    Siphonanthus costulata Hiern in Cat. Afr. Pl. Welw. 1: 843 (1900). Types: Angola, Pungo Andongo, *Welwitsch* 5679 & 5682 (BM, syn.!)

Clerodendrum kentrocaule Bak. in F.T.A. 5: 296, 515 (1900). Type: Angola, Pungo Andongo, *Welwitsch* 5682 (K, holo.!, BM, iso.!)

C. thonneri Gürke in E.J. 28: 292 (1900); Bak in F.T.A. 5: 517 (1900); Thomas in E.J. 68: 68 (1936). Type: Zaire, Boyangi, near Ndobo, *Thonner* 69 (B, holo.†)

var. **nuxioides** (*S. Moore*) *Verdc.*, comb. et stat. nov. Type: Uganda, coast of Lake Victoria, Mutunda, *Bagshawe* 579 (BM*, holo.!)

Calyx-tube ± 5 mm. long. Corolla-tube 0.6–1 cm. long, with lobes mostly at smaller end of range. Fig. 14/9–11, p. 104.

UGANDA. Bunyoro District, Nov 1935, *Eggeling* 2271!; Busoga District: Lake Victoria, Lolui I., 18 June 1953, *G.H.S. Wood* 757!; Mengo District: Mabira Forest, Mulange, Nov. 1920, *Dummer* 4433!
KENYA. N. Kavirondo District: Kakamega Forest, Kakamega–Kaimosi road, near forester's house, 15 Oct. 1953, *Drummond & Hemsley* 4760! & Kakamega Forest, Oct. 1930, *Gardner* in *F.D.* 2482! & ditto, 5 Oct. 1982, *R.M. & D. Polhill* 4856!
TANZANIA. Ngara District: Bushubi, Keza, Buseke, 20 May 1960, *Tanner* 4941! (galled) & 4942!; Buha District: Gombe Reserve, Mkenke valley, 28 Mar. 1964, *Pirozynski* 609!; Ufipa District: Sumbawanga, Chapota, 9 Mar. 1957, *Richards* 8614!
DISTR. U 2–4; K 5; T 1, 4; Zaire
HAB. *Maesopsis* lakeside and riverine forest. *Celtis*, etc., forest, thicket; 1120–1650 m.

SYN. *Siphonanthus nuxioides* S. Moore in J.L.S. 37: 197 (1905)
Clerodendrum validipes S. Moore in J.B. 54: 290 (1916). Type: Uganda, Mengo District, Gaba, *Dummer* 2642 (BM, holo.!)
C. laxicymosum De Wild. in B.J.B.B. 7: 171 (1920); Thomas in E.J. 68: 68 (1936); Moldenke in Phytologia 61: 472 (1987). Type: Zaire, Avakubi, *Bequaert* 1844 (BR, holo.!)
C. nuxioides (S. Moore) Thomas in E.J. 68: 69 (1936); F.P.N.A. 2: 144 (1947); Moldenke in Phytologia 62: 483 (1987)

NOTE. (species as a whole) Baker in F.T.A. 5: 301 (1900) and Thomas in E.J. 68: 91 (1936) suggested that *C. manettii* Vis. (*C. splendens* Manett, *non* G. Don) might be a much earlier name for this species but I have examined the type preserved at Padua and shown it to be the Australian *C. tomentosum* (Vent.) R. Br. (see Verdc. in K.B. 44: 695 (1989)). *C. silvanum* is often galled. P. Jaeger (Marcellia 39: 15–19, figs. 1–6 (1976)) mentions galling of the flowers by a *Paracopium* sp. A claim by Dummer that it was a tree to 18 m. obviously arose as a confusion between the liane and its support.

16. **C. tanganyikense** *Bak.* in K.B. 1895: 71 (1895) & in F.T.A.: 5: 298 (1900); Thomas in E.J. 68: 68 (1936); F.F.N.R.: 367 (1962). Type: Zambia, Lake Tanganyika, Urungu, Fwambo, *Carson* 52 (K, holo.!)

Shrub or small tree 0.9–4.5 m. tall, sometimes scrambling and with spinous persistent petiole-bases 1.5 cm. long on the old stems; stems pale brown to purplish brown, ridged, lenticellate, shortly pubescent and with some longer hairs. Leaves elliptic to ovate, 2–21 cm. long, 1.3–11 cm. wide, acuminate at the apex, rounded to cuneate at the base, margin entire to slightly crenate, venation sometimes impressed when dry, glabrous or sparsely pubescent on the venation beneath; petioles wrinkled and channelled, 0.2–2.5 cm. long, densely hairy above. Inflorescences borne on leafy branches, terminal, forming aggregate clusters 6(–8) cm. long, 7 cm. wide, and a few from lower axils or in some intermediates also borne on leafless lower stems; clusters much shorter than in *C. silvanum* and more pubescent to hairy with adpressed hairs or with denser spreading ± shaggy yellow-brown hairs; peduncles 2–4.5 cm. long, pedicels 4 mm. long. Calyx as in *C. silvanum* but adpressed pubescent (see note); tube cylindrical, 3–5.5 mm. long; lobes triangular, 2 mm. long. Corolla sweet-scented, white; tube 1.4–2 cm. long; lobes 4–5 mm. long, 3–4 mm. wide, ciliolate, ± puberulous on midrib outside and minutely glandular. Style exserted 8 mm. Fruit oblong-ellipsoid, 1.1–1.2 cm. long, 7–9 mm. wide; fruiting calyx fleshy, white, 8–10 mm. wide. Fig. 14/12–14, p. 104.

UGANDA. Kigezi District: Rutenga, Apr. 1948, *Purseglove* 2618! & E. shore of Lake Mutanda, Mushongero [Mushogiri], 9 Jan. 1933, *Rogers & Gardner* 305!; Ankole District: Igara, Feb. 1939, *Purseglove* 572!
TANZANIA. Mpanda District: Kungwe-Mahali Peninsula, NW. slopes of Musenabantu, 13 Aug. 1959, *Harley* 9320!; Ufipa District: ? Kalambo R. area on border, 9 Apr. 1969, *Sanane* 586!
DISTR. U 2; T 4; E. Cameroon, E. Zaire, Burundi and Rwanda
HAB. Forest patches, hillside scrub, *Brachystegia* woodland, *Protea-Pteridium-Hyparrhenia* association; 1200–2100 m.

* Thomas gives K in error; he also treats it as if described in *Clerodendrum*.

SYN. *C. lupakense* S. Moore in J.B. 57: 247 (1919). Type: Zaire, Lupaka, *Kassner* 2458, pro parte (BM, holo.!)
 C. consors S. Moore in J.B. 57: 248 (1919); Thomas in E.J. 68: 68 (1936). Type: Zaire, Lupaka, *Kassner* 2458, pro parte (BM, holo., K, iso.!)

NOTE. This species is subject to the same galling as *C. silvanum. Harley* (9320) reports a smell of chlorine when the stems are cut. *Purseglove* 2804 (Uganda, Kigezi District, Kachwekano Farm, May 1949) has terminal inflorescences to 9.5 cm., petioles to 5 cm. and is described as a scandent shrub to 4.5 m. *Harley* 9566 (Tanzania, below Kungwe Mt, head of Ntali R., 9 Aug. 1959) has the indumentum of *C. tanganyikense* but the 'axillary' inflorescences from old leafless shoots approach those of *C. silvanum* and the plant is a climbing shrub. The two are undoubtedly close and not entirely separable by habit and inflorescence. I have not considered it wise to combine them without much further study in the field. *Exell, Mendonça & Wild* 1294! (Ufipa District, by R. Kalambo, above the Kalambo Falls, 29 Mar. 1955) in *Brachystegia* woodland is a variant with spreading hairs on the calyx. *C. dubium* De Wild. (in F.R. 13: 144 (1914); Thomas in E.J. 68: 67 (1936); type: Zaire, Lubumbashi, *Bequaert* 322 bis (BR, holo.!)) and *C. bequaertii* De Wild. (in F.R. 13: 144 (1914); Thomas in E.J. 68: 67 (1936); type: Zaire, Lubumbashi, *Bequaert* 322 (BR holo.!)) are more or less the same variant of this species with a spreading pubescent calyx 6 mm. long.

17. **C. schweinfurthii** *Gürke* in E.J. 18: 177 (1893); Bak. in F.T.A. 5: 296 (1900); De Wild. in Études Fl. Bas et Moyen Congo 1: 310 (1909) & B.J.B.B. 7: 174 (1920); Thomas in E.J. 68: 70 (1936); F.P.N.A. 2: 144 (1947); T.T.C.L.: 638 (1949); Huber in F.W.T.A., ed. 2, 2: 443 (1963); Verdc. in Ent. Mon. Mag. 98: 272 (1963). Type: Sudan, Niamniam, Nabambisso, *Schweinfurth* 3021 & 3224 (on same sheet) (B, lecto.†, K, isolecto.!)

Liane or scrambling shrub with stems often straggling along the forest floor, 0.45–2 m. long, sometimes forming tangled masses; stems grey or grey-brown, corky and lenticellate, armed with retrorse spinescent old petiole-bases to 1.5 cm., pubescent and with longer hairs, later glabrous. Leaves variable, broadly elliptic, oblong, rhomboid, narrowly obovate to obovate, 5.5–29 cm. long, 5–18 cm. wide, acuminate at the apex, rounded to cuneate at the base, entire to coarsely toothed, glabrous or finely pubescent or puberulent on raised reticulate venation beneath; petiole thick, wrinkled, 0.5–5 cm. long, glabrous, pubescent or hairy. Inflorescences capitate, 5 cm. long, 7–10 cm. wide, usually borne on thick leafless stems but occasionally terminal as well, pubescent; peduncle 1–16 cm. long; bracts and bracteoles linear, 1–3 mm. long; pedicels 2 mm. long. Calyx-tube 3–6 mm. long, widened at throat, puberulous; lobes triangular, 1 mm. long with recurved tips. Corolla white, sweet-scented; tube 2.5–3.6 cm. long, widened at throat; lobes oblong-elliptic, 5–7 × 4–5 mm. Stamens and style exserted 1–2 cm. Fruiting calyx fleshy, white; fruits dark green turning black, ellipsoid, 1 cm. long, 7 mm. wide. Fig. 14/15–17, p. 104.

UGANDA. Bunyoro District: Kyangwali, Jan. 1941, *Purseglove* 1092!; Toro District: Bwamba, Mar. 1957, *Carcasson* in *Bally* 11409!; Mengo District: Busuju, Kasa Forest, 9 Mar. 1950, *Dawkins* 538!
TANZANIA. Buha District: Kakombe valley, 28 Dec. 1963, *Pirozynski* 114!; Mpanda District: Mahali Mts., Kasiha, 26 Sept. 1958, *Jefford & Newbould* 2650! & Kasoge, 19 Aug. 1959, *Harley* 9375!
DISTR. U 2, 4; T 1, 4; Sierra Leone, Ivory Coast to Cameroon, Central African Republic, Zaire, Burundi, Sudan and Angola
HAB. Streamside and swamp-forest, particularly with *Funtumia* and *Pseudospondias*, also in short grassland with bushland relicts; 760–1590(–1800) m.

SYN. *C. congense* Bak. in K.B. 1892: 127 (1892), *non* Engl., *nom. illegit.* Type: Zaire, below Stanley Pool (Pool Malebo) *Johnston* (K, holo.!)
 C. bakeri Gürke in E.J. 18: 175 (1893); Bak. in F.T.A. 5: 296 (1900); De Wild. in B.J.B.B. 7: 165 (1920); F.W.T.A. 2: 273 (1931). Type as for *C. congense* Bak.
 [*C. tanganyikense* sensu Bak. in F.T.A. 5: 299 (1900), quoad *Scott Elliott* 8228, *non* Bak. sensu stricto]
 C. longitubum De Wild. & Th. Dur. in Comptes Rendus Soc. Bot. Belg. 39: 74 (1900); Bak. in F.T.A. 5: 517 (1900). Type: Zaire, Kisantu, *Gillet* [707] (BR, holo.!)
 C. gossweileri Exell in J.B. 68, Suppl. Gamopet.: 142 (1930). Type: Angola, Cuanza Norte, Duque de Braganza, Rianzondo, near R. Lucala, *Gossweiler* 8830 (BM, holo.!)
 C. schweinfurthii Gürke var. *bakeri* (Gürke) Thomas in E.J. 68: 71 (1936); T.T.C.L.: 638 (1949); Everett, Encycl. Hort. 3: 794, figs. (1981)
 C. schweinfurthii Gürke var. *longitubum* (De Wild. & Th. Dur.) Thomas in E.J. 68: 71 (1936); T.T.C.L.: 638 (1949)
 C. schweinfurthii Gürke var. *conradsii* Thomas as "*conradii*" in E.J. 68: 71 (1936); T.T.C.L.: 638 (1949). Type: Musoma District, Ukerewe I., *Conrads* 252 (B, holo. †), *nom. invalid.*

NOTE. The plant is frequently galled by a Tingid bug, *Paracopium glabricorne* Mont. which forms green hollow ovoid ± succulent bodies 1.5 cm. long, 1.1 cm. wide, sitting in the calyces and easily passed over as fruits if one were not familiar with the shape of *Clerodendrum* fruits; these galls are filled with nymphs and adults of the bug (e.g. Buha District, Kasakela Reserve, Gombe Stream, 18

Nov. 1962, *Verdcourt* 3360). Specimens of this species with terminal inflorescences can be confused with *C. tanganyikense* but the calyx is different. The various varieties maintained by Thomas on indumentum characters do not seem very distinctive. His mention of *Schlieben* 4277 (Tanzania, Buha District, Kasulu, 20 Oct. 1933) as the type of var. *longitubum* is of course an error.

18. **C. robecchii** *Chiov.* in Bull. Soc. Bot. Ital. 1917: 53 (1917) & in L'Agric. Colon. 20: 103 (1926); Thomas in E.J. 68: 91 (1936). Types: Ethiopia, Ogaden, without locality, *Robecchi-Bricchetti* 569, 638 (both FT, syn.) & between Imi and Mt. Audo, *Ruspoli & Riva* 1046 (FT, syn.)

Intricately branched shrub 1–3 m. tall, with purplish brown, blackish or yellow-brown stems with pale lenticels; internodes usually short and lateral branches sometimes short and spine-like. Leaves in whorls of 3, narrowly elliptic, oblong-elliptic, ovate, rhomboid or ± triangular, 0.3–1.8 cm. long, 1.5–10 mm. wide (see note), rounded to shortly acute at the apex, attenuate at the base, entire to 1-toothed on either side, rarely distinctly toothed or occasionally triangular and 3-lobed, glabrous (Somalia) to densely pubescent; petiole 1–5 mm. long. Inflorescences few or numerous and scattered along lateral twigs but each is actually terminal on branchlets reduced to nodules, ± 7-flowered, sessile; pedicels 2 mm. long, pubescent and with larger hairs. Calyx cupular, puberulous; tube 1.5 mm. long; teeth shortly triangular, 0.5 mm. long. Corolla sweetly scented, white to cream or pale yellow; tube 1.1–1.3 cm. long, glandular, pubescent outside, slightly widened beneath; limb almost bilabiate with 4 subequal lobes ± 3 × 1.8 mm. and a larger obovate-oblong lobe 4–6 mm. long. Filaments ± 2.5 cm. long, exserted 1.5 cm. Ovary blackish, 1 mm. long, glabrous but glandular; style 2.5 cm. long with bifid apex. Fruit yellow turning black, 7–8 mm. long, 6–6.5 mm. wide, with diverging lobes, glabrous or finely pubescent; fruiting calyx 5–6 mm. wide.

KENYA. Northern Frontier Province: 50 km. W. of Ramu, Lagh Ola, 25 Jan. 1972, *Bally & Smith* 14949!
DISTR. **K** 1; Somalia, Ethiopia
HAB. *Acacia-Commiphora* scrub with *Boscia* and *Anisotes*; 580 m.

SYN. *C. robecchii* Chiov. var. *macrophylla* Chiov. in Bull. Soc. Bot. Ital. 1917: 54 (1917); E.P.A.: 801 (1962). Types: Ethiopia, Ogaden, *Robecchi-Bricchetti* 572, 637 (both FT, syn.) & Somalia, Merehan, *Robecchi-Bricchetti* 558 & Webi, *Robecchi-Bricchetti* 608 (both FT, syn.)
C. somalense Chiov. in L'Agric. Colon. 20: 103 (1926) & Fl. Somala 2: 364, fig. 207 (1932). Type: Somalia, Transjuba, *Gorini* 70 (FT, holo.)
C. microphyllum Thomas in E.J. 68: 71, 103 (1936); Burger, Fam. Fl. Pl. Eth., fig. 60/4 (1967); Moldenke in Phytologia 62: 195 (1987). Type: Somalia, Daodd, *Riva* 1067 coll. *Ruspoli* 964* (B?, holo.†)

NOTE. Thomas emphasised how bad Chiovenda's type material was and that his 1932 figure of *C. somalense* did not match the type or original description. Nevertheless the vernacular name 'dhumot' given for *C. robecchii* is the same as that for *C. microphyllum* and it seems likely the two are the same but further work is needed in Somalia to elucidate the variation. Chiovenda gives the measurements of the leaves of his var. *macrophylla* as 3–6 × 2.5–6 cm.

19. **C. melanocrater** *Gürke* in E.J. 18: 180 (1893) & in P.O.A. C: 341 (1895); Bak. in F.T.A. 5: 299 (1900); Thomas in E.J. 68: 72 (1936); F.P.N.A. 2: 144 (1947); T.T.C.L.: 637 (1949); F.P.S. 3: 194 (1956); Huber in F.W.T.A., ed. 2, 2: 444 (1963); Synnott, Checklist Budongo Forest: 71 (1985); Troupin, Fl. Rwanda 3: 274 (1985). Type: Tanzania, Bukoba, *Stuhlmann* 3322 (B, lecto.†)

Climber or ± scandent shrub 1.8–4.5 m. tall or tall shrub with long thin scrambling branches to 7 m.; stems brown to black, square with keeled pale angles and ridged when dry, ± bifariously adpressed pubescent when young but soon glabrous, sometimes with persistent recurved petiole-bases. Leaves turning quite black on drying, ovate to elliptic or oblong-elliptic, 1.5–16 cm. long, 0.7–9.5 cm. wide, acuminate and mucronate at the apex or tip ± rounded but mucronate, rounded to cuneate at the base, thin, ± pubescent above, glabrous beneath; petiole 0.6–6 cm. long. Inflorescences long terminal thyrses 9–25 cm. long, 8–15 cm. wide, made up of up to 20 lateral many-flowered dichasial cymes, on stalks to 4 cm. long, the lowest from leaf-axils but leaves reduced above to bracts but still with lamina up to 6 × 2 mm. and petiole 4 mm. long; pedicels 0.5–2.3 cm. long; all axes adpressed grey-pubescent or pale ferruginous. Calyx-tube white or mauve-white; tube

* Thomas gives the date as 20 Jan. 1893 but Gillett states that between 15 and 25 Jan. 1893 Ruspoli & Riva were in Ethiopia, on the Webbi Schebelli near Karanle.

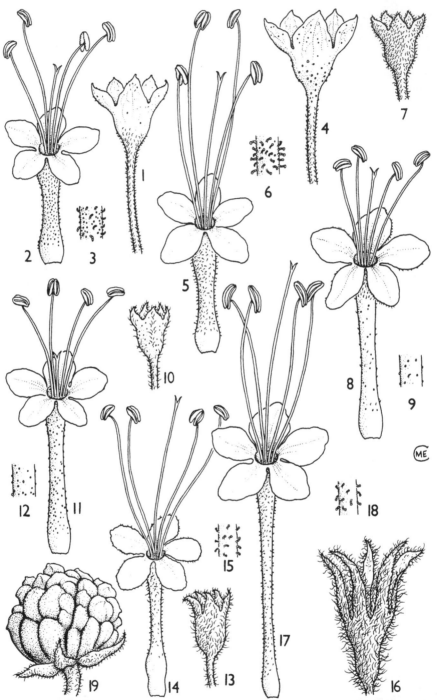

FIG. 16. *CLERODENDRUM MELANOCRATER* — **1**, calyx, × 4; **2**, corolla, × 4; **3**, part of corolla-tube, × 6. *C. VOLUBILE* — **4**, calyx, × 4; **5**, corolla, × 4; **6**, part of corolla-tube, × 6. *C. JOHNSTONII* subsp. *JOHNSTONII* — **7**, calyx, × 4; **8**, corolla, × 4; **9**, part of corolla-tube, × 6. *C. GLABRUM* — **10**, calyx, × 4; **11**, corolla, × 4; **12**, part of corolla-tube, × 6. *C. ERIOPHYLLUM* — **13**, calyx, × 4; **14**, corolla, × 4; **15**, part of corolla-tube, × 6. *C. ACERBIANUM* — **16**, calyx, × 4; **17**, corolla, × 4; **18**, part of corolla-tube, × 6; **19**, fruit, × 2. 1–3, from *Drummond & Hemsley* 4578; 4–6, from *Newbould & Harley* 4263; 7–9, from *A.S. Thomas* 4220; 10–12, from *Tweedie* 961; 13–15, from *Drummond & Hemsley* 2994; 16–18, from *Werner* 997; 19, from *Kibuwa* 2483. Drawn by Mrs. M.E. Church.

cupular, 2–4 mm. long; limb divided into ovate acute lobes 1.5–2 × 2 mm. and with a basal tubular part of up to 1 mm. long. Corolla cream turning yellowish, ink-blue when bruised, drying black and turning black on the plant, finely gland-dotted outside; tube 7 mm. long; lobes ovate, 3 mm. long, 2 mm. wide. Anthers and style exserted 8 mm.; filaments and style green; anthers deep blue. Fruit black, ellipsoid to subglobose, 5–10 mm. long, 5–8 mm. wide, 1–2(–3)-lobed. Fig. 16/1–3.

UGANDA. Bunyoro District: Budongo Forest, Oct. 1932, *Harris* 145 in *F.D.* 1109!; Masaka District: 6.4 km. SW. of Katera, Malabigambo Forest, 2 Oct. 1953, *Drummond & Hemsley* 4578!; Mengo District: Entebbe, lake-shore, Sept. 1926, *Maitland* 142!
KENYA. N. Kavirondo District: Malaba Forest, Kabras, Sept. 1958, *Tweedie* 1706! & Aug. 1965, *Tweedie* 3096! & Nov. 1965, *Tweedie* 3214!; Kakamega Forest Station, 17 Sept. 1949, *Maas Gesteranus* 6264! & 9 Dec. 1956, *Verdcourt* 1668!
TANZANIA. Bukoba, Feb. 1892, *Stuhlmann* 3322 & 20 Mar. 1892, *Stuhlmann* 3650 & 7 Apr. 1892, *Stuhlmann* 3891 & Minziro Forest, Sept. 1958, *Procter* 999!
DISTR. U 2, 4; K 5; T 1; Cameroon, Bioko [Fernando Po], E. Zaire and Rwanda
HAB. Swamp-forest, forest edges, thicket edges, often on termite mounds and rock outcrops; 990–2100 m.

SYN. *Premna melanophylla* S. Moore in J.L.S. 37: 196 (1905). Type: Uganda, Masaka District, Bukora [Bakora], *Bagshawe* 141A (BM, holo.!)
 Clerodendrum melanophyllum (S. Moore) S. Moore in J.B. 45: 93 (1907)
 C. seretii De Wild. in Ann. Mus. Congo Bot., sér. 5, 3: 256 (1909). Type: Zaire, Injolo, *Seret* 996 (BR, holo.)
NOTE. Thomas omitted *C. melanophyllum* from his revision since it had been overlooked by the compilers of the Index Kewensis.

20. **C. volubile** *P. Beauv.*, Fl. Owar. 1: 52, t. 32 (1805); Bak. in F.T.A. 5: 297 (1900); Thomas in E.J. 68: 72 (1936); Huber in F.W.T.A., ed. 2, 2: 444 (1963). Type: Nigeria, Oware, *Palisot de Beauvois* (G, holo., P-JU, iso.)

Twining or less often ± erect shrub (even once described as a tree by a reliable collector in W. Africa) 1–4.5 m. long or tall; stems brown, terete, ridged, glabrous, climbing by means of persistent petiole-bases, these sometimes widened at the apex. Leaves ± oblong, 1.5–15.5 cm. long, 0.6–6 cm. wide, long-acuminate at the apex, rounded to ± cuneate at the base, entire or very coarsely crenate or undulate at the apex, not drying black, glabrous; petiole 0.3–2(–3) cm. long. Inflorescences terminal on lateral branches and probably main branches as well, corymbose, including 2 laterals from first 2 leafy nodes, 6–9(–12) cm. long, 7–10.5 cm. wide; axes very slender, very finely pubescent with short curled hairs; peduncles and secondary axes 0.5–7.5 cm. long; pedicels 0.2–1.5 cm. long; bracts linear to lanceolate, 3–10 mm. long. Calyx-tube subovoid, 2 mm. long, the limb whitish yellow or green, ± funnel-shaped or rather flat, spreading, 5–7 mm. wide, divided into broadly triangular teeth, 1–2 mm. long, 2 mm. wide, glabrous. Corolla yellow-green; tube 5–7 mm. long, shortly glandular-pubescent; lobes oblong-ovate, 2–4 mm. long, 1–2 mm. wide. Stamens and style exserted ± 1.4 cm. Fruit subglobose, 8 mm. long, 8–10 mm. wide, 2–4-lobed; calycine cup ± 9 mm. wide, ridged. Fig. 16/4–6.

TANZANIA. Buha District: 16 km. N. of Kigoma, Kakombe, 7 July 1959, *Newbould & Harley* 4263!; Kigoma District: Gombe Stream National Park, Kahama valley, 25 Aug. 1969, *Clutton-Brock* 235! & Kigoma, July 1963, *Azuma* 622!; Mpanda District: Kungwe-Mahali Peninsula, 19.2 km. N. of Kasoge, Belengi, 2 Aug. 1959, *Harley* 9131!
DISTR. T 4; Senegal to Nigeria, Cameroon, Rio Muni, Gabon, Zaire and Angola
HAB. Gallery forest; 780–900 m.

SYN. *C. multiflorum* G. Don in Edinb. Phil. Journ. 11: 350 (1824). Type: Sierra Leone, Leicester Mt., G. *Don* (K, iso.!)

21. **C. myrianthum** *Mildbr.* in N.B.G.B. 11: 677 (1932); Thomas in E.J. 68: 73 (1936); T.T.C.L.: 638 (1949); Moldenke in Phytologia 62: 328 (1987), pro parte. Type: Tanzania, Ulanga District, near Mahenge, Liando [Liondo], *Schlieben* 2143 (B, holo. †, BM, BR, HBG, iso.!)

Liane reaching the crowns of 15 m. tall trees; young floriferous shoots blackish-violet, glabrous, covered with pale lenticels. Leaves elliptic, 15–23 cm. long, 8–13 cm. wide, fairly long-acuminate to ± caudate-acuminate at the apex, rounded at the base, drying dark olive above, paler and punctate beneath, glabrous or minutely puberulous on venation beneath; petiole thickened above the base, geniculate, persistent, glabrous, 3–5 cm. long;

venation impressed above, prominent beneath. Flowers rather small and very numerous in a large pyramidal or ovoid panicle, 24 cm. long, 12 cm. wide, or subcorymbose; peduncle ± obsolete; secondary axes 3–5 cm. long, pedicels or apparent pedicels 2–7 mm. long; bracts filiform, 3 mm. long; youngest parts puberulous. Calyx rotate-campanulate, glabrous; tube 2.5 mm. long; lobes ovate-triangular, 2–2.5 mm. long, recurved-spreading, acuminate at the apex. Corolla white, stinking (*fide* Schlieben), glabrous; tube ± 1–2 cm. long, slightly narrowed to the throat; lobes elliptic-obovate, 3–4 mm. long, 2.5 mm. wide. Stamens glabrous, exserted 1.5–2.5 cm. Ovary depressed globose, glabrous, 4-sulcate; style glabrous, 3 cm. long. Fruit not known.

TANZANIA.   Ulanga District: near Mahenge, Liando [Liondo], 30 Apr. 1932, *Schlieben* 2143!
DISTR.   **T** 6; not known elsewhere
HAB.   Streamlet valley with rocks and clumps of trees; 900 m.

NOTE.   *Peter* 311 (Lushoto District, Amani to Sigi Valley) cited by Moldenke is sterile and perhaps Combretaceae.

22. **C. formicarum** *Gürke* in E.J. 18: 179 (1893)*; Henriques in Bol. Soc. Brot. 16: 69 (1899); Bak. in F.T.A. 5: 298, 516 (1900); Thonner, Fl. Pl. Afr., t. 133 (1915); De Wild. in B.J.B.B. 7: 167 (1920); F.W.T.A. 2: 273 (1931); Thomas in E.J. 18: 74 (1936); T.T.C.L.: 637 (1949); F.P.S. 194 (1956); Huber in F.W.T.A., ed. 2, 2: 444 (1963); Troupin, Fl. Rwanda 3: 274, fig. 88.2 (1985). Type: Zaire, Khor Kassumbo, R. Kapili [Ghasal-Quellgebiet am Kassumbo in Monbuttu], *Schweinfurth* 3641 (B, lecto.†, K, isolecto.!)

Climber 5–6 m. long, shrub 1.5–4.5 m. tall or small tree with spreading bushy crown to 8 m., rarely (Zambia) a herb 45 cm. tall; stems hollow, purplish brown with 5–8 pale ridges, shortly pubescent when young with short curled hairs but later glabrous; lateral branches with series of bract-like organs at junction with main stems, probably abortive inflorescences; stems usually spiny with reflexed persistent petiole-bases 1–1.8 cm. long (or ?reduced branchlets); sometimes with ant habitations (fide Thonner). Leaves pale green, in whorls of 3–4, oblong, elliptic or ovate-oblong, 1.2–9.5 cm. long, 0.8–6 cm. wide, shortly to ± long-acuminate at the apex, broadly cuneate at the base, entire, glabrous or pubescent on nerves beneath, distinctly 3-nerved from the base; petiole 0.3–2.5 cm. long, often wrinkled. Inflorescences ± 4 cm. long, ± 7 cm. wide, subcorymbose or laxly globose-compound, made up of small dichasial cymes ± 1 cm. long, 2 cm. wide; secondary peduncles up to 3 cm. long; pedicels 0.5–2 mm. long; axes densely shortly pubescent as stem. Calyx pubescent with similar hairs to stem; tube ovoid, 2.5 mm. long; lobes triangular, 1.5–2 mm. long, 1.2 mm. wide. Corolla greenish cream or creamy white, glabrous to pubescent outside; tube (3–)6–8 mm. long; lobes oblong, 2–4 mm. long, 1.5 mm. wide. Anthers dark blue together with style exserted 5 mm. Fruit black with red flesh, 9 mm. long, 7 mm. wide; fruiting calyx 9 mm. wide. Fig. 17.

UGANDA.   NW. Kigezi, Sept. 1947, *Dale* U509!; Masaka District: NW. side of Lake Nabugabo, 9 Oct. 1953, *Drummond & Hemsley* 4701!; Mengo District: Kyagwe [Chagwe], Dec. 1905, *E. Brown* 377!
KENYA.   N. Kavirondo District: Kakamega, Lubiri village, 12 July 1960, *Paulo* 536! & 537! & Kakamega Forest, near Forest Station, 11 Apr. 1973, *Hansen* 904! & Kakamega, May 1944, *Carroll* 17!
TANZANIA.   Lushoto District: Derema bridge, 3 Aug. 1945, *Greenway* 7534!; Kilosa District: Ukaguru Mts., track parallel to road from Lake Marangu westwards, 1 Feb. 1976, *Cribb et al.* 10506!; Morogoro District: near Ruvu, June 1930, *Haarer* 1881!
DISTR.   **U** 2, 4; **K** 5; **T** 1, 3, 4, 6; Guineé to Angola, Zaire, Burundi, Rwanda, Egypt (*fide* F.W.T.A.), Sudan and Zambia
HAB.   Wet evergreen forest with *Celtis* etc., lakeside forest fringes, thorn scrub; 360–1800 m.

SYN.   *C. triplinerve* Rolfe in Bol. Soc. Brot. 11: 87 (?1894) [between 9.12.1893 and 4.8.1894]; F.F.N.R.: 367 (1962). Types: Angola, Malange, S. *Marques* & Golungo Alto, Queta Mts., *Welwitsch* 5661 & 5622 (K, syn.)
     Siphonanthus formicarum (Gürke) Hiern in Cat. Afr. Pl. Welw. 1: 843 (1900)
     Clerodendrum oreadum S. Moore in J.B. 45: 93 (1907); Thomas in E.J. 68: 19 (1936); Moldenke in Phytologia 63: 49 (1987). Type: Uganda, Toro District, Mpanga, *Bagshawe* 1075 & 1123 (BM, syn.!)
     C. formicarum (Gürke) Hiern var. *sulcatum* Thomas in E.J. 68: 74 (1936); F.P.N.A. 2: 146 (1947). Type: Tanzania, Morogoro District, Uluguru Mts., *Schlieben* 3217 (B, holo.†, K, iso.!), *nom. invalid.*

* Actually 22.12.1893 so the balance of probability is that it predates *C. triplinerve* Rolfe published between 9.12.1893 and 4.8.1894; Hiern gives a definite 1894 for this. I have taken Thomas's indication of type as a lectotypification.

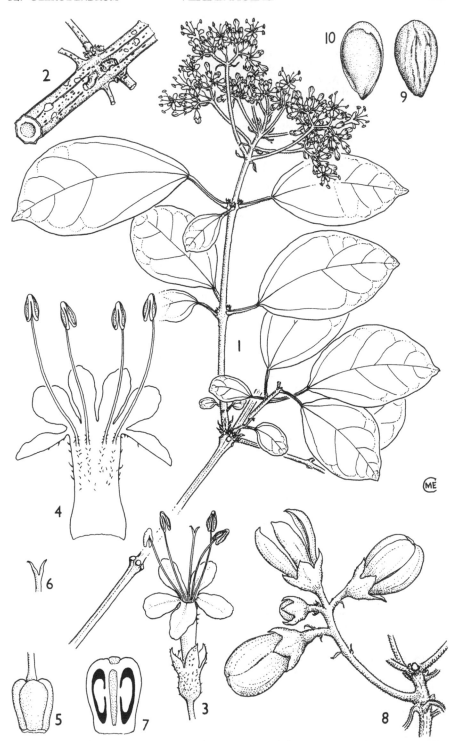

FIG. 17. *CLERODENDRUM FORMICARUM* — **1**, habit, × ⅖; **2**, part of branch, showing hollow centre and persistent petioles, × ⅖; **3**, flower, × 4; **4**, corolla, opened out, × 6; **5**, ovary × 12; **6**, stigma, × 12; **7**, longitudinal section of ovary × 16; **8**, part of fruiting branchlet, × 2; **9, 10**, 2 views of pyrene, × 2. 1, from *Drummond & Hemsley* 4701; 2, from *Hansen* 904; 3–7, from *Paulo* 537; 8–10, from *Snowden* 38. Drawn by Mrs. M.E. Church.

NOTE. Galled corollas have quite campanulate tubes and ovate lobes.

23. **C. johnstonii** *Oliv.* in Trans. Linn. Soc., ser II, 2: 346 (1887); Bak. in F.T.A.: 300 (1900); Thomas in E.J. 68: 75 (1936); F.P.N.A. 2: 142, t. 14 (1947); T.T.C.L.: 638 (1949); K.T.S.: 584 (1961); E.P.A.: 799 (1962); F.F.N.R.: 367, fig. 65 (1962); Troupin, Fl. Rwanda 3: 274, fig. 86.2 (1985); Moldenke in Phytologia 61: 332 (1986). Type: Tanzania, Kilimanjaro at 1500 m., *Johnston* (K, holo.!, BM, iso.!)

Variable in habit; shrub or small tree up to 3–4.5 m. with spreading branches or scrambling shrub 2.5–10.5 m.; older stems square and with stout spine-like curved persistent petiole-bases; younger densely ± adpressed velvety with fawn indumentum or with more orange or ferruginous to bright red-ferruginous ± shaggy hairs. Leaves opposite or ternate (fide F.T.A. & K.T.S.), elliptic, ovate or oblong-ovate, 6.5–19.5 cm. long, 3.5–14.5 cm. wide, acuminate at the apex, broadly cuneate to cordate at the base, discolorous, sparsely pubescent above, pubescent to ± velvety brownish grey pubescent beneath, entire or obscurely crenate; petiole 1.5–10.5 cm. long, jointed above the base. Inflorescences axillary and terminal, forming large complicated dense corymbose panicles of cymes 7–20 cm. wide, scented; peduncle 6–12 cm. long; pedicels 0–3 mm. long; main bracts linear-spathulate, 0.5–1 cm. long, densely fawn-velvety. Calyx ovoid; tube 2 mm. long; lobes triangular, 1.5–2 mm. long. Buds with round limb, pubescent outside. Corolla white or yellowish white; tube (3.5–)9–10 mm. long, ± glabrous; lobes elliptic-oblong, 3.5–4 mm. long, 3–4 mm. wide, ciliate and pubescent outside. Style and stamens exserted 6 mm.; anthers yellowish or blackish. Fruit ovoid, 10 mm. long, 9 mm. wide, 4-lobed, sitting in a cup-like calyx.

subsp. **johnstonii**

Leaves mostly velvety beneath. Corolla-tube (6–)9–10 mm. long. Fig. 16/7–9, p. 114.

UGANDA. Kigezi District: Kabale, 6 July 1945, *A.S. Thomas* 4307!; Mbale District: Bugisu [Bugishu], Bubungi, July 1926, *Maitland* 1241! & Mt. Elgon, Bulago, 19 July 1917, *Snowden* 515!
KENYA. W. Suk District: N. Cherangani Hills, Aug. 1971, *Tweedie* 4085!; S. Nyeri District: Kiandongoro, 23 Sept. 1963, *Verdcourt* 3780!; Teita District: Ngerenyi, below Verbi's House, 16 Sept. 1953, *Drummond & Hemsley* 4366!
TANZANIA. Masai District: Ngorongoro Crater lip, 13 Aug. 1952, *Tanner* 914!; Lushoto District: W. Usambaras, Lushoto–Shume road, 1.6 km. E. of Sungwi [Sungwe], 4 June 1953, *Drummond & Hemsley* 2838!; Mbeya District: Poroto Mts., Kikondo Camp, 20 Jan. 1961, *Richards* 13965!
DISTR. U 2, 3; K 1–5, ?6, 7; T 2–4, 6–8; Zaire, Rwanda, Burundi, ? Ethiopia, Malawi and NE. Zambia
HAB. Submontane and montane forest and derived thicket, scrub, cultivations and wasteland; bamboo forest/grassland margins; 1200–2550 m.

SYN. *C. murigono* Chiov., Racc. Bot. Miss. Consol. Kenya: 99 (1935). Types: Kenya, Fort Hall District, Tuso [Tusu] Forest, *Balbo* 182 (TOM, syn.!) & NE. Meru, *Balbo* 851 (TOM, syn.)
C. johnstonii Oliv. var. *rubrum* Thomas in E.J. 68: 75 (1936); T.T.C.L.: 638 (1949); Moldenke in Phytologia 61: 336 (1986), *nom. invalid*. Type: Tanzania, Morogoro District, Maskati, *Schlieben* 4130 (B, holo.†, EA, iso.!)

NOTE. Material from the W. Usambaras, Ulugurus and other areas in T6 and 7 often has rather markedly longer more reddish ferruginous indumentum and clearly accords with var. *rubrum* Thomas. There are, however, many intermediates particularly in T 2, Mbulu District, etc. and I have not formally recognised it but phytogeographers may take note of a weak subspecific tendency. Material in Uganda, Kenya and most of NW. Tanzania (T2) has short ± fawn-coloured indumentum; in T4 (Buha) the indumentum is fawn-coloured but long and in T8 red but short. *Fyffe* (Uganda, Toro District, Bwamba) is said to be from 2300 ft. (906 m.) but this seems improbably low. Galling can cause what looks like a berry-like fruit 2 cm. long.

subsp. **marsabitense** *Verdc.*, subsp. nov. foliis subtus sparse pubescentibus corollae tubo 3.5–5 mm. longo differt. Typus: Kenya, Northern Frontier Province, Marsabit, *Gillett* 15100 (K, holo.!)

Leaves only sparsely pubescent beneath. Corolla-tube 3.5–5 mm. long.

KENYA. Northern Frontier Province: Marsabit, 14 Feb. 1953, *Gillett* 15100!
DISTR. K 1; not known elsewhere
HAB. Evergreen mist-forest; 1350 m.

NOTE. Material from Tanzania, Mbeya District of subsp. *johnstonii* can have glabrescent leaves.
The sterile specimen *Peter* 1900 (O I 47)! (Tanzania, Arusha District, Mt. Meru, near Arusha forester's house, 26 Feb. 1900, 1800 m.) is a liane with square stems ± 7 mm. wide, with grey-brown bark, buff pubescent when young, ± persistent on old stems, armed with curved spine-like petiole-base remnants up to 2.5 cm. long; innovations on young shoots densely buff velvety; leaves elliptic, 17 cm. long, 9.5 cm. wide, acuminate at the apex, ± rounded to broadly cuneate at the base,

very thin, slightly pubescent above and more densely on the venation beneath; flowers and fruits unknown. Moldenke had named this *C. swynnertonii* S. Moore but it is not that nor does any member of sect. *Capitata* occur in the area. The suggestion of E.W. (?) (annotation on sheet) that it is *C. johnstonii* is likely despite the non-cordate thin glabrous leaves; it could be a young shoot arising from very old main stems.

24. **C. glabrum** E. *Mey.*, Comm. Pl. Afr. Austr.: 273 (1838); Schauer in DC., Prodr. 11: 661 (1847); Bak. in F.T.A. 5: 297, 515 (1900); Pearson in Fl. Cap. 5(1): 219 (1901); Thomas in E.J. 68: 76 (1936); U.O.P.Z.: 198 (1949); T.T.C.L. 637 (1949); K.T.S.: 584 (1961); F.F.N.R.: 367 (1962); Huber in F.W.T.A., ed. 2: 444 (1963); Palmer & Pitman, Trees S. Afr. 3: 1963, figs. p. 1964 (1973); Coates Palgrave, Trees of S. Afr.: 814, fig. 275 (1977); Moldenke in Rev. Fl. Ceylon 4: 457 (1983) & in Phytologia 59: 486 (1986). Type: South Africa, Cape Province, R. Basche, *Drège** (B, holo.†, K, iso.!)

Shrub or small tree 0.9–4.5(–12 *fide* Palmer & Pitman) m. tall, with rough grey-brown bark. Stems erect, lenticellate, brown when young, eventually grey, glabrous or imperceptibly pubescent or rarely (e.g. in T 6) plant pubescent all over. Leaves slightly fleshy (?), with pungent odour, opposite or in whorls of 3, elliptic or elliptic-ovate, 1.5–8.5(–11) cm. long, 0.8–6 cm. wide, obtuse or acute at the apex, cuneate at the base, glabrous or youngest with midrib slightly adpressed pubescent, very densely punctate; petiole 1–1.5(–2.5) cm. long. Inflorescences terminal and with some elements from upper axils, congested but not forming dense heads, 2–9.5 cm. wide; axes ± pubescent; pedicels or apparent pedicels 2–8 mm. long; bracts and bracteoles linear, 1.5–4 mm. long. Calyx glabrescent to ± densely pubescent; tube 2.2 mm. long; lobes triangular to narrowly triangular but not linear, 1–1.5 mm. long. Corolla white or greenish, sweet-scented (or unpleasant according to some collectors), ± regular; tube (4–)6–8 mm. long with sessile glands or ± glabrescent; lobes 3–4 mm. long, 2–3 mm. wide. Stamens mauve or lilac, exserted 5–7 mm. Fruit cream, irregularly globose, 6–10 mm. diameter, strongly wrinkled and ± lobed in dry state, sitting in ± flat strongly ribbed lobed calyx 8 mm. wide. Fig. 16/10–12, p. 114.

var. **glabrum**

Calyx-lobes well developed.

KENYA. Mombasa District: Nyali Beach, 29 May 1934, *Napier* 3303 in *C.M.* 6237!; Kilifi District: Malindi, June 1962, *Tweedie* 2373! & 2374!; Lamu District: 88 km. NE. of Lamu, Kiunga, 9 Aug. 1961, *Gillespie* 174!
TANZANIA. Tanga District: Sawa, 6 June 1955, *Faulkner* 1690! & 29 Aug. 1966, *Faulkner* 3858!; Uzaramo District: 9.6 km. S. of Dar es Salaam, Mjimwema, May 1966, *Procter* 3336!; Mikindani District: Mtwara–Mikindani road, 13 Mar. 1963, *Richards* 17869!; Zanzibar I., Marie Louise road, 7 May 1950, *R.O. Williams* 34!
DISTR. K7; T3, 6, 8; Z; P; Malawi, Mozambique, Zambia, Angola, Botswana, Namibia, South Africa, Seychelles, Comoro Is. and Aldabra; introduced into Sierra Leone
HAB. Littoral sand-dune associations, coastal bushland, evergreen thicket on coral cliffs and rag, grassland; ± 0–50(–480) m.
SYN. *C. glabrum* E. Mey. var. *angustifolium* E. Mey., Comm. Pl. Afr. Austr.: 273 (1838); Moldenke in Phytologia 59: 496 (1986). Type: South Africa, Cape Province, Kei, *Drège* (B, holo.†)
   C. ovale Klotzsch in Peters, Reise Mossamb., Bot. 1: 257 (1861). Type: Mozambique, Rios de Sena, *Peters* (B, holo.†)
   C. ovalifolium Engl., P.O.A. A: 124 (1895), sphalm.
   Siphonanthus glabra (E. Mey.) Hiern in Cat. Afr. Pl. Welw. 1: 842 (1900)
   ?*Clerodendrum glabratum* Gürke in E.J. 28: 295 (1900); Bak. in F.T.A. 5: 516 (1900); Thomas in E.J. 68: 77 (1936); T.T.C.L.: 637 (1949); Moldenke in Phytologia 59: 485 (1986). Type: Tanzania, Uzaramo District, just N. of Dar es Salaam, Kunduchi, *Stuhlmann* 7996 (B, holo.†)
NOTE. *Rawlins* 24! (Kenya, Lamu District, Kui I., June 1956) has much reduced calyx-lobes but is otherwise exceedingly similar and I think no more than a form. *Schlieben* 2635! (Tanzania, Rufiji District, Mafia I., Chole I., 4 Aug. 1932), cited by Thomas as *C. glabrum* and by Moldenke as var. *angustifolium*, has narrower calyx-lobes. I really do not see that *C. glabratum* can be anything but *C. glabrum* but have seen no cited material.
   The Aldabra material with less developed calyx-lobes has been referred to var. *minutiflorum* (Bak.) Fosb. in K.B. 33: 143 (1978) & Fosb. & Renvoize, Fl. Aldabra: 220, fig. 35/3, 4 (1980)

---

* Thomas gives the type as Key (i.e. Kei), *Drège* 3485. This is probably type material of Meyer's var. *angustifolium*.

25. **C. eriophyllum** *Gürke* in E.J. 18: 178 (1893); Bak. in F.T.A. 5: 299 (1900); Thomas in E.J. 68: 77 (1936); T.T.C.L.: 637 (1949); K.T.S.: 584 (1961); Vollesen in Opera Bot. 59: 82 (1980), pro parte. Type: Tanzania, without locality, *Fischer* ser. 1, 331 (B, holo.†)

Shrub or small tree 1–6 m. tall, densely branched and with ± rounded crown; bark very pale brown, corky, irregularly ridged; branchlets finely densely greyish pubescent or velvety; brachyblasts often present in upper axils. Leaves aromatic, opposite or in whorls of 3, ovate, elliptic, oblong or less often ± lanceolate, (1–)2–7(–12) cm. long, (0.5–)1–5(–7) cm. wide, acute or distinctly acuminate, rarely narrowly rounded at the apex, rounded to broadly cuneate at the base, pubescent above, more densely so beneath, ± discolorous, very densely punctate; petiole 0.6–2 cm. long. Inflorescences very variable, ± lax, 4–6 cm. long, with up to 10 component cymes or very condensed heads 2–4.5 cm. diameter; bracts and bracteoles linear-lanceolate, 3–5 mm. long; peduncle 0–7 cm. long; secondary branches up to 1.3 cm. long; pedicels 2–4 mm. long. Calyx densely grey-pubescent; tube 2–2.5 mm. long; lobes linear-lanceolate from a ± triangular base, 2–3 mm. long, up to 1 mm. wide. Corolla sweet scented, white or greenish cream; tube 0.8–1.7 cm. long, sparsely pubescent outside and with sessile or shortly stalked glands; lobes oblong, 3.5 mm. long, 1.5 mm. wide. Style tinged purplish, together with stamens exserted 8 mm. Ripe fruit not seen. Fig. 16/13–15, p. 114.

KENYA. Machakos District: km. 152 on Nairobi–Mombasa road, between Simba and Kiboko, 9 Jan. 1964, *Verdcourt* 3857!; Masai District: between Namanga and Bissel, km. 28.8 on Nairobi road, 31 Jan. 1963, *Greenway* 10865!; Teita District: Voi to Taveta, km. 20.8, 12 Dec. 1961, *Polhill & Paulo* 973!
TANZANIA. Mbulu District: S. slope of Mt. Hanang, Hamit–Katesh, 14 Feb. 1946, *Greenway* 7742!; Lushoto District: W. Usambaras, Mombo–Soni road, 24 June 1953, *Drummond & Hemsley* 2994!; Ufipa District: Namwele, 25 Feb. 1950, *Bullock* 2598!; Zanzibar I., Chukwani, 7 May 1960, *Faulkner* 2546!
DISTR. **K** ?1, 4, 6, 7; **T** 1–8; **Z**; Mozambique, Malawi, Zambia and Zimbabwe
HAB. Grassland, *Acacia-Commiphora-Combretum*, *Apodytes-Rhus-Ozoroa*, etc., bushland and shrub clumps in grassland, also old cultivations; 0–2100 m.

SYN. *Siphonanthus glabra* (E. Mey.) Hiern var. *vaga* Hiern in Cat. Afr. Pl. Welw. 1: 842 (1900). Type: Angola, Benguella, *Welwitsch* 5752 (LISU, holo., BM, K, iso.!)
Clerodendrum eriophylloides Mold. in Phytologia 9: 183 (1963) & in Phytologia 59: 350 (1986). Type: Zanzibar I., Chukwani, *Faulkner* 2785 (S, holo., K, iso.!)
C. glabrum E. Mey. var. *vagum* (Hiern) Mold. in Phytologia 13: 306 (1966) & in Phytologia 59: 499 (1986)

NOTE. The inflorescences are extremely variable, ranging from quite lax to densely capitate but attempts to maintain two varieties based on this have proved unworkable. Judging by Gürke's description the type had ± lax inflorescences. Thomas cited *Schlieben* 3648 with large leaves and lax inflorescences and *Busse* 1043 with condensed inflorescences. *B.D. Burtt* 1312 (Tanzania, Kondoa District, Kolo, 2 Feb. 1928) has small leaves ± 1.5 × 1 cm. and small heads 1–1.5 cm. diameter. *Faulkner* 2546 cited above has leaves to 10.5 × 6 cm., ± obtuse, but is not really separable. *Taylor* 107/2 (Zanzibar, 1929) is even more diverse with lanceolate leaves 3.8 × 0.8 cm. *C. eriophylloides* presumably comes near to typical *C. eriophyllum* but since most named material of the latter had condensed inflorescences it is not surprising another name was proposed. *Peter* 40348 (Tanzania, Pangani District, Hale, R. Pangani [Grosse Brücke] to Ngombezi, 18 May 1926) labelled as *C. tricholobum* is not typically that species but probably a form of *C. eriophyllum*. It is a shrub 1 m. tall, with ± oblong leaves up to 11 × 4.5 cm., lax inflorescences and corolla-tube 1.2 cm. long. It is somewhat intermediate. Moldenke has considered *C. eriophyllum* to be no more than a variant of *C. glabrum* E. Mey. (var. *vagum* (Hiern) Mold.). The two form a difficult complex and I prefer to keep them separate at present.

26. **C. tricholobum** *Gürke* in E.J. 18: 178 (1893) & in P.O.A. C: 341 (1895); Bak. in F.T.A. 5: 303 (1900); Thomas in E.J. 68: 56 (1936); T.T.C.L.: 634 (1949). Type: Tanzania, Pangani, Mauya [Mauja], *Stuhlmann* 587 (B, lecto.†, HBG isolecto.!)

Shrub or scrambler, 1–3 m. tall, sometimes with long trailing branches; stems woolly with very dense long spreading fawn hairs 1–1.5 mm. long, at length glabrescent and nodular. Leaves in whorls of 3, elliptic to ± round, 3–8(–14) cm. long, 2–5(–7.5) cm. wide, acute at the apex, cuneate at the base, entire, pubescent above with long and short hairs, becoming densely papillose with hair bases, paler woolly velvety beneath, rather discolorous; petiole ± 7 mm. long, hairy as stems, often orange-brown at base. Flowers in condensed or capitate dichasial cymes ± 2 cm. wide, axillary from upper nodes and sometimes appearing terminal; peduncle 2–5 cm. long; bracts lanceolate, 9–10(–14) mm. long, 1–3 mm. wide, tapering acuminate, pilose. Calyx white hairy; tube cylindrical, 2–2.2 mm. long; lobes linear from triangular base 2.2–3.5 mm. long, pilose. Corolla strongly scented, white; tube 1–1.8 cm. long, densely covered with stipitate and sessile glands; lobes elliptic to round, unequal, 2.2–2.5 mm. long, 1.2–1.8 mm. wide. Stamens and style exserted 3–8 mm. Ovary smooth. Fruit dull brown, globose, dry, 1–1.3 cm. long, 1–1.5 cm. wide, light and corky, each lobe longitudinally grooved; fruiting calyx-lobes ovate, 5 mm. long, 3 mm. wide, acuminate, ridged.

KENYA. Lamu District: Boni, Kitangani, 27 Dec. 1946, *J. Adamson* 312! & Boni Forest Reserve, 7.7 km. W. of Mararani [Marereni], 29 Nov. 1988, *Robertson & Luke* 5617! & N. of Mombasa to Lamu and Witu, *Whyte*!

TANZANIA. Lushoto District: Pangani R., Magunga–Korogwe, 6 Mar. 1954, *Faulkner* 1370!; Uzaramo District: Utete road, km. 16 [from Dar es Salaam], 13 June 1940, *Vaughan* 2938!; Kilwa District: Selous Game Reserve, Nandembo R., 2 Feb. 1971, *Ludanga* 1198!; Zanzibar, without locality, 1932, *Mrs Taylor* 540!

DISTR. K 7; T 3, 6, 8; Mozambique

HAB. Riverine bushland, light woodland, scrub on coral rock above beach; seasonally flooded forest/woodland/grassland mosaic; 0–300 m.

SYN. [*C. acerbianum* sensu Thomas in E.J. 68: 89 (1936), pro parte quoad *Schlieben* 5866, *non* (Vis.) Benth.]

C. lindiense Mold. in Phytologia 5: 83 (1954) & in Phytologia 61: 492 (1987). Type: Tanzania, Lindi District, Lake Lutamba, *Schlieben* 5866 (BR, holo.!, HBG, K, iso.!)

[*C. eriophyllum* sensu Vollesen in Opera Bot. 59: 82 (1980), quoad *Ludanga* 1198 & *Rodgers* 862, *non* Gürke]

NOTE. There is no doubt this is very close to *C. eriophyllum* although Thomas places it far away in a different section. Gürke included it together with *C. acerbianum* and *C. eriophyllum* in sect. *Cornacchinia* but pointed out that the fruits were not known at that time of either *C. eriophyllum* or *C. tricholobum*.

27. C. acerbianum (*Vis.*) *Benth.* in G.P. 2: 1156 (1876); Boiss., Fl. Orient. 4: 536 (1879); Gürke in P.O.A. C: 341 (1895); Bak. in F.T.A. 5: 295 (1900); Thomas in E.J. 68: 89 (1936); F.P.S. 3: 194 (1956); K.T.S.: 583 (1961); Huber in F.W.T.A., ed. 2, 2: 443 (1963); Berhaut, Fl. Sénégal, ed. 2: 109 (1967); Täckholm, Students' Fl. Egypt, ed. 2: 454 (1974); Moldenke in Phytologia 57: 389 (1985). Type: Upper Egypt, *Acerbi* (ubi?)*

Scandent or erect shrub 1–3 m. long; stems pale straw-coloured, densely spreading velvety grey pubescent, roughened with nodular petiole-bases. Leaves opposite or in whorls of 3–4, oblong to oblong-ovate to oblong-elliptic, 1–18 cm. long, 0.5–5 cm. wide, acute to obtuse or quite rounded at the apex, rounded to subcordate or ± distinctly cordate at the base, pubescent above on the nerves, grey pubescent to velvety beneath; petiole 0.2–1.8 cm. long. Flowers in dense terminal clusters on short lateral axillary branches near the tops of shoots but sometimes appearing to be in long-pedunculate clusters; true peduncles short; apparent peduncles to 12 cm. long, pedicels very short, becoming 7 mm. long in fruit; bract-like leaves up to 2.5 cm. long, 1.5 mm. wide or lacking, in which case the branches are technically peduncles; bracteoles 2–3, narrow, 3 mm. long, 0.3 mm. wide. Calyx-tube 2 mm. long, densely pubescent; lobes 4–5, 2–6 mm. long, 1–2 mm. wide, ± acute to acuminate, very woolly ciliate. Corolla white; tube (0.4–)1.8–3 cm. long, glandular-pubescent outside; lobes oblong-elliptic, 3.5–6 mm. long, 2–3.5 mm. wide. Ovary divided into 4 lobes, deeply grooved between them. Fruit subglobose, 1.2 cm. diameter, hard in life, covered with low spongy-corky processes, resembling a bramble berry in structure; calyx remaining persistent on pedicel with lobes deflexed. Fig. 16/16–19, p. 114.

KENYA. Northern Frontier Province: Tana R., Bura, 1 Oct. 1949, *Hornby* 3103!; Kwale District: Kinango, Jan. 1930, *R.M. Graham* in *F.D.* 2237!; Tana River District: 1°05'S 39°55'E, *Battiscombe* 229!; Lamu District: Ungu [Ngao], on coast near Lamu, *Werner* in *Battiscombe* 997!

TANZANIA. Tanga District: Duga, July 1893, *Holst* 3208!

DISTR. K 1, 7; T 3; Gambia, Guinea Bissau, Egypt, Sudan, Ethiopia and Somalia

HAB. *Acacia* woodland and scrub, *Diospyros-Terminalia-Spirostachys-Dobera-Ficus* association and *Populus-Ficus-Borassus* riverine forest remnants, sand-dunes; ± 0–180 m.

SYN. *Volkameria acerbiana* Vis., Ic. Pl. Quar. Aegypti Nubiae: 23, t. 4/1 (1836); Schauer in DC., Prodr. 11: 656 (1847)

Cornacchinia fragiformis Savi in Mem. Soc. Ital. Mod. 21: 185, t. 7 (1837); Linnaea 14 anh.: 85 (1840). Type: Sudan, W. bank of R. Nile near Abldeharim, *Raddi* (PI, holo.)

Clerodendrum holstii Gürke in Engl., Abh. Preuss. Akad. Wiss.: 21 (1894); Bak. in F.T.A. 5: 303 (1900). Type: Tanzania, Tanga District, Duga, *Holst* 3208 (B, holo. †, K, iso.!)

---

* It could not be found at Padua.

NOTE.   Although this species has been placed in a separate subgen. *Cornacchinia* (Savi) Thomas on account of the curious lobulate fruit, it is so close in facies to *C. eriophyllum* and *C. tricholobum* that I do not believe this single technical character merits such segregation. Thomas gives the type as *Kotschy* 359 from Sudan, Mograd, and Moldenke follows him although he discusses Acerbi later. If the Acerbi specimen is still in existence it is the holotype, if not the Kotschy specimen could be accepted as a neotype.

28. **C. incisum** *Klotzsch* in Peters, Reise Mossamb., Bot. 1: 257 (1861); Vatke in Linnaea 43: 537 (1882); Bak. in F.T.A. 5: 307 (1900); Thomas in E.J. 68: 78 (1936); T.T.C.L.: 636 (1949); Huber in F.W.T.A., ed. 2: 442 (1963); Haerdi in Acta Trop., Suppl. 8: 151 (1964); Vollesen in Opera Bot. 59: 82 (1980); Moldenke in Phytologia 60: 271 (1986). Types: Mozambique, Rios de Sena, Querimba and Boror, *Peters* (B, syn.†)

Shrub 0.6–2.5 m. tall, with straw-coloured stems; youngest parts ± ferruginous pubescent, soon glabrous. Leaves opposite or in whorls of 3, sometimes drying purplish, elliptic to lanceolate, 1.7–13.5 cm. long, 1.2–6 cm. wide, acuminate at the apex, gradually attenuate at the base, usually slightly discolorous, entire or usually coarsely deeply incised-toothed, often only 2–3 teeth on each side, or pinnatilobed, the divisions very acute, with scattered hairs above and on the venation beneath, or quite densely pubescent particularly beneath, rather distinctly gland-dotted, scented (descriptions vary from sweet to highly unpleasant!); petioles 0.5–1.5 cm. long. Cymes terminal and in upper axils and much lower topping short lateral branchlets; peduncles to 2 cm. long; apparent pedicels 2.5–8 mm. long; bracts and bracteoles linear to spathulate, 2–3 mm. long. Calyx cupular, 5 mm. long, the teeth narrowly triangular, 1.5–3 mm. long. Buds with globose limb. Corolla white, opening early in the morning; tube 5–9 cm. long, with sparse to dense gland-tipped hairs; limb 2.5 × 2 cm., the lobes 0.8–1.5 cm. long, 4.5–7 mm. wide. Style purple and white and anthers purple on crimson and purple filaments, exserted 3–5 cm. Fruit depressed-subglobose, 7 mm. long, 9 mm. wide, 3–4-lobed, glabrous, ridged and wrinkled-reticulate in dry state. Fig. 18/1–4.

KENYA.   Kwale District: Lungalunga–Msambweni road, between R. Umba and R. Mwena, 14 Aug. 1953, *Drummond & Hemsley* 3786!; Kilifi District: Arabuko-Sokoke Forest, N. of Sokoke, 9 June 1973, *Musyoki & Hansen* 1016!; Lamu District: 11.2 km. W. of Kiunga, 27 July 1961, *Jarman* in *Gillespie* 70!
TANZANIA.   Tanga District: Ngomeni, 10 July 1967, *Faulkner* 3982!; Bagamoyo District: Kidomole, Apr. 1964, *Semsei* 3742! (densely pubescent form); Kilwa District: Selous Game Reserve, Nunga, 17 Feb. 1971, *Ludanga* 1240!
DISTR.   **K** 7; **T** 3, 6, 8; Somalia, Mozambique; grown in Mombasa gardens; now widely cultivated in the tropics and in greenhouses elsewhere
HAB.   Coastal thicket, forest scrub, *Brachystegia-Hymenaea* woodland with very mixed dense undergrowth, also fixed sand-dunes; 5–450 m.
SYN.   *C. macrosiphon* Hook.f. in Bot. Mag. 109, t. 6695 (1883). Type: living plants sent from Tanzania coast opposite Zanzibar I. to Kew by *Kirk* in 1881 and flowered in 1882 (K, holo.!)
          C. incisum Klotzsch var. *macrosiphon* (Hook.f.) Bak. in F.T.A. 5: 307 (1900); Chiov., Fl. Somala 2: 364 (1932); Moldenke in Rev. Fl. Ceylon 4: 425 (1983) & in Phytologia 60: 277 (1986)
          C. incisum Klotzsch var. *vinosum* Chiov., Fl. Somala 2: 364 (1932); Thomas in E.J. 68: 78 (1936); T.T.C.L.: 636 (1949); E.P.A.: 799 (1962); Moldenke in Phytologia 60: 281 (1986). Type: Somalia, Transjuba, Licchitore, *Senni* 541 (FT, holo.)
          C. incisum Klotzsch var. *longepedunculatum* Thomas in E.J. 68: 78 (1936); Moldenke in Phytologia 60: 276 (1986). Type: Kenya, coast opposite Lamu*, *Hildebrandt* 1911 (B, holo.†)
          C. dalei Mold. in Phytologia 4: 287 (1953) & in Phytologia 59: 240 (1986). Type: Kenya, Kwale District, S. Digo, Majoreni [Maroreni], Ganda Forest, *Dale* in F.D. 3811 (BR, holo., EA, K, iso.!)
NOTE.   Specimens with completely entire leaves are distinctive, e.g. Kenya, Kwale District, Shimoni, 1892, *Whyte* s.n.! & near Majoreni, Ganda Forest, Sept. 1937, *Dale* in F.D. 3811! *Schlieben* 5260, cited by *Thomas* as this species, appears to me in part to belong elsewhere (see p. 124); the calyx is very different. Although very similar, *C. lindemuthianum* Vatke (Linnaea 53: 537 (1882); type: Madagascar, Vavatobé, *Hildebrandt* 3332 (B, holo.†, K, iso.!)) differs in leaf-toothing and even longer flowers and I am not altogether convinced of Moldenke's decision to merge the two (Fl. Madag., Fam. 174: 163 (1956)) and have omitted Madagascar from the range of distribution above. Vatke dealt with *C. incisum* at the same time. Moldenke mentions *C. incisum* var. *parvifolium* Moldenke in Phytologia 3: 407 (1951) & Phytologia 60: 281 (1986) (type: Madagascar, Mt. Vohitrosy, lower Mandrare valley, near Anadabolara, *Humbert* 12471 (P, holo. & iso.)) and *C. incisum* Klotzsch var. *afzelii* Mold. in Amer. Journ. Bot. 38: 325 (1951) & Phytologia 60: 276 (1986) (type: Madagascar, Tuléar, Manasoa Tanosy, *Afzelius* (K, holo.)). Both are I think

* Thomas gives 'Deutsch Ost Afrika, Sansibarkuste bei Lamen', Vatke as ora zanzibariensis prope Lamu.

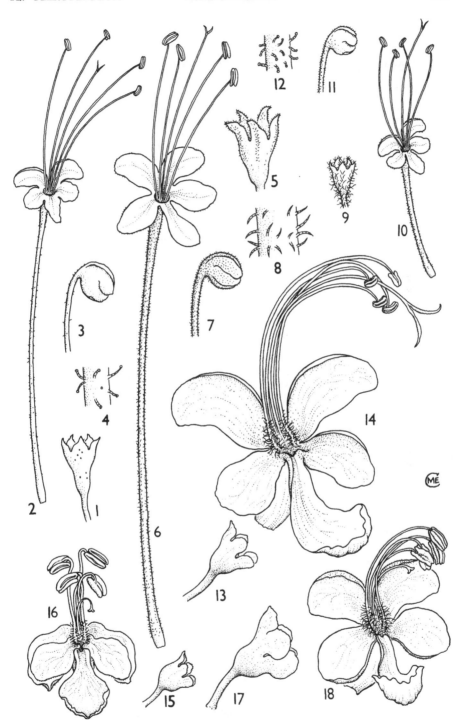

FIG. 18. *CLERODENDRUM INCISUM* — **1**, calyx, × 2; **2**, corolla, × 1; **3**, bud, × 1; **4**, part of corolla-tube, × 6. *C. PUSILLUM* — **5**, calyx, × 2; **6**, corolla, × 1; **7**, bud, × 1; **8**, part of corolla-tube, × 6. *C. TERNATUM* — **9**, calyx, × 2; **10**, corolla, × 1; **11**, bud, × 1; **12**, part of corolla-tube, × 6. *C. SANSIBARENSE* subsp. *CAESIUM* — **13**, calyx, × 2; **14**, corolla, × 2. *C. MAKANJANUM* — **15**, calyx, × 2; **16**, corolla, × 2. *C. KISSAKENSE* — **17**, calyx, × 2; **18**, corolla, × 2. 1–4, from *Magogo & Glover* 926; 5–8, from *Richards* 19226; 9–12, from *Verdcourt* 3095; 13, 14, from *Polhill & Wingfield* 4647; 15, 16, from *Gillett* 20741; 17, 18, from *Gillett* 18021. Drawn by Mrs. M.E. Church.

C. lindemuthianum Vatke. *Boivin* s.n. (BR) is labelled 'Côte orientale d'Afrique, Zanzibar' but I have seen no material from the island.

29. **C. pusillum** *Gürke* in E.J. 30: 390 (1901); Thomas in E.J. 68: 78 (1936); T.T.C.L.: 636 (1949); F.F.N.R.: 367 (1962). Type: Tanzania, Mbeya District, Unyika, Nsangamales [Msangaware's], *Goetze* 1393 (B, holo.†)

Dwarf pyrophytic shrub with very short stems from a thick woody creeping rootstock, the whole plant, including flowers, 13–22 cm. tall. Leaves very congested, lanceolate, oblong, elliptic or oblong-lanceolate, 1.5–4.5(-7) cm. long, 0.7–3.5 cm. wide, truncate to subacute at the apex, ± narrowed at the base, entire or subentire, glabrous above, ciliate and slightly pubescent beneath; petiole 2–3 mm. long. Inflorescence terminal, capitate, few-flowered. Calyx narrowly infundibular-tubular, 9–11 mm. long, including 4 mm. long lanceolate lobes, ciliate and sparsely white-pubescent. Buds with limb globose. Corolla white, densely pubescent outside; tube 10–11.5 cm. long; lobes elliptic, 10 mm. long, 6 mm. wide. Stamens mauve, together with style exserted ± 4 cm. Fruit not seen. Fig. 18/5–8, p. 123.

TANZANIA. Ufipa District: Mbeya–Ivuna road, 2 Nov. 1964, *Richards* 19226!; Mbeya District: Mbozi, Zambi, 19 Nov. 1932, *Davies* 654! & Nsangamales, 30 Oct. 1899, *Goetze* 1393
DISTR. T 4, 7; Malawi, Zambia and Angola
HAB. Presumably *Brachystegia* woodland; 1200–1500 m.

30. **C. lutambense** *Verdc.*, sp. nov. cum *C. inciso* Klotzsch confusa inflorescentiis spicatis, calyce 8–10 mm. longo dense glandulo-pubescenti lobis anguste triangularibus 3 mm. longis incluso, foliis in siccitate pallidis. Type: Tanzania, Lindi District, Lake Lutamba, *Schlieben* 5260 (HBG, holo.!, B, BR, M, NY, S, iso. fide Moldenke*)

Subshrubby herb with several herbaceous or thinly woody shoots 30–80 cm. tall from a woody rootstock; stems ± densely spreading pubescent, later glabrescent and with pale buff-brown cork. Leaves oblanceolate, 3–14 cm. long, 1–4 cm. wide, shortly acuminate at the apex, long attenuate at the base, with 1–5 coarse teeth towards the apex or practically entire, ± densely pubescent above but only on main venation beneath, obscurely pellucid dotted; true petiole 1–2 mm. long. Inflorescences terminal, spicate, ± 5.5–7 cm. long, densely pubescent; peduncle 2.5 cm. long; flowers sessile; larger bracts obovate, 1.2 cm. long, 8 mm. wide, abruptly acuminate; smaller bracts elongate-triangular, ± 9 mm. long, 1.5 mm. wide. Calyx-tube cylindric, 5–7 mm. long, densely pubescent and with sessile glands; lobes narrowly triangular, 3 mm. long. Corolla white, narrowly tubular, ± 10 cm. long, with rather sparse gland-tipped hairs outside; lobes 1.2 cm. long, ± 7 mm. wide, obtuse. Stamens and style probably reddish, exserted 3.5 cm. or more. Fruit not seen.

TANZANIA. Lindi District: Lake Lutamba, 6 Sept. 1934, *Schlieben* 5260! & same locality, [Rutamba], ? Mar. 1943, *Gillman* 1363!
DISTR. T 8; Mozambique
HAB. Presumably *Brachystegia* woodland; 160 m.
SYN. [*C. incisum* possibiliter sensu Thomas in E.J. 68: 78 (1936); T.T.C.L.: 636 (1949) in casibus ambobus quoad *Schlieben* 5260, pro parte, *non* Klotzsch]
        [*C. incisum* Klotzsch var. *macrosiphon* sensu Mold. in Phytologia 60: 275 (1986) adnot., *non* (Hook.f.) Bak.]
NOTE. Out of three specimens cited by Brenan under *C. incisum* in T.T.C.L. *Schlieben* 5260 was the only one he had seen so some English herbarium may have an isotype but it was a mixed gathering since the EA sheet at least is *C. incisum*. I have been unable to examine all parts of the holotype thoroughly and further information on the inflorescence and corolla-limb is needed.

31. **C. ternatum** *Schinz* in Abh. Bot. Ver. Brandenb. 31: 205 (1890); F.T.A. 5: 312 (1900); Thomas in E.J. 68: 79 (1936); T.T.C.L.: 637 (1949); Friedrich-Holzhammer in Prodr. Fl. SW.-Afr. 122: 5 (1967). Type: Namibia, Amboland, Olukunde, *Schinz* 457 (Z, holo.)

Subshrub 15–35(-90 fide Vollesen) cm. tall, usually clump- or carpet-forming with several woody based stems from an extensive creeping woody rootstock up to 25 cm. long or probably more; stems ± pale, pubescent with short curled hairs. Leaves in whorls of 3, linear-lanceolate to oblanceolate, 0.5–7 cm. long, 0.15–1.5 cm. wide, acute at the apex, gradually narrowed to the base, with spaced sharp teeth or almost lobulate or subentire,

* All of these will need examining to decide which species they are or if a mixture (see note).

sometimes born on axillary brachyblasts, slightly pubescent with ± tubercle-based hairs and gland-dotted to densely grey-pubescent. Inflorescence made up of axillary cymes from the upper 4–5 axils; peduncle 1.2–2 cm. long; pedicels 0–1.5 mm. long. Buds with a very excentric globose limb. Calyx cupular, 3 mm. long, including the short broadly triangular teeth 0.5–0.8 mm. long, pubescent. Corolla with a pungent odour usually described as bad-smelling, white or greenish white with white limb; tube 1.3–4.2 cm. long, abruptly funnel-shaped at the apex, glandular and pubescent outside, the glands stalked or subsessile. Filaments purple or reddish purple, white at the base and style reddish purple, both exserted 3–3.5 cm.; stigma bifid; anthers brown. Fruit depressed subglobose, 4 mm. long, 8 mm. wide, deeply 4-lobed. Fig. 18/9–12, p. 123.

KENYA. Teita District: Mwatate, June 1884, *Johnston* s.n.!
TANZANIA. 6.4 km. N. of Old Shinyanga, 15 Dec. 1950, *Welch* 47!; Kilosa District: 40 km. from Mikumi on Iringa–Morogoro road, 26 Feb. 1961, *Verdcourt* 3095!; Rufiji District: S. bank of the Rufiji R., Utete, 2 Dec. 1955, *Milne-Redhead & Taylor* 7468!
DISTR. **K**7; **T**1–3, 5–8; Malawi, Mozambique, Zambia, Zimbabwe, Angola, Namibia and South Africa
HAB. Grassland, *Brachystegia* woodland, *Commiphora-Terminalia* and *Combretum-Sterculia-Grewia* bushland; also old cultivations, often in rocky places; ± 0–1500 m.

SYN. *C. lanceolatum* Gürke in E.J. 18: 181 (1893) & P.O.A. C: 341 (1895); Bak. in F.T.A. 5: 312 (1900); Thomas in E.J. 68: 79 (1936); T.T.C.L.: 636 (1949); F.F.N.R.: 367 (1962), *non* F. Muell. (1863), *nom. illegit.* Type: Malawi, without locality, *Buchanan* 468 (B, holo.†)
C. wilmsii Gürke in E.J. 28: 304 (1900); Pearson in Fl. Cap. 5(1): 224 (1901). Types: South Africa, Transvaal, Lydenberg, *Wilms* 1082 & 1159 (B, syn.†)
C. ternatum Schinz var. *lanceolatum* Mold. in Phytologia 5: 98 (1954); Vollesen in Opera Bot. 59: 82 (1980)

NOTE. Some Uluguru Mt. material has the corolla-tube very short, 1.3–1.8 cm. long, e.g. *Semsei* 1960; it may represent a distinct local race.

32. C. sp. A

Shrub ± 1 m. tall; stems with dense yellowish bristly ± spreading hairs when young, later ± glabrous, closely longitudinally ridged with fine reticulation in the interstices, nodular with petiole-bases of fallen leaves. Leaves oblong-elliptic, 5–9 cm. long, 2.7–6 cm. wide, shortly acuminate at the apex, broadly cuneate or rounded at the base, ± velvety pilose on both surfaces; petioles 1–2 cm. long, densely woolly. Flowers in panicles of dichasial cymes 3 cm. long, 5 cm. wide, the axes hairy like the young stems. Calyx pubescent, cupular; lobes ovate, 1 mm. long, shortly acuminate. Corolla white, funnel-shaped; tube 5 mm. long; lobes ? rounded, 2 mm. long and wide. Stamens exserted ± 2 mm.

TANZANIA. Lindi District: Mlinguru, 17 Dec. 1934, *Schlieben* 5730!
DISTR. **T** 8; not known elsewhere
HAB. Bushland; 275 m.

NOTE. The only specimen seen (HBG) is in bad condition probably rotting through damp when collected and treated in some way, appearing to be covered with white powder (see also sp. 49 which is in precisely the same condition). There appears to be no additional material at either BM or BR. It is not mentioned by Thomas probably arriving too late for inclusion.

33. **C. violaceum** *Gürke* in E.J. 28: 303 (1900); Bak. in F.T.A. 5: 520 (1900); F.W.T.A. 2: 273 (1931); Thomas in E.J. 68: 82 (1936); Huber in F.W.T.A., ed. 2: 441 (1963). Type: Cameroon, Yaoundé, *Zenker* 1428 (B, holo.†)

Liane to 5 m., the old stems sometimes leafless, grey-green or darker, 4-angled or finely ridged, with short pale ferruginous pubescence or glabrescent; lateral leafy shoots with ± condensed internodes at the base. Leaves drying pale or dark, elliptic, 3–13.5 cm. long, 1.5–7(–10) cm. wide, acuminate at the apex, narrowly cuneate at the base into distinct petiole or very attenuate at base and true petiole short, thin, entire or slightly to coarsely serrate, the teeth not spreading, apiculate, ± glabrous to sparsely pubescent above and on venation beneath; petiole 0.2–2.5 cm. long. Inflorescences leafy, few–many-flowered, terminal on main and lateral shoots, 8–20 cm. long, the cymes axillary in sessile ovate acuminate or cordate leaf-like bracts 5 cm. long, 3 cm. wide; upper bracts lanceolate 5–10 mm. long, 1–2 mm. wide; individual cymes 1–2-flowered; axes finely pubescent; peduncle up to 2.5 cm. long; calyx pubescent; tube cupular, 4–6 mm. long; lobes broadly rounded to ovate, up to 5 mm. long and wide, densely gland-dotted within. Corolla violet, violet and white or greenish or, in Ugandan subsp., cream tinged green with posterior petal pale mauve; tube 6–10 mm. long; lobes 0.8–1.6 cm. long, 5–5.5 mm. wide. Stamens exserted 1–3.5 cm. Fruit red, depressed globose, 8 mm. tall, 1.1 cm. wide, 3–4-seeded.

DISTR.   Guineé to Cameroon and Zaire

SYN.   *C. kalbreyeri* Bak. in F.T.A. 5: 311 (1900). Type: Nigeria: Lagos, *Millson* (ubi, syn.) & Cameroon, Cameroon Mt., *Kalbreyer* 94 (K, syn.!)

subsp. **kigeziense** *Verdc.*, subsp. nov. a subsp. *violaceo* planta in siccitate pallide griseo-viridi, ramis vestustioribus aphyllis, foliis basi valde angustatis, petiolis veris ± 2 mm. longis, laminis distincte serratis, corolla viridescenti-alba, lobo posteriore pallide malvino excepto. Typus: Uganda, Kigezi District, Mulole [Murole], *Purseglove* 2690 (K, holo.!, EA, KAW, iso.)

Plant drying a pale grey-green; older stems leafless bearing lateral leafy shoots with ± condensed basal internodes. Leaves long-attenuate and pseudopetiolate at the base, the true petioles short, 2 mm. long; blades coarsely serrate. Corolla greenish white and mauve.

UGANDA.   Kigezi District: Mulole [Murole], Apr. 1948, *Purseglove* 2690!
DISTR.   U 2; not known elsewhere
HAB.   Forest; 2100 m.

NOTE.   The position of this taxon is not clear from the single gathering; it may be specifically distinct from *C. violaceum* as suggested by Gillett (adnot.) when it was first received at Kew. It could be perhaps better treated as one of several forest analogues of *C. myricoides*, but the true climbing habit has influenced my decision.

34. **C. bukobense** *Gürke* in E.J. 18: 182 (1893); Bak. in F.T.A. 5: 311 (1900); Thomas in E.J. 68: 81 (1936); T.T.C.L.: 632 (1949); Fl. Rwanda 3: 272, fig. 87.2 (1985). Type: Tanzania, Bukoba, *Stuhlmann* 3657 (B, lecto.†)

Shrub or ± climbing; young shoots purplish black with pale lenticels and short bifarious pubescence, later grey and glabrous. Leaves elliptic-pandurate to oblanceolate, 7–25 cm. long, 3–21 cm. wide, acuminate at the apex, much narrowed to a square ± cordate base or this so reduced as to appear merely attenuate, subentire to coarsely crenate, thin, with a few scattered hairs above and on nerves beneath but ± glabrous, sessile. Inflorescence a terminal pyramidal leafy panicle of lax dichotomous axillary cymes, the subtending leaves reducing towards apex until elliptic, 7 × 3 mm., peduncles of lowermost cymes up to 10 cm. long; axes finely pubescent; pedicels 2 mm. long. Calyx pubescent at base; tube cupular, 3 mm. long; lobes rounded, ± semicircular, 2 mm. long, 2.5 mm. wide. Corolla blue, very asymmetric; tube 5 mm. long, glabrous above; lobes elliptic, 7 mm. long and wide, slightly ciliate. Fruits red, depressed, 7 mm. long, ± 10 mm. wide, 2–4-lobed.

UGANDA.   Masaka District: NW. side of Lake Nabugabo, 5 Oct. 1953, *Drummond & Hemsley* 4623!
TANZANIA.   Bukoba District: Bushasha, Sept.–Oct. 1935, *Gillman* 470! & Rubare Forest Reserve, Nov. 1958, *Procter* 1064! & Bukoba, 16 Apr. 1892, *Stuhlmann* 4054; Rungwe District: Chivanjee, 29 July 1974, *Leedal* 1891!
DISTR.   U 4; T 1, 7; E. Zaire and Zambia
HAB.   Regenerating secondary rain-forest, woodland; ?1140–1200 m.

SYN.   *C. variifolium* De Wild. in B.J.B.B. 7: 179 (1920). Type: Zaire, Kivu, Rutshuru, *Bequaert* 6121 (BR, holo.)

NOTE.   The original description states leaves attenuate into a winged petiole, upper ones sessile, cordate at base. I suspect the minute cordation was overlooked on the lower leaves.

35. **C. sansibarense** *Gürke* in E.J. 18: 181 (1893); Bak. in F.T.A. 5: 312 (1900); Thomas in E.J. 68: 81 (1936); U.O.P.Z: 198 (1949). Type: Tanzania, Zanzibar I., Tumbatu I., *Stuhlmann* I 511 (B, holo.†, HBG, iso.!)

Shrub (0.4–)2–6 m. tall, erect, scrambling or ? sometimes truly climbing; young stems glabrous to hairy with grey or pale yellow-brown hairs; older stems straw-coloured, glabrous to pubescent. Leaves said to smell offensive, opposite; blades ovate to elliptic-ovate, 4–21(–25) cm. long, 2–11 cm. wide, tapering acuminate to a very sharp apex, attenuate into the petiole, entire to obscurely or more rarely coarsely crenate-serrate, glabrous to sparingly pubescent above and beneath, especially on the nerves, to quite velvety beneath; petiole 0.3–3.5 cm. long. Cymes dichasial, axillary in upper axils and terminal forming leafy pyramidal inflorescences to 20 cm. long, 18 cm. wide; peduncles 4–8 cm. long; pedicels 2 mm. long but apparent pedicels (with bractlets) 1.5 cm. long; all axes glabrous to pubescent; upper leaf-like bracts sessile, ovate, ± cordate. Calyx-tube broadly obconic, 2–3 mm. long, 4 mm. wide; lobes ovate-triangular, 1.5–2.5 mm. long, rounded, soon reflexed, gland-dotted inside. Corolla-tube greenish white or green, asymmetric, 8–10 mm. long; throat hairy; lobes: 4 white, 1 pale blue-green, mauve-blue or purple, 4 oblong, 8–9.5 mm. long, 3.5–5(–9) mm. wide, largest blue one spathulate, 1.3–1.6 cm. long, 3.5–5(–9) mm. wide; limb in bud glabrous to velvety tomentose and gland-dotted, the lobes ciliate. Stamens exserted 2.8–3.5 cm. Fruit scarlet to reddish black, depressed-subglobose, 6–7 mm. long, 7–11 mm. wide, 3–4-lobed; fruiting calyx plate-like, 6 mm. wide, the lobes 3.5 mm. long, 1.2 mm. wide, ciliolate, often strongly reflexed; fruiting pedicels 7 mm. long.

KEY TO SUBSPECIES

Corolla entirely glabrous; lobes often wider than in other
  subspecies, up to 9 mm. wide . . . . subsp. **caesium**
Corolla sparsely to densely pubescent:
  Leaves glabrescent . . . . subsp. **sansibarense**
  Leaves pubescent to velvety . . . . subsp. **occidentale**

subsp. **sansibarense**

Foliage glabrescent. Calyx-lobes usually reflexed. Corolla white with anterior lobe blue or two shades or blue or mauve; limb in bud sparsely to densely pubescent or lobes only ciliate.

KENYA. Kwale District: Shimba Hills, Kivumoni Forest, 13 Apr. 1968, *Magogo & Glover* 855!; Kilifi District: E. & S. of Jilore Forest Station, 12 June 1973, *Musyoki & Hansen* 1037!
TANZANIA. Lushoto District: E. Usambaras, between Ngua and Magunga, 26 June 1953, *Drummond & Hemsley* 3030!; Tanga District: 8 km. SE. of Ngomeni, 31 July 1953, *Drummond & Hemsley* 3555!; Zanzibar I., Chwaka, 'mile 17', 29 Nov. 1963, *Faulkner* 3317! & Mkokotoni, 15 Aug. 1961, *Faulkner* 2885!
DISTR. K 7; T 3, 5(forma), 6–8; Z; Zaire
HAB. Evergreen forest and edges, muddy and sandy seashores; 0–1100 m.

SYN. *C. scheffleri* Gürke in E.J. 28: 301 (1900); Bak. in F.T.A. 5: 519 (1900); Thomas in E.J. 68: 82 (1936); T.T.C.L.: 631 (1949). Type: Tanzania, E. Usambaras, Ngwelo [Nguelo], *Scheffler* 57 (B, holo.†)
  [*C. myricoides* sensu Bak. in F.T.A. 5: 310 (1900) quoad Zanzibar, *Kirk, non* (Hochst.) Vatke]
  C. myricoides (Hochst.) Vatke var. *attenuatum* De Wild. in F.R. 13: 143 (1914). Type: Zaire, Shaba, Kantu, *Kassner* 2381* (BR, syn., BM, K, isosyn.!) & Lubumbashi [Elisabethville], *Homblé* 171 (BR, syn.)
  C. attenuatum (De Wild.) De Wild. in B.J.B.B. 7: 182 (1920), *non* R.Br. (1810), *nom. illegit.*
  C. scheffleri Gürke var. *mahengianum* Thomas in E.J. 68: 82 (1936); T.T.C.L.: 631 (1949), *nomen invalid.* Type: Tanzania, Ulanga District, Mahenge, *Schlieben* 1587 (B, holo.†, BM, K, iso.!)
  C. varium Thomas in E.J. 68: 82, 104 (1936); T.T.C.L.: 633 (1949). Type: Tanzania, Mahenge, *Schlieben* 2250 (B, holo.†, BM, iso.!)
  [?*C. dembianense* sensu Thomas in E.J. 68: 82 (1936); T.T.C.L.: 632 (1949), quoad *Engler* 437, *non* Chiov.]
  C. quadrangulatum Thomas in E.J. 68: 88 (1936). Type: as for *C. attenuatum* De Wild. (1920) *non* R.Br. (1810)
  C. sp. sensu Haerdi in Acta Trop., Suppl. 8: 152 (1964)

NOTE. *Hornby* 522 (Mpwapwa, 9 Nov. 1932) has a woolly calyx with unreflexed lobes and densely hairy bud limb; it is probably a distinct minor local variant. It is difficult to see what else *Engler* 437 (E. Usambaras) can be but subsp. *sansibarense*; *C. dembianense* is said to have verticillate leaves which I have not observed in E. Usambara material. It is interesting to note that almost all material collected by Peter about 50 years ago in the E. Usambaras has glabrous corollas.

subsp. **occidentale** Verdc., subsp. nov. a subsp. *sansibarense* foliis pubescentibus vel subtus dense velutinis, lobis calycis haud reflexis, corolla caerulea, purpurea, alba vel flava, limbo sparse vel dense pubescenti differt. Typus: Tanzania, Handeni District, Kabuku Forest, *Faulkner* 4163 (K, holo.!)

Foliage pubescent or densely velvety beneath. Calyx-lobes not reflexed. Corolla often white or yellow, the limb in bud sparsely to densely pubescent.

TANZANIA. Lushoto District: W. Usambaras, escarpment near Gologolo–Mkumbara [Mkumbala] footpath, 4 June 1953, *Drummond & Hemsley* 2855! & Kwembago, *Shabani* 562! & Lushoto, 3 Jan. 1970, *Archbold* 1161!
DISTR. T 3; not known elsewhere
HAB. Grassland with scrub, forest including clearings and edges; 1000–1750 m.

SYN. [*C. scheffleri* sensu Thomas in E.J. 68: 82 (1936), pro parte, *non* Gürke]

* De Wildeman gives the number as 2382

NOTE. This clearly merges with *C. myricoides* (Hochst.) Vatke var. *discolor* (Klotzsch) Bak. Intermediates with subsp. *caesium* also occur in **T** 6 and **T** 7.

subsp. **caesium** (*Gürke*) *Verdc.*, comb. nov. Type: Tanzania, Iringa District, N. Utschungwe Mts., Muhanga, *Goetze* 633 (B, holo.†)

Leaves glabrous or with very few hairs on venation beneath. Calyx-lobes usually not reflexed. Corolla blue or blue-violet, often large, up to 3 cm. wide; limb entirely glabrous outside, usually not even ciliate; lobes wider than in other varieties, up to 9 mm. wide. Fig. 18/13, 14, p. 123.

TANZANIA. Morogoro District: N. Uluguru Mts., Lupanga Peak, July 1981, *Lovett* 157! & without precise locality, 12 Oct. 1932, *Wallace* 200!; Iringa District: Mufindi Tea Estate, Kibwele Forest, 17 Aug. 1984, *Bridson & Lovett* 524!
DISTR. **T** 6, 7; not known elsewhere
HAB. Upland evergreen forest, woodland, grassland with forest patches; (1350 atypical-) 1600–2100 m.

SYN. *?C. silvicola* Gürke in E.J. 28: 299, 466 (1900); Bak. in F.T.A. 5: 510 (1900); Thomas in E.J. 68: 81 (1936); T.T.C.L.: 631 (1949). Type: Tanzania, Iringa District, N. Utschungwe Mts., Kigula Plateau, *Goetze* 563 (B, holo.†)
C. caesium Gürke in E.J. 28: 300, 466 (1900); Bak. in F.T.A. 5: 519 (1900); Thomas in E.J. 68: 81 (1936); T.T.C.L.: 631 (1949)

NOTE. From the description and locality it does not seem that *C. silvicola* can be anything but a variant of this subspecies but the flowers of that are described as white with middle lobe yellow. The flowers vary considerably in size; *Polhill & Wingfield* 4647 (Morogoro District, Uluguru Mts., above Chenzema towards Lukwangule Plateau, 1 Jan. 1975) has them at the top of the range, corolla 3 cm. wide with lobes 9 mm. wide; *Polhill & Paulo* 1550 (Iringa District, Dabaga Highlands, 21 km. S. of Dabaga, Idewe Forest Reserve, 20 Feb. 1962) on the other hand has the corolla 1.7 mm. wide with lobes ± 4 mm. wide. *Pócs et al.* 88014 (Kilosa District, Ukaguru Mts., Mt. Mamwira, 12 Feb. 1988) has the leaves densely ferruginous pubescent beneath but ± glabrous bud limbs. A few specimens have some calyx-lobes reflexed and others are intermediate with subsp. *sansibarensis* in corolla indumentum, e.g. *Clair-Thompson* 861 (W. slopes of Mt. Rungwe, 11 Mar. 1932)

36. **C. suffruticosum** Gürke in E.J. 28: 302, 466 (1900); Bak. in F.T.A. 5: 519 (1900); Thomas in E.J. 68: 81 (1936); T.T.C.L.: 632 (1949). Type: Tanzania, S. Uluguru Mts., 1200–1800 m., *Goetze* 191 (B, holo.†)

Subshrub ± 1 m. tall; young stems densely pubescent. Leaves opposite, broadly elliptic, 10–13(–17.5) cm. long, 5–7(–10) cm. wide, acuminate at the apex, narrowed at the base into the petiole, coarsely serrate towards the apex, thickly white pubescent on both surfaces; petiole 1–2 mm. long. Flowers in terminal panicles, the axes sparingly covered with multicellular hairs. Supporting leaves broadly ovate, rounded to ± cordate at the base, very shortly petiolate, the lowest 2.5 cm. long and wide the upper becoming smaller; bracts reddish, lanceolate, 3–7 mm. long, 1–2 mm. wide, blunt, entire, narrowing into the stalk, covered with multicellular hairs; pedicels 5–6 mm. long. Calyx broadly campanulate, hairy outside; lobes reddish, broadly ovate, obtuse. Corolla not described. Fruit not described, imperfect.

TANZANIA. Morogoro District: Uluguru Mts., 1200–1800 m., 22 Nov. 1898, *Goetze* 191 & Magado, 28 Dec. 1934, *E.M. Bruce* 383! & Nyandiduma Forest Reserve, Mar. 1955, *Semsei* 1969!
DISTR. **T** 6; not known elsewhere
HAB. Thick grass in forest glades [bushland on lower hillslopes]; (690–)1200–1800 m.

SYN. *C. discolor* Klotzsch var. *verbascifolium* Mold. in Phytologia 7: 79 (1959). Type: Tanzania, Uluguru Mts., above Morogoro, between Silesian [Schlesien] Mission and Lugongo, *Peter* 32288 (B, holo.)

NOTE. No authentic material of this has been seen. Thomas keys it out partly on the stems leaves and calyx being distinctly hairy and the two sheets I have cited in addition to the type seem to agree with this description so far as it goes and I think must be correctly associated save for difference in height. These two sheets are not identical; *Semsei* 1969 has the leaves very narrowed at base but widened and slightly auriculate at point of insertion in fact essentially sessile; *Bruce* 383 appears to have the leaves narrowed into genuine short petioles. Additional information for these two sheets is as follows:-leaves and stems pale ferruginous velvety; older stems (in 383) grey-buff, ± glabrous; leaves as small as 4.5 × 1.5 occur (in 383); inflorescences 15 × 10.5 cm. (383); pedicels 0–6 mm. long; calyx-lobes 3 × 2.8 mm., obtuse; corolla blue, only seen in bud (1969), but glabrous except for ciliate lobes; fruit (383) 7.5 × 8.5 mm.
    With no type available and the only possible sheets I have seen which might belong here being poor and not identical with each other some doubt rests on my interpretation. It is clearly close to *C. sansibarense* Gürke but very different in indumentum from subsp. *caesium* which occurs in the Uluguru Mts.

37. **C. alatum** *Gürke* in E.J. 18: 182 (1893); Bak. in F.T.A. 5: 311 (1900); Thomas in E.J. 68: 83 (1936); F.P.S. 3: 195 (1956); Huber in F.W.T.A., ed. 2, 441 (1963). Type: Sudan, Niamniam, near Tuhamis Seriba, *Schweinfurth* 3796 (B, lecto.†, K, isolecto.!)

Erect herb or subshrub 0.6–1.5 m. tall; stems slender to quite stout, 4–6-angled or narrowly winged, glabrous or pubescent, sometimes hollow and chambered. Leaves opposite or in whorls of 3, elliptic to oblanceolate, oblanceolate-oblong or rarely linear-lanceolate, 5.5–25(–30) cm. long, 1.3–7 cm. wide, acuminate at the apex, cuneate to rounded, truncate or subcordate at the base, entire to sharply toothed in upper third, glabrous or venation pubescent beneath or rarely discolorous and ± pubescent; sessile or petiole ± 2 mm. long. Inflorescence a terminal many-flowered thyrsoid panicle, 12–30 cm. long, composed of many cymes with some lower axillary ones as well; peduncle obsolete; secondary peduncles 2–4.5 cm. long; bracts leafy, ovate, 2.5 cm. long becoming smaller and lanceolate upwards, 7 mm. long; secondary bracts linear-lanceolate, 3–6 mm. long. Calyx glabrous or pubescent; tube broadly cupular, 1.5–2 mm. long; lobes ovate-triangular, obtuse. Corolla white, the limb lilac inside with lower lip blue or violet, sometimes blotched white; tube 6 mm. long; lobes oblong or obovate, 8.5 mm. long, 3.5–4.5 mm. wide, obtuse or truncate. Fruits depressed, 5.2 mm. long, 8 mm. wide.

TANZANIA. Buha District: Kasulu–Kibondo, km. 112, 15 Nov. 1962, *Verdcourt* 3324!
DISTR. T 4; Ivory Coast, N. Nigeria, Zaire, Sudan and Ethiopia
HAB. *Brachystegia-Pterocarpus* woodland; 1400 m.

SYN. *C. myricoides* (Hochst.) Vatke var. *floribundum* Bak. in F.T.A. 5: 310 (1900). Type: Sudan, Niamniam, Gumango Hill, *Schweinfurth* 3887 (K, holo.!)
  C. fleuryi A. Chev. in Bull. Soc. Bot. Fr. 58, Mém. 8: 191 (1912); F.W.T.A. 2: 273 (1931). Type: Ivory Coast, N. of Baoulé, margins of Nzi, Fétékro, *Chevalier* 22160 (P, holo.)
  C. hexagonum De Wild. in B.J.B.B. 7: 169 (1920). Type. Zaire, Uele, *Blommaert* (BR, holo.!)
  C. lelyi Hutch. in K.B. 1921: 395 (1921). Type: N. Nigeria, Mongu, *Lely* 384 (K, holo.!)
  C. alatum Gürke var. *adamauense* Thomas in E.J. 68: 83 (1936), *nomen illegit*. Type: Cameroon, Limbameni, *Ledermann* 4290 & Garoua, *Ledermann* 4935, Baja Buar, *Mildbraed* 9393 & between Kunde and Babua, *Mildbraed* 9251 (B, syn.) & Sudan, Niamniam Gumango, *Schweinfurth* 3887 (B, syn.†, K, isosyn.!)

NOTE. *C. alatum* Gürke var. *pubescens* Thomas (in E.J. 68: 83 (1936); type: Cameroon, Baja, Bouar, *Tessmann* 2 (B, holo.†, K, iso.!), *nomen invalid*.) has distinctly discolorous pubescent leaves and pubescent stems.
  Typical *C. alatum* from the Sudan is extremely distinctive with its very long oblanceolate leaves and extensive inflorescences but in Ethiopia material approaches *C. myricoides*. The single specimen from the Flora area might be considered a distinct subspecies but various characters can be matched elsewhere. See also notes under *C. myricoides* (Hochst.) Vatke subsp. *muenzneri* (Thomas) Verdc. (p. 135). *Kahurananga et al.* 2699 (Kigoma District, Lukoma, 29 May 1975) is intermediate between *C. alatum* and the form of *C. myricoides* represented by *C. phlebodes* C.H. Wright which has the long inflorescence of *C. alatum* but much bigger flowers.

38. **C. sp. B**

Shrub to 1.5 m.; stems ± 8-ridged, pubescent. Leaves (only upper ones seen) scented, narrowly oblong-elliptic, up to 12 cm. long, 3 cm. wide, narrowly acute at the apex, attenuate at the base, ± entire, ± discolorous, shortly closely pubescent on both surfaces. Lateral and terminal inflorescences run together to form a purplish bracteate thyrse 20 cm. long, 10 cm. wide, the lateral elements all subtended by leaves decreasing in size upwards, the branches with bracteoles at each node; all axes densely pubescent; pedicels 1.5–2 mm. long. Calyx purple, 2.5 mm. long, divided into rounded lobes ± 1 mm. long. Corolla purplish with yellow tube, ?4 mm. long.

TANZANIA. ?Kilwa District: Mihumo, 2 May 1956, *B.D. Nicholson* 81B!
DISTR. T 8; not known elsewhere
HAB. "Woodland savanna"; 300 m.

NOTE. This distinctive plant is clearly of the *C. myricoides* aggregate and has a long terminal inflorescence similar to *C. alatum*. Only one corolla has been seen but the collector states 'flowers very small'. If this is confirmed then it must be distinct at some level. The collector gives the native name as 'nunganunga' (Ngindo). The locality has not been traced for certain but is given in U.S. Board on Geogr. Names as = Mkunya, 9°45′S, 38°05′E.

39. **C. prittwitzii** *Thomas* in E.J. 68: 105, 84 (1936); T.T.C.L.: 653 (1949). Type: Tanzania, Dodoma District, Kilimatinde, *von Prittwitz* 62 (B, holo.†)

Perennial subshrubby herb 30–40(–100) cm. tall from thick woody rootstock; shoots few, unbranched, ± densely shortly ± bifariously spreading pubescent. Leaves in whorls of 3, drying ± pale green or brownish, oblanceolate to elliptic-oblong, 4–13(–18) cm. long, 1.2–5(–8) cm. wide, acute at the apex, long attenuate at base, distantly serrate save on narrowed part, shortly pubescent and with longer hairs on the midrib beneath, essentially subsessile or petiole 1–2 mm. long. Cymes arranged in a narrow terminal inflorescence 7–12(–23) cm. long and a few from 1–2 lower leafy nodes; inflorescence-axes sometimes purple, the peduncle up to 4 cm. long and true pedicels 2–3 mm. long; leafy bracts sometimes purplish; bracteoles linear to lanceolate, 2–5 mm. long. Calyx pale green to red in life, vinaceous on drying, obconic, white-pubescent; tube 3 mm. long; lobes rounded, 2 mm. long, 2.5 mm. wide. Corolla white with lower lip blue; tube 6 mm. long; lobes 9 mm. long, 3–4 mm. wide. Style exserted 2 cm. Fruit deeply 4-lobed.

TANZANIA. Chunya District: 152 km. N. of Mbeya on Itigi road, Lupa Forest Reserve, 22 Nov. 1962, *Boaler* 741! & 27 Nov. 1962, *Boaler* 776!; Mbeya District: Ruaha National Park, Magangwe Ranger Post, 19 Dec. 1972, *Bjørnstad* 2169! & same area, May 1970, *Greenway & Kanuri* s.n.!
DISTR. **T** 4, 7; Zambia
HAB. *Terminalia-Combretum* open woodland and *Brachystegia* woodland; 1320–1400 m.

SYN. *C. milne-redheadii* Mold. in Phytologia 3: 264 (1950); F.F.N.R.: 366 (1962); Bjørnstad, Veg. Ruaha Nat. Park 1: 58 (1976). Type: Zambia, Mwinilunga, E. of R. Matonchi, *Milne-Redhead* 3526 (K, holo.!)

NOTE. Clearly closely related to *C. myricoides* but I have accepted it as a peripheral species with a very distinctive habit. Moldenke does not mention *C. milne-redheadii* in his final series of notes on *Clerodendrum*. There are a number of differences between Thomas's description of *prittwitzii* and the available material of *milne-redheadii* but it is very unlikely that they are not the same taxon. In his key (but not in the description) the shoots are said to be branched and the leaves almost entire, the calyx-lobes are described as obovate, 3–4 mm. wide at base, the corolla pale violet with a tube 7 mm. long with 4 lobes 7 mm. long and anticous one ± 1 cm. long. Thomas records two other *von Prittwitz* sheets from Kilimatinde and *Böhm* 13 from Tabora District, Igonda [Gondah] 18 Feb. 1892, the basis of the **T** 4 record.

40. **C. myricoides** (*Hochst.*) *Vatke** in Linnaea 43: 535 (1882); Engl., Hochgebirgsfl. Trop. Afr.: 356 (1892); Gürke in P.O.A. C: 341 (1895) & in E. & P. Pf. 4, 3: 176 (1895); Bak. in F.T.A. 5: 310 (1900), pro parte; Thomas in E.J. 68: 86 (1936); F.P.N.A. 2: 147 (1947); T.T.C.L.: 632 (1949), pro parte; F.P.S. 3: 195 (1956); Jex-Blake, Gard. E. Afr., ed. 4: 109, 225 (1957); K.T.S.: 585 (1961); F.F.N.R.: 366 (1962), pro parte; F.P.U.: 147, fig. 90 (1962); U.K.W.F.: 616 (1974); Vollesen in Opera Bot. 59: 82 (1980); Blundell, Wild Fl. Kenya: 108, t. 44/285 (1982); Fl. Rwanda 3: 276 (1985), pro parte; Blundell, Wild Fl. E. Afr.: 398, t. 615 (1987); Moldenke in Phytologia 62: 328, 452 (1987) (exhaustive bibliography). Type: Ethiopia, lower and middle slopes of Mt. Scholoda, *Schimper* 330 (?TUB, holo., B†, BM!, K!, iso.)

Shrub, subshrub or subshrubby herb, rarely a small tree, usually much branched, 0.9–2.4(–6) m. tall; branches pale brown, ridged or angular, at length with ± corky bark, glabrous to velvety. Leaves opposite or in whorls of 3–4; blades narrowly to broadly elliptic, ovate-elliptic or oblanceolate, oblong or obovate, 2–15(–19.5) cm. long, 0.4–6(–10) cm. wide, usually small but in cultivation can attain large dimensions, acute to acuminate at the apex, cuneate at the base (in one subsp. narrowed to an oblong base), entire to coarsely serrate, glabrous or pubescent to densely velvety, glandular-punctate beneath, ± sessile, or petiole up to 1.5 cm. long. Flowers in few–several-flowered dichasial cymes arranged in unelaborated and lax to quite extensive and elongate panicles 6.5–15(–30) cm. long; peduncles 0–7 cm. long, secondary peduncles up to 4 cm. long; apparent stalks 1–2.5 cm. long but true pedicels 3–5 mm. long. Calyx often entirely purplish or crimson-margined, glabrous to hairy; tube cupular, ± 2.5 mm. long; lobes semicircular to ovate or triangular, 1.2–5 mm. long, quite rounded, obtuse or ± acute. Corolla usually greenish with white to pale blue limb, the anterior lobe dark blue; tube 5–7 mm. long, pubescent at the throat; lobes unequal, (0.6–)1–1.9(–2) cm. long, (1.5–)3.5–7.5 mm. wide, the upper obovate, the lower one spathulate. Stamens and style well exserted. Fruit ± black, subglobose, depressed, 5–6 mm. long, 8–10 mm. wide, mostly deeply 4-lobed, glabrous.

---

* The name of Robert Brown is frequently associated with this species but his mention of it in Salt, Voy. Abyss. App.: lxv (1814) is purely as a name with no description or reference.

KEY TO INFRASPECIFIC VARIANTS

1. Corolla green; subshrubby herb with many unbranched
   stems to ± 40 cm. (T 8) . . . . ` . . . .   a. subsp. **myricoides**
       var. **viridiflorum**

   Corolla at least partly blue or white or if greenish habit
   different 2
2. Leaves glabrous or very sparingly pubescent on venation
   beneath; corolla glabrous outside or lobes ± ciliate 3
   Leaves pubescent to thickly velvety; corolla glabrous
   outside, save for ciliate lobes, to thickly velvety 4
3. Shrub or subshrub or occasionally said to be tree-like,
   usually ± 2(–4.5) m. tall; branches leafy; leaves often
   larger (1–)4–12 × (0.5–)2–7 cm.; corolla-tube 5–6 mm.
   long; lobes 10–20 × 5–7 mm. (widespread) . . . .   subsp. **myricoides**
       var. **myricoides**

   Small tree to 6 m., with leaves ± 3 × 1.2–1.8 cm., congested
   and apparently ± restricted to new apical shoots;
   corolla-tube 6 mm. long; lobes 6 × 1.5–2 mm. (T 6,
   Mafia I.) . . . .   b. subsp. **mafiense**
4. Shrub with very slender branches; leaves small, broadly
   elliptic, 1–4.5 × 0.5–2.5 cm., usually coarsely crenate-
   serrate, with 2–4(–5) lobes on each side; inflorescences
   few, mainly simple, axillary, 1–3-flowered, the peduncle
   very slender and usually much exceeding the leaves (T
   1, Mwanza and Shinyanga) . . . .   c. subsp. **ussukumae**
   Not as above; leaves mostly larger or, if not, inflorescences
   more developed, coarser and more numerous 5
5. Inflorescences denser, leafy, forming often quite long
   terminal thyrses or panicles; axes of individual cymes
   (at up to 12 nodes) mostly shorter; corolla mostly
   pubescent to velvety but sometimes ± glabrous save for
   marginally ciliate lobes (Tanzania, T 4, 7) 6
   Inflorescences laxer and less developed, the axes of the
   cymes longer; corolla glabrous to velvety (widespread) 8
6. Leaves typically obovate, contracted to a narrow rounded
   or even ± subcordate base, often slightly panduriform,
   subsessile, less often narrowed into a short petiole,
   3–10.5 × 1.5–9 cm.; calyx-lobes broadly rounded or
   ovate-triangular, 1–2.5 × 2–3 mm.; corolla-limb densely
   pubescent; shrub or subshrub 0.9–2.4 m. tall (T 7)   e. subsp.
       austromonticola

   Leaves usually elliptic, narrowed to the base (T 4) 7
7. Leaves narrowed at base into very distinct petiole; blades
   3–10 × 2.5–6 cm.; calyx-lobes semicircular, 1.5 mm.
   long, densely hairy; corolla glabrous save for ciliate
   lobes or pubescent; shrub or small tree 3–3.6 m. tall   d. subsp. **napperi**
   Leaves narrowed but ± sessile or very shortly petiolate;
   blades 5–16 × 2–11 cm.; calyx-lobes triangular or ovate,
   usually at least some acute or acuminate and some
   rounded, well developed, 3–5 mm. long, 3–3.5 mm.
   wide; corolla-limb densely pubescent outside; herb or
   subshrub 0.6–1.5 m. tall . . . .   f. subsp. **muenzneri**
8. Corolla-lobes glabrous except for ciliate margin . . .   a. subsp. **myricoides**
       var. **discolor**

   Corolla-lobes pubescent to densely velvety . . . . .   a. subsp. **myricoides**
       var. **kilimandscharense**

a. subsp. **myricoides**

Usually a subshrub or subshrubby herb. Leaves usually rather small (but occasionally attaining large dimensions), elliptic, 1–12(–19.5) cm. long, 0.4–7(–10) cm. wide, toothed or entire, glabrous to velvety. Calyx-lobes usually short, semicircular. Inflorescences mostly lax. Corolla-limb glabrous to densely pubescent outside.

var. **myricoides**

Leaves typically narrowly elliptic, 1.2–8(–12) cm. long, 0.4–4(–6) cm. wide, acute at the apex, attenuate at the base, ± glabrous or sparsely pubescent on nerves above and beneath, entire or slightly or strongly toothed. Corolla-limb glabrous or lobes merely ciliate.

UGANDA. W. Nile District: hill W. of Mt. Eti, 25 July 1953, *Chancellor* 40!; Kigezi District: Ruhinda, Jan. 1951, *Purseglove* 3555!; Busoga District: Butembe Bunya, Nov. 1937, *Webb* 59!
KENYA. Elgon at 1950–2250 m., *T.H.E. Jackson* 364!*; Kiambu District: Kikuyu Escarpment Forest, Lari Forest Reserve, 18 Nov. 1972, *Hansen* 781!; N. Kavirondo District: Kakamega, Lubiri village, 12 July 1960, *Paulo* 535!
TANZANIA. Biharamulo District: Lusahanga, 15 Oct. 1960, *Tanner* 5302!; Kigoma District: Gombe Stream, 27 Jan. 1955, *Benedicto* 29!; Mbeya District: Mbosi Circle, Judyland Farm, 10 Jan. 1961, *Richards* 13814! (intermediate)
DISTR. U 1–4; K 1–7; T 1, 4, 6, 7 (intermediates); Zaire, Rwanda, Ethiopia, Sudan, Somalia
HAB. Grassland, scrub, thicket, open woodland, black cotton soil and rocky outcrops; 900–2250(–2400) m.**

SYN. *Spironema myricoides* Hochst. in Fl. Abyss. exsicc. Un. It. i 1840 no. 330 (?1841)
Cyclonema myricoides (Hochst.) Hochst. in Flora 25: 226 (1842); Schauer in DC., Prodr. 11: 675 (1847); A. Rich., Tent. Fl. Abyss. 2: 171 (1850); Oliv. in Trans. Linn. Soc. 29: 133 (1875)
Siphonanthus myricoides (Hochst.) Hiern in Cat. Afr. Pl. Welw. 1: 844 (1900)
Clerodendrum ugandense Prain in Bot. Mag. 135, t. 8235 (1909); Thomas in E.J. 68: 88 (1936); Everett, Encycl. Hort. 3: 794, 795 (illustr.) (1981). Type: material grown at Kew from seeds collected at Voi, 600 m., *Dawe* (K, holo.!)
C. myricoides (Hochst.) Vatke var. *niansanum* Thomas in E.J. 68: 87 (1936); T.T.C.L.: 632 (1949); Moldenke in Phytologia 62: 469 (1987), *nomen invalid*. Type: Rwanda, Niansa, *Kandt* 91 (B, holo.†, B, EA, iso.!)
C. myricoides (Hochst.) Vatke var. *involutum* Thomas in E.J. 68: 88 (1936); Moldenke in Phytologia 62: 468 (1987), *nomen invalid*. Type: Kenya, Nandi Plateau, *H.H. Johnston* (K, holo.!)

NOTE. Although typically the leaves are small there are populations with larger leaves which are, I think, essentially of the savanna type, e.g. *Pirozynski* 169 (Buha District, Kakombe valley, 7 Jan. 1964) is rather distinctive with rather large entire leaves up to 11 × 4 cm. The badly misnamed *C. ugandense*, named from material cultivated at Kew, has very large leaves, to 12 × 6 cm.; nothing similar has been seen near Voi in Teita District, from whence the seeds supposedly came, despite much collecting in the area. Presumably Dawe was using the railway and on his way to Uganda. Horticulturalists always keep it distinct from *myricoides* and it is best considered a cultivar (see Staples in Baileya 23: 166 (1991))
Moldenke (Phytologia 59: 248 (1986)) cites two specimens under the name *C. dekindtii* Gürke —*Elliot* 422 (Kenya), which I have not seen, and *Peter* 34991! (Tanzania, Ngulu, Goweko W. to Igalula km. 790, 15 Jan. 1926). The latter resembles *C. taborense* closely but has blunt sepals. Both specimens are probably forms of var. *myricoides*.

var. **discolor** (*Klotzsch*) *Bak.* in F.T.A. 5: 310 (1900); Troupin in Fl. Rwanda 3: 276 (1985), adnot. Type: Mozambique, Rios de Sena, *Peters* (B, holo.†)

Similar to var. *myricoides* but stems velvety and foliage densely pubescent to distinctly velvety on both surfaces. Limb of corolla glabrous or lobes merely shortly ciliate at margins.

UGANDA. W. Nile District: R. Oru near Omogo Rest Camp, 8 Aug. 1953, *Chancellor* 140!; Acholi District: Lotuturu [Lututura], June 1950, *Dale* U298! (intermediate); Mengo District: Busiro, 3 km. SE. of Jungo, 11 Feb. 1969, *Lye* 1910!
KENYA. Northern Frontier Province: Mathews Range, Ol Doinyo Lengio, 19 Dec. 1958, *Newbould* 3251!; W. Suk District: Kongelai, July 1961, *Lucas* 191!; Laikipia District: Ngobit Escarpment, 3 Mar. 1933, *Jex-Blake* in *Napier* 2559!
TANZANIA. Morogoro District: Shikurufumi [Chigurufumi] Forest Reserve, May 1955, *Semsei* 2010!; Rungwe District: road to Ipinda [Ipindi], 22 May 1957, *Richards* 9894!; Songea District: Matengo Hills, Miyau, 13 Jan. 1956, *Milne-Redhead & Taylor* 8322!
DISTR. U 1–4; K 1–6; T 1–8; Z; Zaire, Rwanda, Burundi, Ethiopia, Mozambique, Zambia and Zimbabwe
HAB. Grassland, thicket, bushland, woodland including *Brachystegia-Protea-Parinari* and *Juniperus* forest edges, riverbanks in dry country and rocky slopes; 900–2400 m.

SYN. *Cyclonema discolor* Klotzsch in Peters, Reise Mossamb., Bot. 1: 262 (1861)
Clerodendrum discolor (Klotzsch) Vatke in Linnaea 43: 536 (1882); Thomas in E.J. 68: 86 (1936); F.P.N.A. 2: 147 (1947); T.T.C.L.: 632 (1949), pro parte; K.T.S.: 583 (1961); E.P.A.: 798 (1962); F.F.N.R.: 365 (1962); Moldenke in Phytologia 59: 255 (1986)
C. phlebodes C.H. Wright in K.B. 1907: 54 (1907); Thomas in E.J. 68: 84 (1936); T.T.C.L.: 632 (1949). Type: Uganda, Mengo District, Entebbe, *Mahon* (K, holo.!)

---

* This gathering is very similar to the Ethiopian type.
** The highest altitude is based on *Dummer* 3621 from Mt. Elgon.

C. *bequaertii* De Wild. in B.J.B.B. 7: 185 (1920), *non* De Wild. (1914), *nom. illegit.* Type: Zaire, 8 syntypes

C. *bequaertii* De Wild. var. *debeerstii* De Wild. in B.J.B.B. 7: 185 (1920). Type: Zaire, Mpala, *Debeerst* (BR, holo.)

C. *villosulum* De Wild. in Contr. Fl. Katanga: 165 (1921). Type as for C. *bequaertii* De Wild.

C. *villosulum* De Wild. var. *debeerstii* (De Wild.) De Wild. in Contr. Fl. Katanga: 185 (1921)

C. *phlebodes* C.H. Wright var. *pilosocalyx* Thomas in E.J. 68: 84 (1936); T.T.C.L.: 633 (1949), *nomen invalid.* Type: Tanzania, Iringa District, Uhehe, Mgololo, *Goetze* 761 (B, holo.†, BR, iso.!)

C. *discolor* (Klotzsch) Vatke var. *duemmeri* Thomas in E.J. 68: 86 (1936); T.T.C.L.: 632 (1949); Moldenke in Phytologia 59: 262 (1986), pro parte, *nomen invalid.* Type: Uganda, Mengo District, Kirerema and Kipayo, *Dummer* 94 (B, holo.†, BM, K, iso.!)

[C. *discolor* (Klotzsch) Vatke var. *kilimandscharense* Thomas in E.J. 68: 86 (1936), pro parte, quoad *Fries* 247a, etc., *non* Verdc.]

[C. *myricoides* (Hochst.) Vatke var. *niansanum* sensu Thomas in E.J. 68: 37 (1936) quoad Schlieben 5959 *non* Thomas sensu stricto, *nomen invalid.*]

NOTE. Klotzsch mentions that the corolla-lobes of his taxon are pubescent at the margins. Specimens with velvety and ± glabrous leaves sometimes grow together, e.g. *Leippert* 5228 (Baringo District, Kabarnet to Maringat, 7 km., 28 Oct. 1964). C. *phlebodes* is a form with very large leaves up to 19.5 × 10 cm. and long inflorescences.

A number of *Peter* specimens cited as C. *erectum* De Wild. by Moldenke (Phytologia 59: 350 (1986)) are best placed here and De Wildeman's taxon is only a variant of C. *myricoides*. Almost every new collection produces new variants in this complex, many of which may be constant in the areas concerned. *Bidgood et al.* 1936 (Tanzania, Masasi District, Chidya, Kambona Forest Reserve, 12 Mar. 1991) has leaves to 16 × 7 cm. and rather long but still rounded calyx-lobes.

var. **kilimandscharense*** *Verdc.* a var. *discolori* limbo corollae extra satis pubescenti vel velutino differt. Typus: Tanzania, Kilimanjaro, Marangu, *Volkens* 226 (K, holo.!, BM, HBG, iso.!)

Similar to var. *discolor* but corolla-lobes pubescent to thickly velvety hairy outside.

TANZANIA. Biharamulo District: Lusahanga, 15 Oct. 1960, *Tanner* 5331!; Arusha District: Arusha National Park, track to Sakila [Sakile] Swamp, near Ngurdoto Crater, 14 Dec. 1968, *Richards* 23329!; Moshi District: Lyamungu, 19 Aug. 1932, *Greenway* 3041!; Zanzibar I., Kizimkazi [Kisimkazi] 'mile 36', 19 July 1963, *Faulkner* 3221!

DISTR. T 1–3, 4 (intermediate), 5, 6 (intermediate), 7, 8; Z (see note)

HAB. Grassland, bushland and forest edges, dry stony hillsides, etc.; (0–)600–2350 m.

SYN. C. *discolor* (Klotzsch) Vatke var. *kilimandscharense* Thomas in E.J. 68: 85 (1936); T.T.C.L.: 632 (1949); Moldenke in Phytologia 59: 263 (1986), pro parte, *nomen invalid.* Types: Tanzania, Kilimanjaro, Marangu, *Volkens* 226 and 252 (both B, syn.†)

C. *discolor* (Klotzsch) Vatke var. *eudiscolor* forma sensu Thomas in E.J. 68: 85 (1936), pro parte quoad *Uhlig* 261, etc.

[C. *discolor* sensu Brenan, T.T.C.L.: 632 (1949), quoad *Greenway* 3041, *non* (Klotzsch) Vatke sensu stricto]

NOTE. In Zanzibar material approaches C. *sansibarense* with leaves larger and broader, to 9.5 × 7 cm. and in truth here there may be only one taxon. Such is the variation in the *myricoides* complex that total illogicalities occur in some areas. Thomas records var. *kilimandscharense* from Kenya (Machakos District, Kibwezi [Kibwesi], *Kassner* 683 and N. Nyeri District, Rongai R., *Fries* 74a (actually 274a)) but these are var. *discolor*.

var. **viridiflorum** *Verdc.*, var. nov. inter taxa affinia ob flores virides et habitum ± herbaceum multicaulem differt. Typus: Tanzania, Songea District, *Bidgood et al.* 2091 (K, holo.!, EA, iso.!)

Subshrubby herb with many unbranched stems to 40 cm. from a woody rootstock. Leaves ± sparsely pubescent beneath. Corolla pale green, the petals glabrous except for ciliate margins.

TANZANIA. Songea District: Kwamponjore valley, 7 Feb. 1956, *Milne-Redhead & Taylor* 8629! & between Peramiho & R. Wuwawesi, 21 Feb. 1956, *Milne-Redhead & Taylor* 8629A! & 40 km. on Songea–Njombe road, 22 Mar. 1991, *Bidgood et al.* 2091!

DISTR. T 8; not known elsewhere

HAB. *Brachystegia-Uapaca* woodland and derived overgrazed grassland; 960–1000 m.

NOTE. Vollesen attests to the distinctness of this variant in the field.

b. subsp. **mafiense** *Verdc.*, subsp. nov. ob folios parvos subglabrescentes subsp. *myricoidis* var. *myricoidis* affinis sed habitu arborescenti ramis vetustioribus aphyllis, corollis parvis differt. Typus: Tanzania, Rufiji District, Mafia I., *Greenway* 5251 (K, holo.!, EA, K, iso.!)

---

* I have maintained Thomas's spelling to avoid confusion.

Much-branched tree 6 m. tall with rather slender pale stems; youngest parts pubescent, soon glabrous. Leaves and inflorescences clustered at first node of old growth, the rest of the stems leafless (and probably so in nature); blades obovate-elliptic, ± 3 cm. long, 1.2–1.8 cm. wide, very shortly acuminate at the apex, cuneate at the base, entire, glabrescent above, pubescent on the nervation beneath; petiole 5 mm. long. Inflorescences ± 5 cm. long; peduncle and axes slender, finely pubescent, some at least in dry state distinctly purple. Calyx slightly pubescent at base. Corolla pale blue, ± 1 cm. long, the limb entirely glabrous save for ciliate margins.

TANZANIA.   Rufiji District: Mafia I., Kilindoni, 13 Sept. 1937, *Greenway* 5251!
DISTR.   T 6; not known elsewhere
HAB.   *Hymenaea* (*Trachylobium*), Rosaceae woodland, thicket; ± 50 m.

NOTE.   This has a very distinctive appearance. A specimen, *Harris et al.* 5376 (Tanzania, Uzaramo District, Pugu Hills, near Kiserawe, 8 Nov. 1970), is very similar in general habit with very pale bark and leaves and foliage tufted at shoot apices but is described as a shrub (?scandent) up to 3 m. *Peter* 31361 and 31416 also from the Pugu area are the same but also without habit data.

c. subsp. **ussukumae** *Verdc.*, subsp. nov., subsp. *myricoidis* valde similis et sine charactere singulari distinguenda sed permixtione foliorum late ellipticorum 1–4.5 × 0.5–2.5 cm. subtus velutinorum plerumque grosse crenato-serratorum distincte petiolatorum, inflorescentiarum plerumque axillarium gracilium 1–3-florium, pedunculorum quam laminorum distincte longiorum distinguenda. Typus: Tanzania, Shinyanga District, Mwantine [Mantini] Hills, Igaramhuri, *B.D. Burtt* 5105 (K, holo.!, BR, EA!, iso.)

Shrub 1.8–2.4 m. tall with slender stems; young parts square, spreading pubescent. Leaves opposite, broadly elliptic, 1–4.5 cm. long, 0.5–2.5 cm. wide, obtuse to acute at the apex, cuneate at the base, entire to usually coarsely crenate-serrate with 2–4(–5) lobes on each side, mostly in upper half, discolorous and velvety beneath; petioles distinct, up to 7 mm. long. Inflorescences mainly simple axillary, 1–3-flowered dichasial cymes, the peduncle slender and usually much exceeding the leaves. Calyx glabrescent to sparsely pubescent. Corolla pale to bright blue or mustard yellow with red-purple centre (*fide* Tanner); tube glabrous to slightly pubescent, 7 mm. long.

TANZANIA.   Mwanza District: *Rounce* 233! & Buiru, 19 Apr. 1952, *Tanner* 668! & shores of Speke Gulf, 1 June 1931, *B.D. Burtt* 2473! & Nassa, Igalukiro, Ngasamo, 22 Feb. 1950, *Tanner* 549!
DISTR.   T 1; not known elsewhere
HAB.   Thicket of *Commiphora eminii*, *Tinnea*, *Combretum*, etc. including along lakeshores on hills of great granite rocks; also in old cultivations on black cotton soil; 1120–1260 m.

NOTE.   Burtt remarks on the highly pungent odour of the leaves and Tanner (668) states whole plant unpleasantly aromatic but the label of his 549 states 'leaves not aromatic'. The Wasukuma use the split stems for basket making and another note mentioned that the leaves are put on fires to keep away mosquitos and the roots used for constipation. It seems possible that *C. discolor* (Klotzsch) Vatke var. *crenatum* Thomas in E.J. 68: 85, 86 (1936) — type: 'Uganda', Bukumbi, 1190 m., *Conrads* 8 (B, holo.†), *nomen invalid.* — is this taxon but the brief key entry is inadequate for certainty. I believe the Bukumbi in Mwanza District must be the locality meant; there is another in Toro District in Uganda, but Conrads never went there. Moldenke in Phytologia 59: 262 (1986) also cites *Tanner* 668 and *B.D. Burtt* 5105 as possibly being var. *crenatum* together with specimens from Zimbabwe and the Transvaal; he had not seen the type nor does he mention that the name is invalid.

If only the 'Flora area' is considered this taxon appears highly distinctive but forms of var. *myricoides* in Ethiopia approach it; since it differs strongly from other populations in NW. Tanzania it needs a name.

d. subsp. **napperi** *Verdc.*, subsp. nov. a subsp. *myricoide* var. *discolori* (Klotzsch) Bak. habitu subarborescente, ramis crassis juventute dense villosis, foliis late ellipticis vel ± rotundatis subtus dense velutino floccoso-pubescentibus differt; affinis subsp. *austromonticola* Verdc. sed foliis basi in petiolo valde angustatis distincta. Typus: Tanzania, Ufipa District, Mbizi Forest, *Napper* 1103 (K, holo.!, EA, iso.)

Shrub (or even described as 'small tree') 3–3.6 m. tall, with thick ± square stems 3–5 mm. wide in dry state; young parts densely floccose-villous; older stems leafless; bark rough and fissured. Leaves opposite and probably also in whorls of 3 on some shoots, broadly elliptic to almost round, 3–10 cm. long, 2.5–7 cm. wide, acute to rounded at the apex, narrowed at the base into a distinct petiole, subentire to shallowly crenate-serrate in upper half, discolorous, densely velvety floccose-pubescent beneath; petiole up to 1.5 cm. long. Inflorescences terminal and axillary at first 4–6 nodes, usually 5–7-flowered, 4–11.5 cm. long, the peduncles up to 7 cm. long, mostly exceeding or equalling the leaves. Calyx densely hairy; lobes semicircular, 1.5 mm. long. Corolla green with lip blue or mauve, ± 2 cm. long; limb glabrous, save for marginal ciliae, or pubescent.

TANZANIA.   Ufipa District: uplands above Sumbawanga, 8 Nov. 1956, *Richards* 6920! & Sumbawanga, 15 Jan. 1950, *Bullock* 2228 & 2228A! & near Sumbawanga, 15 Dec. 1934, *Michelmore* 1062!
DISTR.   T 4; not known elsewhere
HAB.   Unburnt *Themeda-Hyparrhenia* grassland below lower edge of forest; 1800–2400 m.

e. subsp. **austromonticola** *Verdc.*, affinis *C. myricoidis* (Hochst.) Vatke var. *discoloris* (Klotzsch) Bak. inflorescentiis longioribus densioribus thyrsoideis, foliis obovatis saepe ± leviter panduriformibus basi angustatis basi ipsa saepe anguste oblonga interdum subcordata, limbo corolla extra dense pubescenti differt. Typus: Tanzania, Rungwe, Kiwira Forest Station, *Richards* 14342 (K, holo.!)

Shrub or subshrubby herb 0.9–2.4 m. tall. Leaves mostly rather large, obovate, ± round or broadly elliptic, often slightly panduriform, 3–10.5 cm. long, 1.5–9 cm. wide, acute to shortly acuminate at the apex, narrowly cuneate or narrowed to an oblong base, sometimes slightly amplexicaul or subcordate, sessile or sometimes pseudopetiolate but slight widening at extreme base present, coarsely bluntly crenate to sharply serrate but lower narrowed part entire, discolorous, slightly to densely pubescent beneath. Inflorescences many-flowered with terminal and axillary ± much branched cymes at up to 12 nodes forming a ± dense narrow leafy thyrse, the axes, calyces, bracts and indumentum often brilliant violet, at least when dry. Calyx-lobes usually broadly rounded, less often triangular, acute, 1–2.5 mm. long, 2–3 mm. wide. Corolla-limb densely pubescent, upper lip purple, lower mauve or blue.

TANZANIA. Mbeya District: Mbosi Circle, Mbimba [Mbimbe] Agricultural Station, 12 Jan. 1961, *Richards* 13890! & World's End View, Ipinda, 6 Feb. 1976, *Grey-Wilson & Mwasumbi* 10564! & Mbeya Mt., 'Catchment B', 11 Jan. 1963, *Napper* 1702!; Njombe District: Matamba, 7 Jan. 1957, *Richards* 7527!

DISTR. **T** 7; not known elsewhere
HAB. Plateau grassland; 1500–2415 m.

NOTE. The flowers are variously described as having the upper lip pale blue, blue or purple and the lower royal blue, blue or mauve. *C. myricoides* (Hochst.) Vatke var. *stolzei* Thomas (in E.J. 68: 88 (1936), *nomen invalid*; type: Tanzania, Rungwe District, Kyimbila, *Stolz* 317 (B, holo.†, K, iso.!)), probably belongs here, at least partly, although Thomas cites other material including a specimen from Ethiopia ('Gallahochland').

I have seen none of the material cited by Thomas under *C. discolor* (Klotzsch) Vatke var. *pluriflorum* Gürke (in E.J. 30: 391 (1901); Thomas in E.J. 68: 85 (1936); T.T.C.L.: 632 (1949); type: Tanzania, Rungwe District, Kananda, 1500 m., *Goetze* 1437 (B, holo.†)). The material comes from **T** 7–8 and is described as having inflorescences 30 cm. long with violet hairs but the leaves are entire. It might be a form of subsp. *austromonticola*. After describing this variety Gürke comments on the difficulties of keeping *C. discolor* and *C. myricoides* separate. Thomas records var. *pluriflorum* from Mozambique, Ethiopia and Angola also.

Leedal 2012 (Mbeya District, Ilembo, 2100 m.) is a peculiar plant with sharply square stem with brown peeling bark, sessile glabrous leaves, at least to 12 × 11 cm., narrowed to base, peduncle to 6 cm., few-flowered cymes, glabrescent calyces and tomentose buds. It may be form of subsp. *austromonticola* but has not been matched with any material. It may not be normal.

f. subsp. **muenzneri** (*Thomas*) *Verdc.*, comb. et stat. nov. Type: Tanzania, S. Ufipa, Msamvia [Msambia], *Fromm & Muenzner* 108 (B, holo. †)

Herb or subshrub 0.6–1.5*(-?) m. tall. Leaves mostly rather large, paired or in whorls of 3–4; blades ovate, elliptic- or oblanceolate-ovate, 5–16 cm. long, 2–11 cm. wide, acute to acuminate and mucronulate at the apex, rounded to cuneate at the base, coarsely crenate-toothed or ± lobulate, ± discolorous, adpressed pubescent above, ± densely pubescent and gland-dotted beneath. Inflorescence many-flowered with numerous axillary dichasial cymes, up to 4 per node arranged in a thyrse up to 30 cm. long, 12 cm. wide, the leaves gradually becoming bracts towards the apex; true bracts ovate to lanceolate, 0.5–1.5 cm. long, 2–7 mm. wide. Calyx densely pubescent; 3 upper lobes ovate-triangular, 5 mm. long, 3–3.5 mm. wide; 2 lower often rounded, 3–4 mm. long, 3 mm. wide; or all lobes narrowly triangular, 1–1.5 mm. wide. Corolla-limb mostly densely pubescent in bud.

TANZANIA. Mpanda District: S. Rukwa, 23 Dec. 1969, *Sanane* 973!; Ufipa District: Kipili road, 15 Feb. 1971, *Sanane* 1559! & Kipili, overlooking lake, 1 Feb. 1950, *Bullock* 2377! & Old Sumbawanga road near Mpui, 3 Jan. 1962, *Richards* 15899!

DISTR. **T** 4, ?7; see note
HAB. Open *Brachystegia* woodland on red sand; 780–1800 m.

SYN. *C. muenzneri* Thomas in E.J. 68: 105, 83 (1936); T.T.C.L.: 632 (1949); Moldenke in Phytologia 62: 327 (1987)

NOTE. East African material had mostly been determined as a form of *C. alatum* Gürke which typically has narrowly oblanceolate leaves ± 25 × 4.5 cm. There is no doubt that the taxon treated above is very closely allied and even in the Sudan and Ethiopia rather similar plants do occur which are similarly intermediate with *C. myricoides*. The merging of *C. myricoides* and *C. alatum* is really not acceptable but if wholesale lumping is done as by Persson then it is difficult to avoid adding *alatum* to the lengthy list of synonyms. Several sheets, e.g. *Richards* 7243 (Ufipa District, Chala Mt., 13 Dec. 1956), have distinctly acute or acuminate calyx-lobes thus approaching *C. taborense* but the facies is entirely that of subsp. *muenzneri*; they completely break down the calyx-lobe distinction between Thomas' sections *Pleurocymosa* and *Chaunocymosa* (i.e. *Cyclonema*)

---

* Thomas states 'ramis floriferis tantum notis' so perhaps collectors' notes provided this figure.

although they are separable on other grounds. *Grant 203* (Biharamulo District, Usui) has small calyx-lobes on one specimen and longer ones on the other; an identical-looking specimen from 5°S 33°E, i.e. Tabora, has larger calyx-lobes. Oliver (Trans. Linn. Soc. 29: 133 (1875)) treats all these as a variety of *Cyclonema myricoides* but does not mention the Tabora locality although reproducing a description on the field label. Baker (in F.T.A. 5: 310 (1900)) treats them as *C. myricoides* sensu stricto. It is clear the size and shape of the calyx-lobes varies like all the other characters. *Robinson 4805* (Rukwa Escarpment, Muse Gap, 29 Dec. 1961, river bank) clearly belongs near subsp. *muenzneri*, with similar long triangular calyx-lobes but the inflorescence is sparse and the thin entire sparsely pubescent leaves up to 17 × 7.5 cm render it intermediate with *C. sansibarense*. Yet another variant with the long calyx-lobes of subsp. *muenzneri* but ± glabrous buds, e.g. *Richards 22051* (Mbeya District, 32 km. Mbeya–Iringa, 25 Jan. 1967) and *Boaler 397* (Mbeya District, Chimala, 3 Jan. 1962) links with *C. taborensis*.

NOTE.   (on *myricoides* complex as a whole). One of the most difficult complexes in eastern Africa it is beset with the usual savanna/forest problems. I have maintained several species whereas E. Persson (Uppsala) has reduced them all to *myricoides*. Logically this may be inescapable but the result is a horrendously variable taxon obscuring all local evolutionary trends and ecology. Undoubtedly the distributions between forest and grassland populations break down in disturbed areas, e.g. the Korogwe area of the W. Usambaras and in northern Kenya on Mt. Kulal and Mt. Marsabit, where what I have referred to *C. myricoides* comes very close to *C. sansibarense*. Typical *C. alatum* Gürke in the Sudan and NE. Zaire is very distinctive but in Ethiopia even this merges with *C. myricoides*. The ultimate result is a monotypic sect. *Cyclonema* (sect. *Oligocymosa* Thomas). In a group of great horticultural potential, recognition of several taxa seems the wisest course.

41. C. sp. C

Shrub 2–3 m. tall with thin striate grey-brown branches, the youngest parts adpressed pubescent but soon glabrous. Leaves at apices of shoots and on very abbreviated branchlets in the axils of the fallen leaves on previous year's stems, ovate or ± rounded, 1.2–4 cm. long, 0.8–3.4 cm. wide, ± acute at the apex, cuneate at the base, coarsely crenate, sparsely shortly pubescent above and on main venation beneath; petioles 1–10 mm. long. Flowers in axillary 2–3-flowered cymes; peduncles ± 2.5 cm. long and secondary axes 1.5 cm. long, both very slender; pedicels 3–4 mm. long; all axes puberulous; bracts very narrowly triangular, 2.2 mm. long, attenuate. Calyx-tube cupular, 2 mm. long, venose, sparsely pubescent; lobes ovate, 1.5–2.5 mm. long. Buds glabrous. Corolla-tube 7–10 mm. long, hairy inside; limb blue and green or purplish white, with one lip 1 cm. long, the other lobes 8 mm. long, 5 mm. wide. Style and stamens exserted about 2.7 cm. Fruits not known.

KENYA.   Meru District: N. end of Nyambeni Range, Lagadema [Locadema] Hill, 20 Dec. 1971, *Bally & Smith* 14711! & Meru National Park, Kiolo R. crossing, 12 May 1972, *Ament & Magogo* 262!; Kitui District: 69 km. from Mwingi on Kora road, 26 Nov. 1983, *Mutangah & Muasya* 323!
DISTR.   **K** 4; not known elsewhere
HAB.   Unrecorded except outwash gully; 600–1000 m.

NOTE.   Probably no more than a form of *C. myricoides* but of distinctive habit. Leaves tufted on very short shoots occur on ± typical *myricoides* in seasonally very dry areas.

42. C. sp. D

Probably a scrambler; stems slender, brownish straw-coloured, densely softly spreading pubescent. Stem-leaves not known. Inflorescence-leaves only preserved, much shorter than the peduncles, rounded ovate, 1–1.5 cm. long, 1–1.3 cm. wide, acuminate at the apex, cordate at the base, entire, pubescent beneath, sessile. Inflorescence very lax, the leafy bracts becoming small towards apex, 3 mm. long and wide. Apparent peduncles (lateral axillary axes) slender, ± 3 cm. long; secondary elements 1.5–2 cm. long, pubescent; secondary bracts and bracteoles linear, 1–3 mm. long. Calyx-tube 1.5–2 mm. long; lobes broadly rounded to triangular, ± 1 mm. long. Corolla-tube 6–7 mm. long; lobes: short one ± round, ± 5 mm. long and wide, rest obliquely elliptic-oblong, 7 mm. long, 5 mm. wide. Stamens exserted ± 2 cm.

KENYA.   Lake Turkana [Rudolf], *Wellby*!
DISTR.   **K** 1 / Ethiopia
HAB.   Probably bushland

NOTE.   A plant of distinctive habit, possibly not from Kenya. Clearly closely related to *C. myricoides* but until it has been matched with new material I have left it separate.

43. **C. tanneri** *Verdc.*, sp. nov. affinis *C. myricoidis* (Hochst.) Vatke foliis ± uniformiter minoribus, inflorescentiis axillaribus numerosis ab apice rami ad nodos successivos sex dispositis, calycis lobis triangularibus, corolla minore ± 5–6 mm. longa differt. Typus: Tanzania, between Biharamulo and Kahama, *Verdcourt* 3451 (K, holo.!, EA, iso.!)

Shrub 75–90 cm. tall with slender angular stems, the nodes sometimes purplish, with many axillary branchlets in 3s above; young shoots with short adpressed ± sparse brown or whitish pubescence. Leaves in whorls of three, elliptic, 1–2.5(–5) cm. long, 0.4–1.5 cm. wide, acute at the apex, cuneate at the base, entire or with 1–2 acute teeth, often only on one side, ciliate or with few very short hairs on the midrib, ± glabrescent. Inflorescences axillary 1–3-flowered cymes, 1–3 per node, often 2 per axil; peduncles 1.5–3 cm. long, pedicels 1–3 mm. long; axes very sparsely shortly white pubescent; bracts lanceolate, 2.5 mm. long, 0.5–0.8 mm. wide. Buds asymmetric. Calyx purple; tube cupular, 2 mm. long; lobes triangular, 2 mm. long. Corolla-tube greenish or brownish yellow, 5 mm. long, glabrous outside but with pubescent throat; lip blue-mauve towards throat and side-lobes yellow, 5 mm. long, 3 mm. wide. Filaments and style greenish, exserted 12 mm. Fruit depressed, 8 mm. long, 5 mm. wide, ± 3-lobed. Fig. 19, p. 138.

TANZANIA. Biharamulo District: 44.8 km. from Biharamulo on the Kahama road, 24 Nov. 1962, *Verdcourt* 3451!; Buha District: about 6.4 km. from Kibondo, 15 Nov. 1962, *Verdcourt* 3319! & Kibondo, Nyabibuye, 5 Nov. 1960, *Tanner* 5600A!
DISTR. T 1, 4; not known elsewhere
HAB. *Combretum-Terminalia-Piliostigma-Ozoroa* woodland, 'light forest' and grassy hillsides; 1335–1350 m.
NOTE. *Tanner* 5905 (Kibondo, Kakonko, 20 Mar. 1961) is a white-flowered form. Named for Mr. Tanner who has collected extensively in **T** 1 and **T** 3.

44. **C. taborense** *Verdc.* a speciebus sectionis *Cyclonematis* (*Chaunocymosa*) calycis lobis anguste triangularibus vel lanceolatis differt; speciebus sectionis austroafricanae *Pleurocymosae* Verdc.* fallaciter similaris sed fructo leviter haud profundo lobulato corollae lobis ciliatis habitu fruticoso ramoso differt. Typus: Tanzania, Tabora District, Simbo Forest Reserve, *Ruffo* 953 (K, holo.!, EA, K, iso.!)

Shrub 1–2 m. tall with ± slender branches, at first with rather woolly spreading fawn hairs, later glabrous. Leaves opposite or in whorls of 3, aromatic, elliptic to oblanceolate or in one variety sometimes almost round, 1.3–7.5 cm. long, 0.7–4.5 cm. wide, acute to rounded but minutely mucronulate at the apex, cuneate to rounded or subcordate at the base, discolorous, glabrous, puberulous or velvety pubescent beneath, entire or with 1–few teeth towards apex; petiole 0–3 mm. long. Flowers few, in simple axillary and terminal dichasia, primary and secondary peduncles 1–4 cm. long; pedicels 3–5 mm. long; bracts linear, 3–4 mm. long. Calyx pubescent; tube 2.5–3.5 mm. long; lobes narrowly triangular to lanceolate-triangular, 2–4 mm. long, mostly ± equalling the tube, narrowly acute. Corolla green or blue; tube 5–6 mm. long, glabrous or asperulous-pubescent; limb glabrous save for ciliate margins, the lobes narrowly oblong-obovate, 8 mm. long, 3 mm. wide. Fruit 6 mm. long, 7–9 mm. wide, 4-lobed.

var. **taborense**

Leaves narrowly elliptic or oblanceolate, up to 4.5 cm. long, 2.5 cm. wide, mostly velvety pubescent beneath.

TANZANIA. Tabora District: Beekeeping Institute, 20 Dec. 1977, *Shabani* 1223! & Tabora, 9 Mar. 1937, *Lindeman* 271!; Mpwapwa, 30 Jan. 1931, *Hornby* 360!
DISTR. T 4, 5; not known elsewhere
HAB. *Brachystegia* woodland; 1200 m.
NOTE. Field notes record that it is called 'mpugambu' in Kinyamwesi, is evil-smelling and used as an insect repellant.
C. corbisieri De Wild. (in F.R. 13: 144 (1914); type: Zaire, Shaba, Welgelegen, *Corbisier* in Homblé 592 (BR, holo.!)) is very similar but has the inflorescences more distinctly terminal and without leaves in the upper part and shorter more broadly triangular calyx-lobes.

---

* Sect. *Pleurocymosae* Verdc., sect. nov. a sectione *Cyclonema* (Hochst.) Gürke (*Chaunocymosa* Thomas) calycis lobis semper anguste triangularibus vel lanceolatis, fructu profunde, fere perfecte 4-lobato, corollae lobis haud ciliatis differt. Typus: *C. triphyllum* (Harv.) H.H.W. Pearson. I have chosen *Cyclonema myricoides* as the lectotype of *Cyclonema* Hochst.

Fig. 19. *CLERODENDRUM TANNERI* — **1**, habit, × ⅔; **2**, flower, × 4; **3**, calyx, × 4; **4**, longitudinal
section of corolla, × 4; **5**, ovary, × 10; **6**, stigma, × 10; **7**, longitudinal section of ovary, × 14; **8**, part
of fruiting branchlet, × 2; **9**, fruit, × 3; **10**, pyrene, × 4. 1, from *Tanner* 5905; 2–7, from *Tanner*
5600A; 8–10, from *Verdcourt* 3319. Drawn by Mrs. M.E. Church.

var. **latifolium** *Verdc.*, var. nov. a var. *taborensi* foliis latioribus late ellipticis vel rotundato-cordatis, usque 7.5 × 4.5 cm. differt. Typus: Tanzania, 18 km. S. of Tabora, *Boaler* 499 (K, holo.!, EA, K, iso.!)

Leaves broadly elliptic, the upper broadly ovate to almost round, subcordate at the base, up to 7.5 cm. long and 4.5 cm. wide, puberulous beneath.

TANZANIA. Tabora District: 18 km. S. of Tabora on Sikonge road, 11 Feb. 1962, *Boaler* 499!
DISTR. **T** 4; not known elsewhere
HAB. Regenerating *Brachystegia* on old cultivations on granite hillslopes; 1200 m.

45. **C. rupicola** *Verdc.*, sp. nov. in subgen. *Cyclonemate* (Klotzsch) Gürke ponenda affinis *C. myricoidis* (Hochst.) Vatke ramulis folia inflorescentiasque ferentibus valde reductis in axillis foliorum primariorum delapsorum insidentibus, calycis lobis longioribus intus dense glandulosis differt. Typus: Kenya, Tana River District, Kora Reserve, *Hemming* 83/111 (K, holo.!, EA, iso.!)

Subshrub 0.5–3 m. tall, of *Commiphora*-like habit; stems branched, pale dull ± purplish brown, finely pubescent when young, later glabrous and wrinkled. Leaves tufted on very short pubescent nodular shoots, oblanceolate-obovate, 3.5–9 cm. long, 1.2–5 cm. wide, rounded at the apex, very narrowly cuneate at the base forming an apparent petiole, entire, very sparingly pubescent at base and on nerves beneath; true petiole ± obsolete. Inflorescences apparently axillary but probably terminal on the nodules, up to 5 cm. long, 1–3-flowered, puberulous; peduncle 5 mm. long; secondary axis 8 mm. long; pedicels flattened, 6 mm. long; bracts oblong, 7.5 mm. × 3 mm.; bracteoles linear, 2–5 mm. long. Calyx puberulous; tube 2.5–3 mm. long; lobes ovate to narrowly triangular, 3–4 mm. long, rounded at apex, densely glandular inside. Corolla dull wine-red or lateral lobes green and central one pale violet; tube curved, ± 1 cm. long; lobes oblong, 1.1–1.3 cm. long, 6–7 mm. wide, rounded at apex. Fruit globose, 1 cm. diameter, one seen 3-lobed.

KENYA. Northern Frontier Province: 5 km. W. of Kula Mawe, 8 Dec. 1971, *Bally & Smith* 14468! & Garissa area, NE. of Dadaab, 28 Nov. 1978, *Brenan et al.* 14791!; Tana River District: Kora Reserve, 0°21'S, 38°46'E, 3 May 1983, *Hemming* 83/111!
DISTR. **K** 1, 7; not known elsewhere
HAB. *Acacia-Cordia* open bushland with lava boulders and tussock grassland on rich volcanic soil; 160–800 m.

NOTE. *Gillett* 12739 (Northern Frontier Province, Dandu, 10 Apr. 1952) has shorter calyx-lobes as in *C. myricoides*, more pubescent foliage and inflorescences and smaller flowers; it is very similar in habit and I am temporarily referring it to *C. rupicola*. Although *C. rupicola* is clearly close to *C. myricoides*, and dry country populations of that species can have reduced lateral branchlets, it has a distinctive habit and a rather succulent look to the stems. To leave it as a mere form would be stretching the variability of *C. myricoides* too far.

46. **C. commiphoroides** *Verdc.*, sp. nov. affinis *C. myricoidis* (Hochst.) Vatke foliis inflorescentiisque in ramulis lateralibus valde congestis insidentibus, pedunculis gracilioribus 1–2-floris differt. Typus: Tanzania, Iringa District, 55 km. on Mafinga–Madibira road, *Bidgood et al.* 1297 (K, holo.!, EA, NHT, iso.!)

Shrub or small tree 2–3.6 m. tall; innovations sometimes ± pale ferruginous pubescent but stems otherwise glabrous, blackish and wrinkled, ± slender. Leaves clustered on very short abbreviated side shoots, 3–5 per shoot, the internodes suppressed. Leaves oblanceolate, obcuneate or narrowly elliptic, 1.2–6 cm. long, 0.6–2.2 cm. wide, acute to rounded at the apex, narrowed to the base, drying ± pale green, sessile, coarsely serrate towards the apex, glabrous but minutely gland-dotted. Peduncles very slender, apparently from the apices of short shoots but actually lateral, 2.5–4.5 cm. long, straw-coloured or purple, 1–2-flowered, secondary axes 1.5 cm. long, true pedicels 7 mm. long; inflorescences very lax; bracts narrowly elliptic, 1.5–2.5(–5) mm. long, 1–2 mm. wide. Calyx glabrous; tube 2–3 mm. long; lobes 1–2 mm. long, 3 mm. wide, rounded, glabrous. Corolla pale greenish outside, whitish inside with lower lip pale bluish purple, glabrous outside; tube ± 10 mm. long, lobes up to 1.5 cm. long, 5 mm. wide; stamens exserted 2 cm. Fruit black, 7 mm. long, 1–1.5 cm. wide, lobed.

TANZANIA. Iringa District: 36 km. W. of Iringa towards Idodi, 14 Feb. 1972, *Bjørnstad* 1388! & 57.5 km. N. of Iringa on Dodoma road, Nyangolo [Nyangoro], top of escarpment, 16 Jan. 1954, *Wigg* 988!; Njombe District: 48 km. N. of Njombe on Mbeya road, S. of Ilembula, 25 Jan. 1952, *Wigg* 1004! & 17.5 km. E. of Iringa–Dodoma road, Kitapilimwa, 21 July 1963, *Mathias & Taylor* A 145!
DISTR. **T** 7; not known elsewhere
HAB. *Brachystegia* woodland and *Markhamia, Piliostigma, Strychnos*; 1200–1500 m.

SYN. *Clerodendrum* (sect. *Chaunocymosa*) *sp*.; Mathias in Taxon 31: 491 (1982)
NOTE. Mathias records the use of this plant, 'munyludeke', for epilepsy.

47. C. sp. E

Stem very graceful, pale brown, densely shortly spreading pubescent. Leaves oblong-elliptic, 1–3.8 cm. long, 0.2–1.4 cm. wide, sharply acuminate at the apex, rounded at the base, ± discolorous, very shortly pubescent above and beneath; petiole 1 cm. long. Cymes axillary, 2–3-flowered, up to 2.5 cm. long; peduncles up to 1.3 cm. long, secondary peduncles 7 mm. long; pedicels 2.5 mm. long, all axes pubescent; bracts tinged vinous, linear, up to 6 mm. long, ciliate; bracteoles 2.5 mm. long. Calyx cupular, 2 mm. long, 3 mm. wide, pubescent; lobes vinous, oblong, 2.5–3 mm. long, 1.5 mm. wide, rounded or somewhat acute. No corolla nor fruit preserved.

TANZANIA. Shinyanga District: Msalala, Nov. 1882, *Hannington* s.n.!
DISTR. T 1; not known elsewhere
HAB. Not known

SYN. [*C. myricoides* sensu Bak. in F.T.A. 5: 310 (1900), pro parte, *non* (Hochst.) Vatke]

NOTE. Two fragmentary shoot-tips do not seem to match any other material I have seen; the sheet had been placed with the indeterminate specimens where it rested for very many years. Baker's identification is most certainly wrong. It is strange no other material has been collected in what was one of B.D. Burtt's collecting areas. It comes somewhat close to *C. tanneri* Verdc., which has adpressed indumentum and shorter not vinous calyx-lobes.

48. C. sp. F

Subshrubby herb of unknown height; very young stems densely adpressed pubescent, soon glabrescent, pale brown, 4-angled. Leaves elliptic, 1–2.5 cm. long, 4–8 mm. wide, acute at the apex, cuneate at the base, subsessile, pubescent above, densely pubescent beneath. Dichasia short, axillary in upper axils, the branches approximate, 2–2.5 cm. long; peduncle 1.5 cm. long; pedicels very short, ± 1–1.5 mm. long; all axes pubescent; bracts and bracteoles linear, 1–2 mm. long. Calyx-tube cupular, ± 2 mm. long and wide, densely pubescent; lobes probably purple, semicircular, ± 1 mm. long; sparsely pubescent or glabrescent. Corolla pale green, tinged mauve; tube ± 6 mm. long; lip-lobe narrowly oblong, 4 mm. long, others 2.5 mm. long. Stamens and style exserted up to 1.5 cm.

TANZANIA. Dodoma District: ± 16 km. Itigi–Chunya, 6 Nov. 1960, *Richards* 13527!
DISTR. T 5; not known elsewhere
HAB. "Woodland scrub possibly after burning"; 900 m.

NOTE. This is clearly closely related to *C. sp. E*, but the flowers have much shorter pedicels and the calyx-lobes are shorter. More material is required to check the constancy of the characters which might, to some extent, be due to the time of flowering and environmental factors. The Itigi area is known for many near-endemic plants.

49. C. sp. G

Shrub to 3 m.; stems angular above, densely adpressed hairy with matted hairs. Leaves opposite, narrowly elliptic, 4.5–14 cm. long, 2–4.5 cm. wide, acuminate at the apex, attenuate to a very narrowly rounded base or cuneate, densely pubescent above, with denser longer adpressed hairs beneath but scarcely velvety. Inflorescences axillary and pseudoterminal, 3–6 cm. long. Calyx-tube cupular, 2 mm. long; lobes rounded, 1.5 mm. long. Corolla blue; tube ± 8 mm. long; lobes ?rounded, ± 6 mm. long and wide, very sparsely hairy outside and ciliate. Style exserted at least 2 mm.

TANZANIA. Lindi District: Lake Lutamba, 9 Feb. 1935, *Schlieben* 5959!
DISTR. T 8; not known elsewhere
HAB. Bushland; 240 m.

NOTE. Clearly belongs to subgen. *Cyclonema* but not *C. myricoides*. The note attached to species 32 also refers to this species. Both specimens appear to have been treated in the same way.

50. C. sp. H

Shrub 2–3 m. tall with pale grey-brown wrinkled stems, the youngest parts together with the young foliage minutely spreading puberulous. Leaves in whorls of 3, drying pale yellow-green, ± triangular in outline, 6 cm. long, 5.5 cm. wide, obtuse at the apex, broadly cuneate then tenuously decurrent into a whitish 5 cm. long petiole, coarsely crenate; petiole-bases leaving a prominent scar. Flowers and fruits unknown.

KENYA. Northern Frontier Province; 15 km. from Wajir on El Wak road, 31 May 1977, *Gillett* 21285! DISTR. **K** 1; not known elsewhere HAB. *Acacia-Commiphora* bush with *Delonix* on red sand; 250 m.

NOTE. This sterile fragmentary specimen would not have been collected nor mentioned here but for the distinctive foliage. It seems almost certain it is a *Clerodendrum* and probably of the subgen. *Cyclonema*, although there is a slight possibility it could belong to some other Verbenaceous genus.

51. **C. makanjanum** *H. Winkler* in F.R. 18: 124 (1922); Thomas in E.J. 68: 88 (1936); T.T.C.L.: 631 (1949); Gillett in K.B. 14: 342–343 (1960); K.T.S.: 585 (1962); Moldenke in Phytologia 62: 145 (1987). Type*: Tanzania, S. Pare Mts., Makanya [Makanja]–Same, *Winkler* 3797 (BRSL, holo.)

Scrambling or ± erect shrub with silvery grey bark, usually flowering when leafless or with leaves very undeveloped (but occasionally with developed leaves), 0.6–3 m. tall; branches yellow-brown, fleshy, later darker and ± brittle, wrinkled when dry; youngest parts pubescent but soon glabrous. Leaves slightly fleshy, ovate, oblong or elliptic, 1.5–6 cm. long, 1.2–4 cm. wide, obtuse or subacute at the apex, attenuate into the petiole, glabrous, irregularly and coarsely serrate or crenate-dentate; petioles 3–5 cm. long. Inflorescences 5–10(–13) cm. long on short or very short axillary shoots nodulated by petiole-bases, resembling a labiate with dense whorls of flowers, the nodes ± 7 mm. apart; rhachis rich purple and ± fleshy; pedicels purple, 2–6 mm. long. Buds glandular-pubescent outside. Calyx greenish to purple, 2–4 mm. long; lobes broadly triangular, 1–1.5 mm. long. Corolla white and blue, pale green or dull yellow-green with pinkish purple lip; tube curved, 3–4 mm. long; lobes declinate, 8 mm. long, 3.5 mm. wide, with mauve lip darker near the tip, other lobes white or all pale mauve. Filaments and style white or greenish with purple anthers and mauve stigma. Stamens erect and arcuate, exserted 1.2–1.3 cm. Fruit subglobose-obovoid, 9 mm. long, 8 mm. wide, (2–)3–4-lobed when dry. Fig. 18/15, 16, p. 123.

KENYA. Machakos District: near Mtito Andei, about 240 km. from Mombasa, 29 Aug. 1959, *Verdcourt* 2380!; Masai District: Nairobi–Magadi road, 9.5 km. E. of turn-off for Olorgesailie, 19 Oct. 1974, *Gillett* 20741!; Teita District: Mwatate–Voi road, 6.4 km. from Mwatate, 18 Sept. 1953, *Drummond & Hemsley* 4416! TANZANIA. Pare District: N. Pare Hills, Lembeni–Same, 23 Jan. 1957, *Bally* 11369!; Lushoto District: Mazinde, 1 May 1953, *Drummond & Hemsley* 2330!; Kilosa District: at Ruaha R., 5 km. W. of junction with Yovi R., 1 Sept. 1970, *Thulin & Mhoro* 820! DISTR. **K** 4, 6, 7; **T** 3, 6; not known elsewhere HAB. *Acacia* desert thornbush, dry semideciduous thicket and succulent savanna, *Adansonia-Cordia-Sterculia-Commiphora-Cassia-Platycelyphium-Combretum*, etc., bushland and also forest and *Euphorbia* /grassland margins; 450–1150 m.

NOTE. Thomas places this in his sect. *Stacheocymosa* together with *C. kissakense* Gürke. *C. wildii* Mold. is very closely related but Gillett gives reasons for keeping them separate.

52. **C. kissakense** *Gürke* in E.J. 28: 304, 466 (1900); Bak. in F.T.A. 5: 520 (1900); Thomas in E.J. 68: 89 (1936); T.T.C.L.: 631 (1949); Vollesen in Opera Bot. 59: 82 (1980); Moldenke in Phytologia 61: 407 (1986). Type: Tanzania, Morogoro District, Kissaki Steppe, *Goetze* 42 (B, holo.†)

Erect perennial herb or subshrub 20–60 cm. tall from a woody stock; stems glabrous, lenticellate, drying dark and herbaceous above, only corky and striate at the extreme base, rather labiate-like in aspect. Leaves in whorls of 3, lanceolate to oblong-oblanceolate, 4–28 cm. long, 1–7 cm. wide, shortly acuminate at the apex, narrowed to the base, actually sessile, glabrous, irregularly serrate, perhaps somewhat fleshy; apparent winged petiole up to 5 cm. long. Flowers in cymes at separated leafless nodes forming a terminal inflorescence 8–12(–30) cm. long and in the axils of upper leaves, the upper nodes with reduced leaves forming linear-oblong bracts 0.8–1.5 cm. long, 2–5 mm. wide; pedicels 5–7 mm. long. Calyx-tube cupular, 3 mm. long; lobes broad and obtuse, rounded-

* Thomas gives *Engler* 1517 as 'typus' but Winkler clearly designates the specimen I have cited as 'original'; Gillett has seen *Winkler* 3766 (BRSL).

oblong, 3 mm. long, 4 mm. wide, imbricate. Corolla bluish violet (Thomas), greenish yellow with lower lip bluish (Gürke), nearly white with lower lobe deep blue (Gillett); tube a little longer than the calyx; lower lip cymbiform. Fruit depressed, 7 mm. tall, 11 mm. wide; pyrenes with spaced ribs showing on outside of dry fruit. Fig. 18/17, 18, p. 123.

TANZANIA. Kilwa District: 53 km. S. of Kilwa Kivinje turn-off on Dar es Salaam–Lindi road, 25 Nov. 1966, *Gillett* 18021! & 17 km. SW. of Kingupira, 16 Dec. 1976, *Vollesen* in *M.R.C.* 4227!; Lindi District: 100 km. NW. of Lindi, 2 Feb. 1935, *Schlieben* 5997; Tunduru District: Ruvuma [Rovuma] R., Kwa Mtira, 9 Feb. 1901, *Busse* 1007!
DISTR. T 6, 8; Mozambique*
HAB. Fireswept grassland with *Combretum* and other scattered trees, *Brachystegia* woodland; 114–175 m.

SYN. *C. kissakense* Gürke var. *rovumense* Thomas in E.J. 68: 89 (1936); T.T.C.L.: 631 (1949); Moldenke in Phytologia 61: 408 (1986), *nomen invalid.* Type: Tanzania, Tunduru District, Ruvuma [Rovuma] R., Kwa Mtira, *Busse* 1007 (B, holo., EA, iso.!)

NOTE. Var. *rovumense* was described as differing from the type in its larger leaves and red-violet flowers but there is too little material to assess —probably all the cited material belongs to the variety. Although Thomas (op. cit.: 89) puts var. *rovumense* Gürke l.c. there is no mention of the variety by Gürke and Thomas's phrase in German is not adequate to validate it.

53. **C. uncinatum** *Schinz* in Verh. Bot. Ver. Brandenb. 31: 206 (1890); T.T.C.L.: 633 (1949); F.F.N.R.: 366 (1962); Friedrich-Holzhammer in Prodr. Fl. SW.-Afr. 122: 5 (1967). Type: Namibia, NW. Kalahari, Gorochas, *Schinz* 456 (Z, holo.)

Rambling shrub or subshrub 0.3–1.5 m. tall, forming tangled masses to 2 m. wide; branches pale, densely covered with pale spreading long hairs, shorter hairs and glands; also with curved ± orange-brown spines and straight spines derived from peduncles after flowers have fallen; root large and tuberous. Leaves broadly elliptic or almost round, or ovate, 0.7–4(–4.7) cm. long, 0.5–2.5 cm. wide, subacute to mostly rounded at the apex but usually shortly apiculate, rounded to broadly cuneate at the base, entire, densely pubescent and glandular all over; some leaves near base of stem can be obovate and up to 13 × 8 cm.; petiole 2–6 mm. long. Flowers solitary, axillary in numerous axils along the stems; peduncles 0.8–1.2 cm. long; true pedicels 1–4 mm. long; bracts paired, linear, 4–8 mm. long. Calyx densely hairy; tube cupular, 5 mm. long; lobes narrowly triangular, 3–4 mm. long, 2 mm. wide, acuminate. Corolla scarlet or orange-red with yellow throat, densely hairy outside; tube 0.6–1.3 cm. long; lobes oblong or obovate-elliptic, 1–2.5 cm. long, 0.6–1 cm. wide. Style and stamens exserted 1.5–2 cm., curved above. Fruit ovoid or didymous, 1.5 cm. long, 0.8–1.5 cm. wide, pubescent, sitting in the persistent scarcely accrescent calyx. Fig. 20.

TANZANIA. Tabora District: Tabora, near police lines, 29 July 1948, *Semsei* 66 in *F.H.* 2498! & near Urambo, 10 Oct. 1949, *Bally* 7641!; Ufipa District: Muse, 21 July 1950, *Bullock* 3007!; Dodoma District: Manyoni, Kazikazi, 14 July 1932, *B.D. Burtt* 3806!
DISTR. T 4, 5; Zaire, Malawi, Zambia, Zimbabwe, Botswana, Angola and Namibia
HAB. Bushland, *Brachystegia* and *Isoberlinia* woodland, cultivations, sometimes as a weed; 780–1260(–2100 *fide* Bullock) m.

SYN. *Cyclonema spinescens* Oliv. in J.L.S. 15: 96 (1876) & in Hook., Ic. Pl. 13, t. 1221 (1877) *non* Klotzsch (1861), *nom. illegit.* Type: Tanzania, Kigoma District, S. of Kawele, *Cameron*** (K, holo.!)
   Kalaharia spinipes Baillon, Hist. Pl. 11: 111 (1892). Type: S.Africa, no specimen cited
   Clerodendrum spinescens (Oliv.) Gürke in E.J. 18: 180 (1893); Bak. in F.T.A. 5: 313 (1900); De Wild. in Ann. Mus. Congo, Bot., sér. V, 3: 136 (1909); Thomas in E.J. 68: 89 (1936)
   Kalaharia spinescens (Oliv.) Gürke in P.O.A. C: 340 (1895); Warb., Kunene-Sambesi-Exped.: 350 (1903)
   K. spinescens (Oliv.) Gürke var. *hirsuta* Mold. in Phytologia 3: 418 (1951). Type: Tanzania, Dodoma District, Kazikazi, *B.D. Burtt* 3806 (BR, holo., K, iso.!)
   K. uncinata (Schinz) Mold. in Phytologia 5: 132 (1955)
   K. uncinata (Schinz) Mold. var. *hirsuta* (Mold.) Mold. in Phytologia 5: 132 (1955)

NOTE. Under *Holst* 2798 (said to have been collected at Tabora, Limbo, 7 Mar. 1912) Thomas mentions 3000 m. but this is an error due to some label mix up; Holst did not collect near Tabora nor as late as 1912 since he died in 1894.

---

* *Stuhlmann* 713 cited as without locality by Thomas collected in January 1889 was almost certainly collected on Mozambique I. or in Quelimane.
** See Gillett in K.B. 14: 319 (1960).

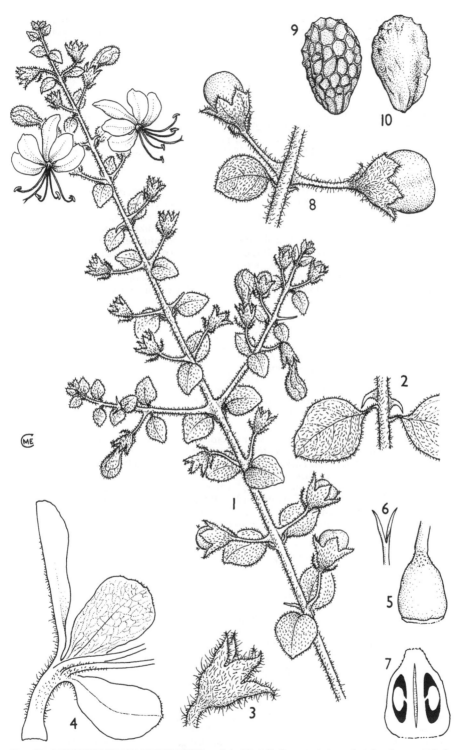

FIG. 20. *CLERODENDRUM UNCINATUM* — **1**, habit, × ⅔; **2**, part ot stem, showing thorns, × 1; **3**, calyx, × 2; **4**, longitudinal section of corolla, × 2; **5**, ovary, × 6; **6**, stigma, × 6; **7**, longitudinal section of ovary, × 8; **8**, fruits, × 1; **9**, **10**, 2 views of pyrene, × 2. 1, 2, from *Bullock* 3007; 3–7, from *Sanane* 212; 8–10, from *B.D. Burtt* 3333. Drawn by Mrs. M.E. Church.

Thomas also mentions a *C. uncinatum* Schinz var. *parviflorum* Schinz supposedly published with the species but I am unable to find it. B.D. Burtt mentioned that he considered the plant would make a good dry-season ornamental but it does not appear to have been cultivated.

13. AVICENNIA

L., Sp. Pl.: 110 (1753) & Gen. Pl., ed. 5: 49 (1754); Bakhuizen van den Brink in Bull. Jard. Bot. Buitenz., sér. 3, 3: 199–226, tt. 14–22 (1921); Biswas in Notes Roy. Bot. Gard. Edin. 18: 159–166, tt. 243–246 (1934); Jafri, Fl. W. Pakistan 49, Avicenniac. (1973); Moldenke in Ann. Missouri Bot. Gard. 60: 149–154 (1973) & in Rev. Fl. Ceylon 4: 126 (1983) (very full references)

Shrubs or small trees of the mangrove zone; pneumatophores present; nodes swollen. Leaves simple, lanceolate to elliptic, decussate, coriaceous, entire; stipules absent. Flowers sessile, yellow to orange in terminal and axillary often capitate cymes, regular, subtended by an involucre of 1 scale-like bract and 2 scale-like bracteoles. Calyx deeply 5-lobed; lobes ovate, imbricate. Corolla-tube shortly campanulate; lobes 4, ovate. Stamens inserted at the throat, equal or slightly didynamous, alternating with the lobes. Ovary with free central placentation; ovules 4, pendulous; style shortly divided. Capsule broadly ellipsoid, compressed, opening by 2(–4) thick valves; usually only 1 erect seed; embryo with large longitudinally folded cotyledons and villous radicle, viviparous; endosperm fleshy.

A genus of 3–6(–12) species according to delimitation often, and with some justification, placed in a separate family Avicenniaceae Endl. Distinctive characters are the involucres, free central placentation with ± 4-winged placenta, orthotropous ovules, viviparous embryo and fleshy endosperm; also there are important anatomical differences. Nevertheless, for the purpose of this Flora I have followed Hutchinson and Cronquist in treating it as a distinctive subfamily of Verbenaceae. Lam (Verben. Malay Arch.: 343 (1919)) and A.F.W. Schimper (Bot. Mitt. Trop. Heft 3: 98, 129 (1891)) take a very conservative view of the species and recognise only one in the Old World and 2 in the New World, but I have more or less followed Moldenke's narrower views. Admittedly there are many intermediates but the East African material is rather uniform.

A. marina (*Forssk.*) *Vierh.* in Denkschr. Akad. Wiss., Wien, Math.-Nat. 71: 435 (1907); Lam & Bakh. in Bull. Jard. Bot. Buitenz., sér. 3, 3: 102 (1921), pro parte; U.O.P.Z.: 137 (1949); T.T.C.L.: 629 (1949); Täckholm, Students' Fl. Egypt: 155 (1956); K.T.S.: 581, fig. 106 (1961); E.P.A.: 803 (1962); Jafri, Fl. W. Pakistan 49, Avicenniac.: 2, fig. 1 (1973); Ju-Ying Hsiao in Fl. Taiwan 4: 411, t. 1055 (1978); Fosberg & Renvoize, Fl. Aldabra: 219, fig. 35/1 (1980); Moldenke in Rev. Fl. Ceylon 4: 127 (1983) (very full synonymy and description etc.); Collenette, Fl. Saudi Arabia: 494 (1985). Type: 'islands and shores of Red Sea' (Al Luhaygah [Lohaya]), *Forsskål* (C, syn., photo.!, BM, isosyn.)

Shrub or small tree 1–9 m. tall or even attaining 15 m. when left in favourable conditions; bark brownish yellow-green, smooth. Pneumatophores 10–25(–40) cm. long. Stems finely grey-tomentose. Leaf-blades elliptic or ovate-elliptic to elliptic-lanceolate, 3–12 cm. long, 1.5–5 cm. wide, acute to acuminate or even obtuse at the apex, cuneate at the base, green or yellow-green and glabrous above, minutely silvery grey or whitish tomentose beneath, glaucous, sometimes blackening on drying; petiole 3–8(–14) mm. long, gradually passing into the lamina. Flowers in small dense heads 0.7–1.2 cm. diameter, with 3 heads per terminal inflorescence but lateral branches originating from lower leaflets or leafy nodes; sometimes a pair of additional opposite flowers borne on central peduncle well below the head; bracts and bracteoles ovate or ± round, concave, addressed to calyx, 2–4 mm. long, 1.5–3 mm. wide, acute, ciliate. Sepals ovate, elliptic or ± round, 3.5–4 mm. long, 2.5–3 mm. wide, obtuse, ± densely ciliate and tomentose outside. Corolla yellow, apricot or dark orange, turning black; tube 2–3 mm. long, glabrous; lobes ovate, 2.5–4 mm. long, 2–3 mm. wide, pubescent outside. Anthers sulphur-yellow turning black. Ovary yellow-green, narrowly conical, 2.5 mm. long, pubescent above, glabrous below; style 0.8 mm. long, glabrous, 2-fid. Capsule subglobose, broadly ellipsoid or ovoid, usually not beaked when mature, 1.2–3 cm. long, 0.7–2.5 cm. wide, velvety scaly-tomentose. Seed usually single, compressed. Fig. 21.

KENYA. Mombasa, Nov. 1884, *Wakefield*!; N. Kilifi, *R.M. Graham* 251!; W. side of Lamu town, 22 Feb. 1956, *Greenway* 8933!

FIG. 21. *AVICENNIA MARINA* — **1**, habit, × ⅔; **2**, flower, × 6; **3**, outer bract, × 6; **4**, middle bract, × 6; **5**, inner bract, × 6; **6**, corolla opened out, × 6; **7**, ovary, × 6; **8**, fruit, × ⅔; **9**, seed, × ⅔. 1–7, from *Semsei* 1813; 8,9, from *Renvoize* 762. Drawn by P. Halliday.

TANZANIA.   Tanga Bay, 4 Nov. 1929, *Greenway* 1850!; Uzaramo District: 16 km. N. of Dar es Salaam, 26
Apr. 1933, *B.D. Burtt* 4468!; Lindi, 9 Dec. 1955, *Milne-Redhead & Taylor* 7593!; Zanzibar I.,
Marahubi, 17 Apr. 1962, *Faulkner* 3031!
DISTR.   **K** 7; **T** 3, 6, 8; **Z**; **P**; N. and S. Yemen, Oman, Saudi Arabia, Bahrain, Socotra, Egypt to South
Africa (Transkei, Kentani), Madagascar, Seychelles, Aldabra, Comoro Is., Persian Gulf to Pakistan,
S. India, Sri Lanka, Andaman Is., Malay Peninsula, Indochina, Philippines, N. Borneo, Sarawak,
China (incl. Hainan), Taiwan, Japan and a variety in Australia, New Zealand, New Caledonia and
Solomon Is. (detailed Malesian distribution omitted as very confused).
HAB.   Locally dominant in sandier parts and inland fringes of mangrove associations, sandy dunes,
mud of tidal rivers and salty creeks, colonizes new mud banks; sea-level

SYN.   *Sceura marina* Forssk., Fl. Aegypt.-Arab.: 37 (1775)
        [*Avicennia officinalis* sensu Klotzsch in Peters, Reise Mossamb., Bot. 1: 266 (1861); Gürke in
        P.O.A. C: 342 (1895); Bak. in F.T.A. 5: 332 (1900); V.E. 1(1): 233, fig. 202 (1910); Lam, Verben.
        Malay Arch.: 340 (1919), pro parte, *non* L.]
        A. officinalis L. var. *ovalifolia* forma *tomentosa* + forma *flaviflora* Kuntze (sic), Rev. Gen. Pl. 3(2):
        249 (1898). Type: 'Sansibar' (? Zanzibar I.), *O. Kuntze* (K, iso.! [only one specimen on the
        sheet])

NOTE.   *A. officinalis* L., *A. alba* Blume and *A. marina* are closely related species and have often been
considered to be conspecific. Bakhuizen van der Brink considered *A. marina* to extend to Asia and
Australia and maintained var. *alba* (Blume) Bakh. and three other varieties apart from the typical
one and gives immensely detailed synonymies and references. Biswas considered that *A. marina*
did not occur anywhere on the Indian Coast. Lam (1919) considered there was only one variable
Old World species and did not even recognise varieties but this is an extreme view not supported
by the material. I have followed Moldenke's determinations to a large extent since he has
examined a vast amount of material throughout the world. The wood is used in joinery for ribs,
legs, etc., as a fuel, for lime burning and a brown dye is obtained from the bark.

New names validated by B. Verdcourt in this Part

Clerodendrum L.
  subgen. Clerodendrum
    sect. **Capitata** *Verdc.*, p. 85
    sect. **Cylindrocalyx** *Verdc.*, p. 85
    sect. **Eurycalyx** *Verdc.*, p. 85
    sect. **Konocalyx** *Verdc.*, p. 86
    sect. **Macrocalyx** *Verdc.*, p. 85
    sect. **Microcalyx** *Verdc.*, p. 86
    sect. **Odontocalyx** *Verdc.*, p. 86
    sect. **Oxycalyx** *Verdc.*, p. 85
      subsect. **Apiculata** *Verdc.*, p. 85
      subsect. **Fallax** *Verdc.*, p. 85
    sect. **Siphonocalyx** *Verdc.*, p. 85
  subgen. Cyclonema (Hochst.) Thomas
    sect. **Pleurocymosa** *Verdc.*, p. 137
    sect. **Stacheocymosa** *Verdc.*, p. 86
Clerodendrum cephalanthum Oliv.
  subsp. **impensum** (*Thomas*) *Verdc.*, 108
  subsp. **mashariki** *Verdc.*, p. 108
  subsp. **montanum** (*Thomas*) *Verdc.*, p. 109
  subsp. **swynnertonii** (*S. Moore*) *Verdc.*, p. 108
    var. **schliebenii** (*Mildbr.*) *Verdc.*, p. 109
Clerodendrum commiphoroides *Verdc.*, p. 139
Clerodendrum hildebrandtii Vatke
  var. **puberula** *Verdc.*, p. 110
Clerodendrum johnstonii Oliv.
  subsp. **marsabitense** *Verdc.*, p. 118
Clerodendrum lutambense *Verdc.*, p. 124
Clerodendrum myricoides (Hochst.) Vatke
  subsp. **austromonticola** *Verdc.*, p. 135
  subsp. **mafiense** *Verdc.*, p. 133
  subsp. **muenzneri** (*Thomas*) *Verdc.*, p. 135
  subsp. myricoides
    var. **kilimandscharense** *Verdc.*, p. 133
    var. **viridiflorum** *Verdc.*, p. 133

INDEX TO VERBENACEAE

GEOGRAPHICAL DIVISIONS OF THE FLORA

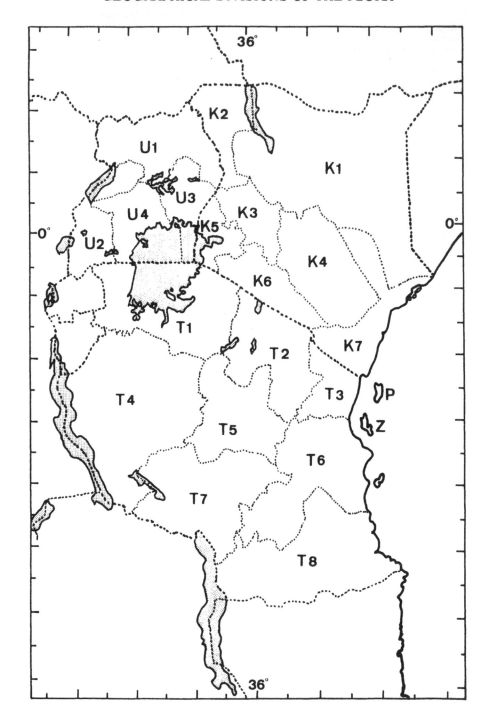

Milton Keynes UK
Ingram Content Group UK Ltd.
UKHW031152141024
449569UK00024B/871